ANALYSIS OF AIR POLLUTANTS
Peter O. Warner

ENVIRONMENTAL INDICES
Herbert Inhaber

URBAN COSTS OF CLIMATE MODIFICATION
Terry A. Ferrar, Editor

CHEMICAL CONTROL OF INSECT BEHAVIOR: THEORY AND APPLICATION
H. H. Shorey and John J. McKelvey, Jr.

MERCURY CONTAMINATION: A HUMAN TRAGEDY
Patricia A. D'Itri and Frank M. D'Itri

POLLUTANTS AND HIGH RISK GROUPS
Edward J. Calabrese

SULFUR IN THE ENVIRONMENT, Parts I and II
Jerome O. Nriagu, Editor

ENERGY UTILIZATION AND ENVIRONMENTAL HEALTH
Richard A. Wadden, Editor

METHODOLOGICAL APPROACHES TO DERIVING ENVIRONMENTAL AND
OCCUPATIONAL HEALTH STANDARDS
Edward J. Calabrese

FOOD, CLIMATE AND MAN
Margaret R. Biswas and Asit K. Biswas, Editors

CHEMICAL CONCEPTS IN POLLUTANT BEHAVIOR
Ian J. Tinsley

RESOURCE RECOVERY AND RECYCLING
A. F. M. Barton

ATMOSPHERIC MOTION AND AIR POLLUTION
Richard A. Dobbins

INDUSTRIAL POLLUTION CONTROL—Volume I: Agro-Industries
E. Joe Middlebrooks

BREEDING PLANTS RESISTANT TO INSECTS
Fowden G. Maxwell and Peter Jennings, Editors

COPPER IN THE ENVIRONMENT, Parts I and II
Jerome O. Nriagu, Editor

ZINC IN THE ENVIRONMENT, Parts I and II
Jerome O. Nriagu, Editor

NEW TECHNOLOGY OF PEST CONTROL
Carl B. Huffaker, Editor

NEW TECHNOLOGY OF
PEST CONTROL

NEW TECHNOLOGY OF PEST CONTROL

EDITED BY

CARL B. HUFFAKER

Professor of Entomology and Entomologist in
The Experiment Station, Division of Biological Control
Department of Entomology and Parasitology
The University of California, Berkeley

and

Director, The International Center for Integrated and Biological Control
The University of California, Berkeley and Riverside

SPONSORED BY

THE INTERNATIONAL CENTER FOR INTEGRATED AND BIOLOGICAL CONTROL THE UNIVERSITY OF CALIFORNIA

A WILEY-INTERSCIENCE PUBLICATION

JOHN WILEY & SONS

NEW YORK • CHICHESTER • BRISBANE • TORONTO

This publication was prepared under National Science
Foundation Grant DEB 75-04223, with funds therein
provided by the National Science Foundation (NSF) and
the Environmental Protection Agency (EPA). Any opinions,
findings, conclusions, and/or recommendations contained
in this publication are those of the authors and do not
necessarily reflect the views of the participating
institutions or agencies, the National Science Foundation
or the Environmental Protection Agency.

Volume 1 in a series on the NSF/EPA-funded integrated
pest management project titled: "The principles, strategies
and tactics of pest population regulation and control in
major crop ecosystems."

Library of Congress Cataloging in Publication Data

Main entry under title:
New technology of pest control.

(Environmental science and technology)
"A Wiley-Interscience publication."
Includes index.
1. Pest control, Integrated.   2. Insect control.
I. Huffaker, C. B., 1914–   II. International
Center for Integrated and Biological Control.
SB950.N48   632'.7   79-4369
ISBN 0-471-05336-8

Printed in the United States of America

10 9 8 7 6 5 4 3 2

CONTRIBUTORS

P. L. Adkisson
Department of Entomology
Texas A & M University
College Station, Texas

J. C. Allen
Agricultural Research and Education Center
University of Florida
Lake Alfred, Florida

E. J. Armbrust
Illinois Natural History Survey and Illinois
    Agricultural Experiment Station
Urbana, Illinois

D. Asquith
Pennsylvania State University Fruit
    Research Laboratory
Biglerville, Pennsylvania

D. Baasch
Division of Biological Control
The University of California
Riverside, California

R. F. Brooks
Agricultural Research and Education Center
University of Florida
Lake Alfred, Florida

L. G. Brown
Department of Agricultural Engineering
Mississippi State University
Mississippi State, Mississippi

R. N. Coulson
Department of Entomology
Texas A & M University
College Station, Texas

B. A. Croft
Pesticide Research Center
Michigan State University
East Lansing, Michigan

G. L. Curry
Department of Industrial Engineering
Texas A & M University
College Station, Texas

D. W. Davis
Department of Biology
Utah State University
Logan, Utah

H. A. Dean
Texas A & M University Agricultural
    Research and Extension Center
Weslaco, Texas

D. W. De Michele
Department of Industrial Engineering
Texas A & M University
College Station, Texas

G. W. Fick
Department of Agronomy
Cornell University
Ithaca, New York

T. W. Fisher
Department of Entomology
The University of California
Riverside, California

J. L. Foltz
Department of Entomology
Texas A & M University
College Station, Texas

E. H. Glass
Department of Entomology
New York State Agricultural Experiment
  Station
Geneva, New York

A. P. Gutierrez
Division of Biological Control
The University of California
Berkeley, California

F. P. Hain
Department of Entomology
North Carolina State University
Raleigh, North Carolina

R. G. Helgesen
Department of Entomology
Cornell University
Ithaca, New York

D. C. Herzog
Department of Entomology
Louisiana State University
Baton Rouge, Louisiana

S. C. Hoyt
Tree Fruit Research Center
Washington State University
Wenatchee, Washington

C. B. Huffaker
International Center for Integrated and
  Biological Control
The University of California
Berkeley and Riverside and Division of
  Biological Control
Department of Entomological Sciences
The University of California
Berkeley, California

R. L. Jensen
Department of Entomology
Louisiana State University
Baton Rouge, Louisiana

M. Kogan
Department of Agricultural Entomology
University of Illinois
Urbana, Illinois

W. A. Leuschner
Department of Forestry and Forest Products
Virginia Polytechnic Institute and State
  University
Blacksburg, Virginia

R. F. Luck
Division of Biological Control
The University of California
Riverside, California

G. R. Manglitz
Agricultural Research Service
USDA and University of Nebraska
Lincoln, Nebraska

N. F. Marolsan
Department of Chemical Engineering
Louisiana State University
Baton Rouge, Louisiana

C. W. McCoy
Agricultural Research and Education Center
University of Florida
Lake Alfred, Florida

F. D. Miner
Department of Entomology
University of Arkansas
Fayetteville, Arkansas

L. D. Newsom
Department of Entomology
Louisiana State University
Baton Rouge, Louisiana

B. C. Pass
Department of Entomology
University of Kentucky
Lexington, Kentucky

T. L. Payne
Department of Entomology
Texas A & M University
College Station, Texas

J. R. Phillips
Department of Entomology
University of Arkansas
Fayetteville, Arkansas

R. L. Pienkowski
Department of Entomology
Virginia Polytechnic Institute and State
  University
Blacksburg, Virginia

P. E. Pulley
Data Processing Center
Texas A & M University
College Station, Texas

R. L. Rabb
Department of Entomology
North Carolina State University
Raleigh, North Carolina

R. E. Rice
San Joaquin Agricultural Research and
  Extension Center
The University of California
Parlier, California

L. A. Riehl
Department of Entomology
The University of California
Riverside, California

W. G. Rudd
Departments of Computer Science and
  Entomology
Louisiana State University
Baton Rouge, Louisiana

W. G. Ruesink
Illinois Natural History Survey and Illinois
  Agricultural Experiment Station
Urbana, Illinois

C. A. Shoemaker
Department of Environmental Engineering
Cornell University
Ithaca, New York

R. Skeith
Department of Industrial Engineering
University of Arkansas
Fayetteville, Arkansas

R. F. Smith
Department of Entomological Sciences
The University of California
Berkeley, California

R. W. Stark
Coordinator of Research
University of Idaho
Moscow, Idaho

C. G. Summers
Department of Entomological Sciences
The University of California
Berkeley, California

S. G. Turnipseed
Department of Entomology and Zoology
Clemson University
Clemson, South Carolina

Y. Wang
Division of Biological Control
The University of California
Berkeley, California

W. E. Waters
Department of Entomological Sciences and
  Department of Forestry
The University of California
Berkeley, California

W. H. Whitcomb
Department of Entomology and Nematology
University of Florida
Gainesville, Florida

D. L. Wood
Division of Entomology and Parasitology
The University of California
Berkeley, California

SERIES PREFACE

Environmental Science and Technology

The Environmental Science and Technology Series of Monographs, Textbooks, and Advances is devoted to the study of the quality of the environment and to the technology of its conservation. Environmental science therefore relates to the chemical, physical, and biological changes in the environment through contamination or modification, to the physical nature and biological behavior of air, water, soil, food, and waste as they are affected by man's agricultural, industrial, and social activities, and to the application of science and technology to the control and improvement of environmental quality.

The deterioration of environmental quality, which began when man first collected into villages and utilized fire, has existed as a serious problem under the ever-increasing impacts of exponentially increasing population and of industrializing society. Environmental contamination of air, water, soil, and food has become a threat to the continued existence of many plant and animal communities of the ecosystem and may ultimately threaten the very survival of the human race.

It seems clear that if we are to preserve for future generations some semblance of the biological order of the world of the past and hope to improve on the deteriorating standards of urban health, environmental science and technology must quickly come to play a dominant role in designing our social and industrial structure for tomorrow. Scientifically rigorous criteria of environmental quality must be developed. Based in part on these criteria, realistic standards must be established and our technological progress must be tailored to meet them. It is obvious that civilization will continue to require increasing amounts of fuel, transportation, industrial chemicals, fertilizers, pesticides, and countless other products; and that it will continue to produce waste products of all descriptions. What is urgently needed is a total systems approach to modern civilization through which the pooled talents of scientists and engineers, in cooperation with social scientists and the medical profession; can be focused on the development of order and equilibrium in the presently disparate segments of the human environment. Most of the skills and tools that are needed are already in existence. We surely have a right to hope a technology that has created such manifold environmental problems is also capable of solving them. It is our hope that this Series in Environmental Sciences and Technology will not only serve to make this challenge more explicit to the established professionals, but that it also will help to stimulate the student toward the career opportunities in this vital area.

Robert L. Metcalf
Werner Stumm

PREFACE

This volume is a summary report of the progress toward development of integrated pest management systems accomplished in recent years by the NSF/EPA Integrated Pest Management Project. This considerable progress is of great significance, but of even greater import is the tremendous impact this project has had on the ways of conducting and administrating pest management research, particularly multidisciplinary research. This report describes a major landmark in the evolution and unification of the plant protection discipline.

The history of integrated pest management goes back much farther than the mere coining of the term *integrated control* in the early 1950s and the subsequent elaboration and clarification of the concept. The origins are deeply rooted in the evolution of pest control practices as developed by entomologists and plant pathologists in the 19th century.

The history of man is a series of attempts to gain increasing control over his environment. At first this control was minimal; then the gradual gain in man's capacity to control his environment paralleled the gradual rise of civilization. As man aggregated in villages near rivers, and as he planted crops nearby, he encountered increasingly severe pest attacks against himself and his crops. For thousands of years, man could do little about these pests but appeal to the power of magic and a variety of gods. For the most part, he had to live with and tolerate the ravages of plant diseases and insects; then gradually early man learned how to improve his condition through "trial and error" experiences. Among these improvements were the beginnings of pest management.

Prior to the emergence of the crop protection sciences in the 18th century and even before the broad outlines of the biology of insect pests and the causality of plant diseases were understood, man evolved many cultural and physical control practices for the protection of his crops. The appropriate use of these methods reduced the crop damage potential of essentially all classes of pests and provided satisfactory economic control of some. But there were many pests of high damage potential that could not be controlled adequately by early agriculturists through any combination of known cultural control methods. As biological knowledge grew during the 18th and 19th centuries, and pest problems became more severe as the result of an intensification of agriculture and the introduction of pests into new areas, man became increasingly preoccupied with the search for more effective pest control measures.

Although a few early entomologists and plant pathologists strongly advocated a fundamental ecological approach to plant protection, their pleas were mostly ignored as more and better chemicals were sought to control these pests. There were optimistic expectations as early as the turn of the 20th century that all diseases and insects would ultimately be controlled by chemical pesticides alone. Despite the occasional warnings about the hazards of unilateral approaches to pest control, pest protection in the United States since about 1920 gradually shifted toward heavy dependence on chemical pesticides and, though to a much lesser extent, on resistant varieties for plant diseases.

The results of these trends are well known; and they culminated in a series of crises in pest control in the mid 1950s. The limited base of pest control tactics then in use was no longer sufficient to control the target pests; and it interfered with the control of other pests, and in releasing other species from existing natural control enabled them to become pests. In some cases, the chemicals unfavorably modified the physiology of the crop plants, created hazards to man's health, destroyed pollinators and other desirable wildlife, and in other ways produced undesirable effects.

The introduction of new pest species into the agroecosystems also placed stress on an already overburdened pest control technology. Furthermore, the pressure over the years toward greater intensity of production rapidly forced changes in the agroecosystems and created new environments for the pests, with the result that agroecosystems often became more vulnerable to them. Changes in tillage, water management, crop varieties, fertilization, and other agronomic practices greatly influenced pest incidence, very often in favor of the proliferation of many pest species. The increased complexity and intensity of agricultural production practices, along with reduced genetic diversity in many agricultural crop species, combined to produce a new magnitude of crop hazards from pests.

The intensification and increasing complexity of crop protection problems, coupled with the associated environmental, financial, and health hazards of heavy chemical usage, have combined to stimulate great interest in the importance of crop protection and in the broad ecological approach as the appropriate avenue to sound and acceptable solutions. It was in the climate of this interest that the International Biological Programme, National Science Foundation, Environmental Protection Agency (IBP/NSF/EPA) Integrated Pest Management Project was established in the early 1970s.

The concept of integrated pest management has not arisen from disjunct developments in crop protection; it is simply a more holistic, and still evolutionary, stage in the development of an ecologically based pest control strategy. It is significantly different in that its new conceptual approach sets crop protection in a new context within a crop production system. Many components of the integrated pest management concept were developed in the late 19th and early 20th centuries, but integrated pest management as now conceived is unique because it is based on ecological principles and integrates multidisciplinary methodologies in developing agroecosystem management strategies that are practical, effective, economical, and protective of both public health and the environment. The early efforts of crop protectionists to control pests with ecologically based cultural methods were not satisfactory; consequently, entomologists, plant pathologists, and, later, weed scientists, became preoccupied with the search for pesticides that were both economical and effective. Often, unfortunately, chemical methods were used not to supplement cultural methods but to supplant them.

Our state of technology and understanding of host–pest interactions has now evolved to the point that an integration of pest control tactics for the control of a given class of pest—for example insects, or plant pathogens—and for multiple classes of pests is not only feasible; it is necessary, given the inadequacies of single-method, single-discipline approaches and their potential for undesirable effects on nontarget beneficial and pest species.

The NSF/EPA-supported Pest Management Project has taken the lead role in providing the mechanisms—the way of doing business—in the complex milieu of multidisciplinary plant protection as a component of crop production. The *Huffaker Project*, as it has been dubbed, has shown the way toward a new era in plant protection. This is what this book is about.

RAY F. SMITH

Berkeley, California
January, 1980

CONTENTS

xv

NEW TECHNOLOGY OF
PEST CONTROL

1

RATIONALE, ORGANIZATION, AND DEVELOPMENT OF A NATIONAL INTEGRATED PEST MANAGEMENT PROJECT

C. B. Huffaker

International Center for Integrated and Biological Control
The University of California, Berkeley and Riverside

and

Division of Biological Control, Department of Entomological Sciences
The University of California, Berkeley

Ray F. Smith

Department of Entomological Sciences
The University of California, Berkeley

INTRODUCTION

The importance of pest control in general, and insect control specifically, and the increasing public concern for the adverse side-effects of pesticides on public health and the environment generally, have called for agriculture and forestry to devise ways of controlling the pests which are more economical, more lasting, and less harmful to public health and the environment, including the soil itself. The accumulation of some pesticides or their ingredients in crop soils [e.g., heavy metals (Brown & Jones, 1975)] are rendering such soils unsuitable for growth of some crops (in the future, for many crops?). The U.S. International Biological Program (IBP) initiated such an effort in 1971, with funding supplied by the National Science Foundation (NSF), Environmental Protection Agency (EPA), the U.S. Department of Agriculture (USDA) (initially) and 18 universities with their agricultural experiment stations (Fig. 1.1). The program has been continued with NSF-EPA support following the termination of IBP. It has engaged some 250 or so scientists for the past 5 years, most of whom have been supported by regular funds of their experiment stations. A central and unifying force has been the gradual shifting of the program toward concentration on the crops themselves, rather than on the pests specifically (mostly insect pests), by employment of systems analysis and computer technology. This approach has helped furnish guidance of the research, analysis of processes, and development of delivery systems appropriate to widescale commercial or public adoption. Only this catalytic element has enabled us to attract and hold participants from a wide spectrum of disciplines. Yet, we emphasize that major improvements have been made by altering strategy and utilizing more ecological *tactics* in various combinations, and many others are possible by grassroots ad hoc biological studies, *entirely* aside from the employment of systems analysis in the formal mathematical or modeling sense. In fact, we expect that much of the finer aspects of full sophisticated modeling will not be used in the implementation programs that the use of such models are helping us to devise, by giving us better insights into the roles of, and importance of various factors—so that we can more confidently concentrate on only the *key* ones in our scouting and necessary monitoring for practical control programs.

The systems chosen for this effort were alfalfa, cotton, soybean, citrus, pome and stone fruits, and pine forests (concentrating on bark beetles). The

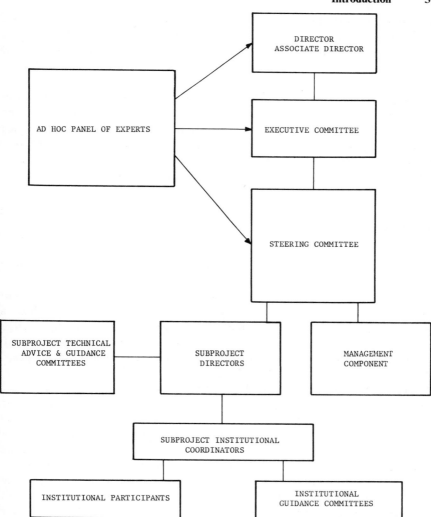

Figure 1.1. Organization chart of IPM Project.

rationale used in selection of these systems was based on a number of factors: importance of the crop, the problems posed by insect pests, the quantity of insecticides being used on them or likely to be used soon (if better programs are not developed), the prospects of developing and using more economical, more lasting ecologically based tactics (alternatives) and improved use and reductions in use of the more disruptive chemical pesticides, the availability of staff having a background and interest in developing integrated and systems analysis approaches to pest control, a distribution of each problem over a wide belt of the country, and overall, so as to represent all regional areas of the country. The rationale for selection of each individual crop (or forest) will be given in Chapters 2 to 14.

We need, however, to deal here with the general rationale and justification for establishing such a program. This requires a brief discussion of the dilemma being faced by society at all levels with respect to the need to produce increasing quantities of cheap food and fiber in ways that are not in the long run counterproductive through breakdown of effectiveness, escalating costs, or hazards to public health and the environment (above).

RATIONALE AND NEED FOR A NEW APPROACH

Shortly after World War II pest control shifted largely from a biological discipline to a chemical one. This era of unilateral dependence on pesticides (particularly insecticides) provided, indeed, good disease control, spectacular insect control, and, subsequently, weed control. There has also been a rather unilateral effort to develop crop varieties concentrating only on high yields with consequent disregard for loss of, or possible incorporation of, characters for tolerance of or resistance to pests, particularly insect pests. Both of these major "advances" have with time come up short of early hopes. Neither has rested on the broad ecological dictum that the whole interacting system must be considered, and neither rested on an appreciation of nature's two basic systems of containing excessive abundance—natural enemies and the plants' own factors for resisting pest attack. One reason for their failure to consider the whole system is that scientists are by nature specialists and individualists; they like to tinker with their own specific thing. To a regrettable degree, individuals, departments of research, and extension, in the same university and, to a greater degree, those in different universities, have concerned themselves very little with what the others were doing. A notable example of this has come to light since this program was initiated. Plant pathologists discovered that use of the fungicide benomyl will reduce effects of several soybean diseases and increase yields. Its use has been widely recommended despite the fact that it disrupts natural control of a number of major insect pests of soybean. Worse still, it is now being suggested that insecticides be added in the sprays and used in prophylactic manner. This would drastically disturb or essentially destroy the excellent progress already made to develop ecologically and economically sound integrated methods for control of soybean insects. Investigations to establish the utility of other fungicides that are less disruptive require collaborative research between the two disciplines.

A major objective has been to bring diverse expertise and institutions to bear on the common problem(s). In this the program has had a major success. Thus, it was the hope, and it has been fulfilled, that crop growth specialists, entomologists, economists, plant pathologists, weed specialists, systems analysts, mathematicians, engineers, ecologists, foresters, extension workers, and computer specialists, among others, would pool their talents in this effort.

The southern corn blight outbreak in 1970 represents the era of seeking only maximum yields. There was a consequent extreme narrowing of the genetic base, until essentially 90 percent of all the corn grown in the United States had a common genetic background (Smith, 1972). Moreover, troubles arose in the Green Revolution programs for the same reason of neglecting resistance to pests. There is beginning to be a broader program of plant breeding in general, and resistance to pests is being incorporated into high yield lines of a number of crop plants (Maxwell et al., 1972; Gallun et al., 1975; Chapters 3, 5, 7). Plant breeding for resistance to pests (insects and diseases) is taken as one of the major alternative tactics for exploration in the three short-term crops of this program—soybean, cotton, and alfalfa—and substantial progress has been made in all three (Chapters 3, 5, 7). At the same time, this program has confirmed again that no single tactic can be relied upon alone. The same characters in cotton that confer resistance to bollworms and budworms, for example, confer greater susceptibility to fleahoppers and other plant bugs. Thus the plant breeders provide certain components of the overall solution to a pest complex, while other components are provided through other agencies. All components must be interdigited and structured into a model, not necessarily mathematical, for a sound pest control system for a crop.

In the era of nearly total reliance on pesticides, there was general disregard of the fact that even the simplest agroecosystem presents a complicated maze of delicately balanced ecological interactions. (Elaborate calendar date schedules of insecticide and fungicide treatments on a prophylactic basis were developed.) This program, too, ignored one of nature's two basic systems of containment of pest abundance—the natural enemies.

The system ignored the enormous plasticity and potential of the insects to counterattack. Resistance to the insecticides became commonplace until now some insect populations cannot be controlled by any insecticide [e.g., tobacco budworm, *Heliothis virescens* (F.), in the area of the Rio Grande River of Mexico and Texas, and of *Heliothis armigera* (Hübner) in the Ord River Valley of West Australia]. The target insects in any event resurged quickly, most of them eventually developed resistance, natural enemies were destroyed, and previously innocuous or minor species became major pests. Increasing dosages and more kinds of, and more frequent, pesticide treatments led to a sort of pesticide addiction from which it is difficult to disengage. The cost of pest control became a major threat to economic production of many crops, even as there seemed to be little chance of alleviating the problem. At the same time, a series of governmental and public developments (since then intensified in the United States) have, in effect, commanded that we draw back from the pesticides addiction that had dominated our research and actions in pest control for the last three decades (Newsom, 1970). A basic need existed to revive an earlier era of research on agricultural pests that centered on the pests' biology and ecological, cultural, and biological ways of handling them, not to the abandonment, however, of insecticides, which when properly used remain our most reliable *immediate* solution to a problem.

Prior to this program, there had been little effort to tie pest control in its various phases to crop growth and production, and economic gain or loss in any explicit way. The analysis of the whole complex of factors and processes in the culture and growth of a crop, control of the pests, and in determining loss/benefit relationships is often too complex for intuitive or purely empirically derived solutions. What was needed was a new technology relative to crop production and decision-making that would utilize the power of computers and systems analysis in the various forms that have been used in engineering, industry, and commerce. Heretofore, the farmer has had to be his own systems analyst. Logically, scientists and technologists in these various disciplines of crop production and crop protection could be brought into these areas, and should be able to develop a structured, analytical approach to the problems and decisions to be made that would make the farmer's decisions easier for him, with reference to both short-term and long-term interest.

OBJECTIVES, STRATEGY, AND TACTICS OF THE PROGRAM

The general objective has been the development of improved, ecologically oriented pest management systems that optimize, on a long-term basis, costs and benefits of crop protection. Thus, the project has strived to develop systems of insect pest control (and to an extent crop production as a whole) which will return greater profits to the farmer. At the same time, these systems, by reducing use of expensive and counter-productive (disturbing) insecticides would produce large benefits accruing to society as a whole and to our endangered environment.

Inherent to this general objective is recognition that most crop pests are tremendously adaptive, highly reproductive organisms, and that most of them in nature are not likely to be eradicated—moreover, that they do not cause catastrophic damage to their host plants in natural systems. Thus, we seek to maximize their great natural control forces: the weather, host plant resistance, and natural enemies. *Containment* of their populations rather than *prevention* or *eradication* is the logical strategy. This at least is the strategy or philosophy employed for integrated control of insect pests and, where feasible, for control of other pests [i.e., in integrated pest management (IPM)]. The various tools, methods, and details of how a strategy is to be carried out constitute the tactics. The terms should not be used interchangeably. Since tactics are used in any strategy, they are interrelated, but a tactic is not a strategy.

In 1967, the FAO Panel of Experts on Integrated Pest Control defined "integrated control" as follows:

A pest management system that in the context of the associated environment and the population dynamics of the pest species, utilizes all suitable techniques and methods in as compatible a manner as possible and maintains the pest populations at levels below those causing economic injury.

While the weather cannot be manipulated directly, we can intensify its harmful effects on the pests and ameliorate those on the natural enemies, or to favor host plant resistance to the pests by various cultural or habitat management practices. The other two factors of nature's great triumvirate (above), plant resistance and natural enemies, have been taken as the cornerstone of the effort. In addition, efforts have been intensified to find better ways of using chemical pesticides—to an extent by seeking more physiologically selective ones, but primarily by using nonselective ones in ecologically selective ways (see especially Chapters 3, 5, 7, 9, 10).

The objective of the strategy is to *optimize* pest control in the long-term economic, social, and environmental spheres. While an integrated pest management system found to be optimal for control of one pest, or the whole complex of pests, may be associated with a resulting maximum crop yield allowable for a given unit of land to which applied, the goal of the IPM strategy is *not* specifically to maximize yields. Pest control strategies that seek to effect maximum crop yields are not necessarily the most optimal form of pest control from the standpoint of both costs and benefits to the farmer or to society as a whole. Thus, we utilize the concept of the cost/benefit ratio— that is, the costs required for pest control as compared to the value of the benefits. If pest control is overly costly, maximum yield may be associated with financial loss. Total benefits are measured in terms of overall economic, social, and environmental gains. Thus, there is a fundamental difference in "optimal" pest control and pest control that strives solely for "maximum yields." One of the most difficult obstacles to farmer adoption of integrated pest management is the farmer's insistent adherence to the concept of seeking maximum yields. This erroneous concept was not only advocated but was also often widely adopted when the post-World War II synthetic pesticides were still providing highly satisfactory results. Despite the world need for higher yields, the goal of attaining absolutely maximum yields is no longer practical in many field crops due to the problem of pesticide resistance, excessive costs of pest control, and governmental bans on some pesticides. Other countermanding factors resulting from agroecosystem and environmental disturbance have largely refuted the value of the "preventive" or "insurance" pest treatment strategy.

The more specific objectives relating to the various tactics that may be employed in the overall strategy are as follows:

1. Development of a scientific understanding of the significant biological, ecological, and economic processes in the growth of the crops and the population dynamics of the pests and their natural enemies, and other factors affecting them, and of the interactions among these processes and factors.
2. Development of alternative tactics, especially cultural, biological, and host resistance factors, which are ecologically compatible and which can be expected to reduce the use of broad spectrum biocides and lessen the adverse effects of their use.
3. Development of better methods of collecting, handling, and interpreting relevant biological, meteorological and crop production data.
4. Utilization of systems analysis in the general and specific sense, including modeling as a central unifying and research-guiding tool in the pursuit of the main goal and its subsidiary goals.
5. Building of models of the crop production and pest systems, integrating these with economic analysis (impact, etc.), and conducting pilot tests for each crop system.

The basic steps (adapted from guidelines of F. R. Lawson and associates, USDA, Columbia, Missouri) for developing a program have been as follows (not followed specifically by each of the crop subprojects, but all features dealt with by most of them):

1. Separate the *real* pests from those *induced* by insecticides in the different regions involved.
2. Establish realistic economic injury levels for the real pests with appropriate attention to the hidden costs of controls.

3. Separate the real pests into those causing intolerable losses (i.e., *key* pests) and those causing only light or sporadic damage controllable by occasional or limited use of pesticides.

4. Identify the key factors controlling, or of great potential value for controlling (e.g., a key resistant variety, natural enemy, or a cultural measure), populations of the key pests and measure their effects.

5. Design and test control systems based upon these guidelines in each of the areas where the key pests and/or factors are different.

6. Modify control systems according to time and area conditions and new inputs as the program develops.

Obviously these steps are not easily satisfied, and not necessarily in the order given. Some of them imply complex and multifaceted studies, the nature of which could relate to any of the component tactics of the overall strategy of insect containment. Continuing extensive research is mandatory. The specific objectives relating to development of measures or tactics to be used to satisfy the overall strategy can be divided into (1) the search for the development of *direct pest control measures* and (2) development of *supportive* tactics which, although they do not exert any control effect in themselves, are required in the application of the total pest management system.

DIRECT CONTROL TACTICS

The direct tactics for integrated insect pest management that may be used and to some extent have been explored in the IBP project are (1) crop plant resistance, (2) biological control, (3) cultural control, (4) pesticides, (5) attractants and repellents, and (6) growth regulators. Quarantine measures and eradication procedures, while they may, if successful, ease the problem of developing successful integrated control systems, are not in themselves tactics of *containment* of existing populations (the integrated control strategy) but rather of *prevention* or *eradication* strategy. They are therefore not dealt with here.

Crop Plant Resistance

Over long periods of time plants, both natural and cultivated ones, have developed a great diversity of methods for warding off or tolerating attack by plant-feeding organisms. Our cultivated plants have been inadvertently selected for characters offering such resistance. An extensive pool of genetic factors for such resistance exists in our domestic and in wild plant varieties which can be used to develop crop varieties more resistant to pests. Use of plant resistance should be a first consideration in the long-term development of integrated pest management programs. It has become the most important method of control for the most important plant diseases (Glass et al., 1975). Glass et al. (1975) noted as follows:

> Precise knowledge of the genetic basis for pest resistance in plants is not essential to the breeding of resistant plants; however, a general understanding of the genetics of inheritance and plant breeding is necessary. Pest resistance may be due to a simple dominant gene or it may be complicated and polygenic in nature. In the past, emphasis was placed on utilizing dominant, major gene resistance to pests, especially to plant pathogens. Using standard backcross procedures, such resistance can be readily transferred to outstanding cultivars without disrupting their favorable agronomic qualities. Furthermore, the gene usually remains stable across a wide range of environmental conditions. Current emphasis, however, is on the development of cultivars possessing polygenic resistance, because this tends to retard the development of pest races or biotypes able to overcome the resistance.
>
> Resistant plant material is derived from many sources. Originally, most cultivated plant species possessed a high level of genetic variability; in some areas this is still the case. Such high genetic variability offers promising opportunities for the discovery of genotypes with useful levels of resistance. Useful resistance has also been obtained by hybridizing resistant wild species with closely related cultivated types to effect an interspecific transfer of the useful traits. In other situations, mutation breeding using ionizing radiation, chemicals or physical treatments has been successful in inducing resistance in cultivated species. Verticillium wilt resistant cultivars of Mitcham peppermint represent such a case.

Plant pathologists have extensively used breeding for pest resistance as a primary method of plant disease control. Work has embraced diseases caused by fungi, viruses, mycoplasma, nematodes, and bacteria. Early success with wilt diseases of cotton, cowpea and cucurbits in the United States, and with downy mildew and powdery mildew of grape in France, and with wheat stem rust and crown rust of oats in various countries led on to other successes of variable duration. For the rusts and for tobacco mosaic virus on tomato, and various other cases, continued success has been dependent upon continued breeding to counter the resistance-breaking new biotypes that arise (Coons, 1953; Day, 1972). Borlaug (1965) noted that the useful life of a stem rust variety of wheat is about 5 years. Nevertheless, a large percentage of our major crop plants are now comprised of varieties possessing some disease resistance (Coons, 1953; Glass et al, 1975; Maxwell et al. 1972). Cultivars resistant to a complex of pathogens are now available [e.g., tobacco resistant to six major diseases (Glass et al., 1975)].

The development of and maintenance of resistance to a complex of such diseases is, however, complicated, and the method is not looked upon as a solution for all pest problems. For one thing, resistance to one type of pest may render susceptibility to another (Zink & Duffus, 1969; Beck & Maxwell, 1976). Complex breeding procedures are required. Sequential screening procedures for each pathogen and each segregating generation are required. It is difficult or impossible to arrange for appropriate intensities of selection pressure for each of a complex of pathogens in the screening program. Earlier infection by another pathogen may render the test plants either more or less susceptible to a given pathogen being tested. The soil biota in the field may in various ways influence the disease susceptibility quite differently than what would occur in a pure culture (Baker & Cook, 1974).

Entomologists have lagged far behind the plant pathologists in developing resistant cultivars despite there having been several early notable successes with the method for insect control—for example, wheat resistant to the Hessian fly (still successful), woolly apple aphid resistant apple (Winter Majestic), and use of resistant American grape rootstocks in Europe for grape phylloxera (Painter, 1951). The inattention has derived in part from the originally high effectiveness and simplicity of chemical insect control. The problems arising from the unilateral use of such insecticides (above), however, have focused new and increased attention on development of insect resistant cultivars.

This increased recent effort has resulted in substantial advances. It has been noted that some 25 vegetable crops have been developed that are resistant to 35 species of insects, and some resistant to both insects and diseases are being developed [e.g., in the cucurbits, wheat, corn, soybean, alfalfa, and cotton (Horber, 1972; Smith et al., 1974; Huffaker & Smith, 1976)].

It will be seen in Chapters 3, 5, and 7 that some progress has been made in this project in the development of resistance to insect pests of alfalfa, soybean, and cotton, and in combining such resistant factors with others for disease resistance in agronomically acceptable varieties in some instances. Particularly noteworthy has been the degree of resistance to the Mexican bean beetle and in some lines to the bean leaf beetle, soybean looper, velvetbean caterpillar, and *Heliothis* spp. This accelerated program of development was made possible by cooperative breeding and screening in Brazil and Puerto Rico and several states in the southern United States, centered in South Carolina.

In California, development of alfalfa resistant to the spotted alfalfa aphid in the early 1960s became a significant feature of insect control of this pest, and together with actions of introduced and native natural enemies made it largely unnecessary to use insecticides, thus preserving the high degree of natural control of other insect pests of alfalfa that normally exists (Stern et al., 1959). Now, work is beginning to develop cultivars resistant to a recent invader, the Egyptian alfalfa weevil.

Biological Control

For control of insects, in contrast to plant diseases, biological control has been the more central core tactic around which integrated control has been based, rather than host plant resistance. Biological control occurs when the action of parasites, predators, or pathogens temporarily control or continually regulate a host (or prey) population at densities below what they would be in the absence of these natural enemies. In practice, such control is not of importance to man unless the degree of control results in economic benefits. When man manipulates the natural enemies, the pest population, or other factors of the environment to achieve biological control, this is referred to as *applied biological control*. If biological control occurs without manipulation this is referred to as *naturally occurring biological control*. By far the greatest activity in biological control, however, has been in the conduct of *classical biological control*, which involves the importation of natural enemies from exotic regions, usually the native home of a pest exotic to the target area, and their establishment and successful control of the target pest species. Predatory and parasitic (parasitoid) insects have predominated as agents in such biological control. The premise upon which the practice rests is as follows (from Huffaker et al., 1976):

> The premise of biological control is that organisms have natural enemies, that in certain circumstances of place, time, and combination of species, strains, etc., many organisms are held at low, noninjurious levels by these natural enemies, and that in the cases of specific pest species or potential pest species these natural enemies, or natural enemies of their relatives, may provide a control solution. Experience has proved the soundness of the premise. Applied biological control has had many successes, mainly through the introduction of exotic natural enemies to control exotic pests, but more recently by manipulating indigenous natural enemies. Hagen et al. (1971), Rabb (1971), and MacPhee and MacLellan (1971) have recently detailed the value that can result from indigenous natural enemies. Naturally-occurring biological control exists all around us and the possibilities of its conservation and augmentation are many. Yet the spectacular examples of manipulated or applied biological control as achieved through introductions of exotics to control exotics have constituted most of the world effort until recent years.

DeBach (1971, 1974) claims that of all the prospective alternative tactics (to chemicals) for use in integrated control of insect pests, biological control has exceeded all others in proven effectiveness, with the possible exception of cultural controls. Huffaker (1970) noted that examples include biological control of insects, mites, and weeds, and to a degree of vertebrate pests, vectors of disease, and of dung accumulations in pastures. It has been successful, he notes ". . . on land and in water, from the tropics to cold temperate regions, in forests and ranges, and in cultivated crops and ornamental plantings."

In some situations where detrimental pesticides are not used for control of other pests on the crop, biological control alone has provided continuing, essentially permanent control. Examples of such classical biological control are commonly referred to as complete, substantial, or partial in degree of success. Laing and Hamai (1976) extended the data provided by DeBach (1964) and noted that for the insect pests and weeds combined there have been 385 successful attempts made in the various countries or island provinces of the world, with 327 of these being control of insect pests (102 complete, 144 substantial, and 81 partial) and 57 being weeds (13 complete, 26 substantial, and 18 partial). In terms of actual species of pests involved, these examples represent 157 insect species and 29 species of weeds, with 104 of the insect pests being completely or substantially controlled and 24 of the weeds being completely or substantially controlled.

An extensive literature exists on the theory and practice of biological control and its role in the balance of nature (see, e.g., DeBach, 1974; Huffaker & Messenger, 1976), the behavior and performance features of various natural enemies, including elaborate modeling of these relationships (see, e.g., reviews by Hassell, 1976; May, 1976; Huffaker et al., 1977). Space does not permit discussion of these features here. It must suffice here to say that we seek to obtain those natural enemies that exist, predominate, and have a high actual searching effectiveness in the field at very low host densities in their native environments which are climatically close to the target area. This will most nearly assure obtaining a natural enemy with the desired characteristics (Huffaker et al., 1977)—that is, one capable of and self-sufficient in regulating its host(s) population(s) at noneconomic densities, continuingly.

Lastly, various manipulative procedures have been investigated for increasing the effectiveness of natural enemies beyond what they can achieve in a self-sufficient manner unaided (above). This has been termed *augmentation* and *conservation*. Various methods are used in such augmentative and conservation practices, including for example, use of favorable cultural practices, strip cropping or cutting, addition of habitat resources, alternate hosts, refuges or subsidiary foods, or even the pest itself at critical times, and of course protection by use of modified spray programs, crop residue management, and so on (Ridgway & Vinson, 1977). This area has received far less emphasis than the program's director first envisioned. Part of this is because the granting agencies considered that the central focus should be placed in an entirely new innovative sphere of activities—the complete analysis and simulation description (modeling) of the respective ecosystems, with emphasis to be placed on gaining a better understanding of all the "resident" factors and their possible use as appropriate tactics, without substantial effort on importation of new natural enemies. Another factor is that the vast majority of the personnel had only limited interest and enthusiasm for classical biological control. Nevertheless, in the citrus, soybean, cotton, and the pome and stone fruit subprojects some effort was made in this area, with the one major success being the successful establishment of a form of *Aphytis lingnanensis* Compere from Hong Kong into Florida for control of snow scale, *Unaspis citri* (Comstock). This success has been so striking that about $8 million in savings to the citrus industry are credited to it each year. This alone is making it easier also to establish an integrated pest management program for citrus in Florida.

A considerable effort has been made in the soybean subproject to develop microbial agents as highly specific and environmentally sound controls for use in integrated pest management of soybean pests. Limited efforts were also made in the cotton and pome and stone fruit subprojects (Chapters 5, 9).

Moreover, in all the programs there was a solid effort to establish the impact of the principal parasites and predators that were active in the respective ecosystems and to make allowance for their actions in the development of integrated systems for crop pest management. Of real significance, it has been found that predators as a complex seem to be more important than the parasitoids in control of insect pests in such row crops as soybean and cotton.

These developments will be treated in more detail in the respective chapters, as relevant.

Cultural Control

Cultural control methods are among the most economical of all methods and can be widely applicable. Cultural and physical methods provided good control of many important insect pests for hundreds of years. Timing of planting and harvesting to escape infestation or infection or periods of intense exposure, rotating of crops or varieties, managing water, manuring, controlled burning, cultivation, sanitation (crop residue destruction, etc.), use of crop-free periods, and destruction of bordering or other host plants or volunteer plants are well known, widely practiced examples of cultural and physical control of insect pests and of some plant diseases. Their use interferes with pest development, commonly through exposure to rigors of the environment (another of nature's main controls but only indirectly manipulatable), but also in fostering natural enemy or antagonist action or plant resistance.

The development of synthetic broad-spectrum insecticides following World War II released growers from these rather unspectacular methods, and many of them were largely abandoned. In forests silvicultural methods are still widely used and their use appears to be most indicated by current research. These techniques seldom give complete control of a complex of agricultural pests, but they present much promise as supplementary components in integrated control systems. In certain situations they may serve as the principal means of control as is commonly so in the People's Republic of China (NAS, 1976).

Two of the major pests of cotton in Texas are the boll weevil and pink bollworm. Effective control of these species was developed by a series of largely cultural measures: destruction of cotton stalks soon after harvest and adoption of cultural practices preventing sufficient food to develop healthy diapausing individuals. A defoliant or desiccant is used at crop maturity plus an organophosphorus insecticide to facilitate early harvesting and to reduce the boll weevil populations. All stalk residues are shredded and plowed under. These measures provide about 90 percent control of these two key pests.

Use of an early planted small area trap crop of soybeans to attract the entering bean leaf beetles has been successful in preventing damage by this species and bean pod mottle virus which it transmits on the extensive larger acreage planted a little later in Louisiana.

Crop rotation was once used extensively to control corn rootworms in the corn belt, and is still a useful approach in an integrated pest management program, with perhaps soybean used as a valuable nitrogen fixing and high protein diet food crop in rotation with corn. [However, heavy use of the herbicide atrazine in corn has rendered substantial areas unfit for growing soybean in rotation. This suggests the hazards faced from single-discipline efforts and neglect of the long-view ecological holistic approach to pest control. So also does the accumulative residue of heavy metals in soils of apple orchards in the Northwest and citrus in Florida, for example, wherein crop plant physiology can be seriously affected (Brown & Jones, 1975).]

Various other features of cultural or silvicultural measures have been explored in the project, and these will be dealt with in the respective chapters.

Conventional Pesticides

For integrated insect pest management we seek minimal use of pesticides, applied so as to give an optimal effect in controlling the specific target pest population, while conserving natural enemies. Sometimes, perhaps usually, we do not need, or want, complete elimination of the target pest, and certainly not of all the pests and potential pests in the crop, for this would indirectly render the environment sterile for support of natural enemies, even if these natural enemies were not killed directly by the materials applied. This is why it is unjustified, indeed counterproductive, for EPA to set a fixed very high killing efficiency for a product to be registered. A highly specific, environmentally sound microbial material that gives a lower initial kill may be better than a conventional insecticide that kills 100 percent of a target population. Moreover, the microbial product's beneficial effects are extended in two ways: by its conservation of all nontarget organisms, including other natural enemies, and by its own multiplicative capacity after application such that endemic or epidemic mortality from the microbial pathogen itself will continue further into the season.

Of much significance is the question, "What can be done about outbreaks of secondary pests?" What we need is selective use of pesticides. All pesticides have some selectivity, but the degree of selectivity varies greatly. We have sought mainly materials of high toxicity to invertebrates and low toxicity to mammals. But many of our materials are simply biocides. We need differential toxicity within the insects, not necessarily species-specific specificity, but specificity for groups such as thrips, aphids, lepidopterous larvae, and so on. Gasser (1966) and Metcalf (1966, 1972) discussed the future possibilities for developing selective chemical insecticides. Gasser, however, thinks that industry will not make the necessary effort because of the great costs in developing a chemical which may not return much profit. This suggests government subsidization, or their development by government. We cannot afford to neglect this area, for insecticides remain our most reliable immediate solution to a problem at hand.

In the absence of physiological specificity, we are " making do " to a marked degree by obtaining ecological selectivity using physiologically nonselective materials in a special way: (1) by selective placement on the plant, (2) by specific timing, and (3) by careful dosage and formulation selection (Smith, 1970; Newsom, 1976). A number of materials used in such ways have had considerable utility in various programs.

Significant employment of conventional insecticides and acaricides in ecologically more selective ways than in common commercial usage has been explored in the program (Chapters 3, 5, 7, 9, 10) except in the pine bark beetle program, and in that program use of pheromones, both for monitoring and for possible direct control, is being investigated. The full ecological consequences of *trap out* or *confusion* techniques for control of bark beetles are as yet unknown (e.g., the effects on natural enemies or the development of resistance to trapping out or confusion).

Attractants

Insect repellents have been enormously useful in protecting people and domestic livestock from attacks of blood sucking insects, but no practical use of repellents to protect crops has been developed (Glass et al., 1975), so no effort was made in this program on repellents. Attractants, however, have long been used in insect control (e.g., in various poisoned baits). The possibilities of using attractants in baits or on trap crops for the purpose of concentrating pests where they will do little damage or can be destroyed, or for the purposes of attracting natural enemies into the habitat have received some attention.

Opportunities have been much expanded recently with discovery and characterization of several powerful insect attractants, notably the sex pheromones. This whole category of chemicals might be better referred to as *behavior modifying chemicals*. They are in general exceedingly powerful in their abilities to incite a behavioral response. Potential uses include (1) monitoring for insect presence and density so as to determine critical timing of events and (2) use as a direct control tactic.

Monitoring by the new chemicals being studied offers a great improvement in ascertaining the best time to apply an insecticide. By use of precise timing it is possible to reduce treatments for pests in apple orchards (Chapter 9). Studies on the use of California red scale female traps (natural pheromones released) in citrus trees in the San Joaquin Valley suggest that in such a case the damage potential from the pest can also be directly measured by monitoring and thus the need (or not) to make a treatment can be ascertained (Chapter 10). They are products normally present in the environment, though in very small amounts. The influence of such materials on the complex of natural enemies, however, and, as well, other possible unknown side-effects if to be used in large amounts, requires study.

These pheromones are not as strictly specific as originally believed (Kaae et al., 1972) but they do seem to have a very narrow range of specificity, although some of the natural enemies associated with their plant feeding hosts seem to have evolved behavior responsive to them even though they and their hosts are taxonomically unrelated. Available evidence suggests that, otherwise, they are environmentally safe. Newsom et al. (1976) noted:

Much additional research remains to be done before their promise can be realized. Answers must be obtained to such basic questions as number of applications required for control, most effective formulations, time of application and dosage rates, and mechanisms of mating disruption, i.e., confusion, habituation, arrestment, and repellency. Taschenberg et al. (1974) have called attention to the possibility that habituation of males to the chemicals, as well as confusion, may be a mechanism responsible for disruption. Finally, the important question of their possible carcinogenic and teratogenic activity must be resolved.

The excellent results obtained with these chemicals in large-scale field trials for control of such important pests as *Anthonomus grandis*, *Argyrotaenia velutinana*, and *P. gossypiella* is reason for optimism that synthetic pheromones will become important components of integrated pest management systems of the future. It should be emphasized, however, that these chemicals will be most useful in such systems, and not as the panacea that they were envisioned to become by some workers.

Among the attractants that have been used and are being studied are the following: methyl eugenol in baits for oriental fruit flies (*Dacus dorsalis* Hendel), protein hydrolyze for various insects, *Ri Bliere* for red-banded leaf roller [*Argyrotaenia velutinana* (Walker)], *Codlemone* for codling moth [*Laspeyresia pomonella* (L.)], *Grandlure* for cotton boll weevil (*Anthonomus grandis* Boheman), *Gossyphere* for pink bollworm [*Pectinophora gossypiella* (Saunders)] several attractants for bark beetles (Chapter 13), and so on (see especially Birch, 1974).

SUPPORTIVE TACTICS, INCLUDING BACKGROUND INFORMATION

In addition to the direct pest control tactics (above), a number of supportive tactics, including securement of the necessary information for modeling pest population dynamics and assessing pest damage, are required for making pest management decisions in keeping with the overall strategy, and in assessing alternative strategies and tactics.

Securement of the necessary information can best be obtained by monitoring or sampling the pest population and its principal natural enemies at appropriate times of the year. Systems science and modeling (Chapter 2) require rather accurate data on a variety of parameters. Equations or graphic guidelines must be developed to represent the interrelationships among the principal pest species, their natural enemies, soil and weather factors, the various alternative crop production procedures, including insecticide or other pesticide treatment, and economic factors bearing on the profit or loss situation, with due consideration for the hidden costs of control. Through monitoring and inclusion of short-term weather predictions (or average data of longer term historic basis for the locality), we must make predictions on the probable consequences of various management decisions (no action is implied). It will be seen in Chapter 5, for example, that by the use of short-term prediction (5–7 days) of probable losses from *Heliothis* in cotton in Arkansas, it was possible in 1976 to reduce the insecticide treatments of cotton from the conventional 10 to 2 for the season and in 1977 from 10 to 1.

Establishing economic thresholds and real need to use an insecticide, or to intervene in any way, is of foremost importance. This is the number one priority. The economic threshold, however, is not a fixed level, but a dynamic concept, the density level satisfying the concept depending upon a variety of circumstances which may vary markedly with the location and as the season progresses. We need to determine the relationships of various levels of economic damage to the size and stage of development of the crop and of each major pest species, for the complex of pest species, and the probability of alleviation by natural enemies or by the crop's own capacity to compensate for pest damage at a particular time. The project has made real progress in this area, but much more information is still needed, particularly with regard to the effects of the whole complex of pests, including also the plant pathogens, nematodes, and weeds.

Glass et al. (1975) noted:

The growing of a crop and the management of its pests are ecological endeavors. There are strong biological and ecological interactions among the various basic factors affecting: crop productivity (plant variety, soil type, fertility, water supply, temperature, wind, sunlight, etc.), the potential for damage by pests of the crop, the potential of the pests' natural enemies to circumvent such damage and the various practices used by man to control the pests. Because of these interactions, often involving strong compensations, the end results of various actions are difficult to predict. There is a great need to characterize, both quantitatively and qualitatively, the population dynamics and mobility of our major crop pests, the damage they cause and the various interactions among the many factors mentioned above.

Far too little attention has been given the crop and the means by which it produces its particular item(s) of yield. We must know the effects on yield of various environmental and agronomic factors, as well as of pests and the various pest control tactics, whether these operate alone or in a combination. Moreover, the level of acceptability of the cost of controlling a pest is dependent upon the market outlook. Thus, economic relationships must receive attention.

For both the pests and their natural enemies, much basic information is required. This, however, will vary greatly with each situation and type of pest. Long-term life table studies are perhaps generally prohibitive. But short-term population dynamics studies in which the causes and extent of mortality are tabulated in a life table will often prove useful relative to insect pests. Particularly essential is basic information on the potential of natural enemies to fully or adequately control various actual or potential pests in the absence of disruptive effects of agricultural practices (especially pesticides, but may also include, for example, the manner of crop refuse destruction). This information can be obtained by experimentation. These studies will require extensive, detailed and often frequent monitoring and sampling of populations. For insects, the influence of pesticides on natural enemy populations and on the pest species must be known. The methods of application as well as the kinds of materials used may greatly affect both the pest and the natural enemies. Much basic information is needed on the differential effects on pests and natural enemies that can be achieved in this way. More information on differential toxicity with respect to other groups of pests and their natural enemies is also needed.

PROMISE AND OBSTACLES

Pest control in the United States has reached a crossroads. It will now go backward or forward, depending upon whether or not government at all levels, pest control scientists, and private industry will accept and implement the outlined possibilities inherent in this new approach to pest control. This will require a new approach to research and to adoption or commercial use of new research findings and implementation technology on a broad scale. It also must include changes in the way research and research application programs are funded, how they are organized and managed, and how the priorities are set with respect to allocation of effort and funds to different areas of research and implementation testing and demonstration, and the communication of the benefits of such new programs, particularly to the farmer, and to the consumer, government officials, and other levels of society.

The IPM project which this book describes, was organized in an unusual manner and its funding appeared to be the result, in part at least, of an unusual series of concommitant events—scientific, social, and political—which resulted in establishment of the project on a sound financial and organizational basis such that the 18 universities involved could effectively pool their efforts to address the broad objectives of the project. It is hoped that similar efforts in the future will be favored by support.

There are perhaps three areas which stand out in importance in preventing, up to now, a much broader attack on the problems confronting us in achieving effective pest control, while preserving or improving benefits to society in other ways.

The first is that the chemical industry has for too long dominated the pest control scene, and this has resulted in an almost complete departure from some of the older more ecologically based methods of pest control. A virtual army of insecticide salesmen have in many parts of the country practically replaced the traditional dependence of the farmer upon his university, their researchers, and extension farm advisors for advice on pest control. There seems to be no way that this can be corrected unless the government, by law, removes the conflict of interest that is inherent in a chemical industry representative selling both *insecticides* and the *advice to use them*. This would simply be the bringing of toxic pesticides into the same pattern of dispensation as that of toxic medicinal drugs, wherein the person who advises their use is not the same person who sells them. Rather, advice is sold only by real professionals.

The second need is that the method of funding and managing research programs to develop improved pest control (i.e., integrated pest management) must allow for some changes. Existing routes of funding, through small individual research grants on small pieces of basic or applied research, or through the USDA which (as with ARS) either fails to bring in a broad-based vigorous input from the many competent research scientists in the universities, or when it does so (as with CSRS) the monies are channelled and filtered down to State Experiment Station Directors, Deans, Heads of Departments and Sections, and by "formula funding," such that the money getting to researchers is usually a little here and a little there. It is thus very difficult, if not impossible, to fund a major cropbelt-wide coordinated integrated pest management program, such as this IPM effort, with the strength required for success.

Thirdly, the management and research priorities of such programs must be revamped. At present, the management is automatically subject to the cross-currents, opposing viewpoints, and, yes, parochialism and personal interests or special background of the administrators of the various levels in the several universities usually involved in such coordinated programs as now exist. A program of appropriate scope and technical depth, centered on use of systems science and modeling as a means of setting research priorities, guiding the research, evaluation of direct results on pest abundance, and on the economic and social benefits to the farmer and society, requires a strong centralized management largely independent of the domination by these administrators. The large IPM program supported jointly by NSF and EPA, with USDA assistance, became possible because Government realized the need for such a centrally managed and block-funded effort. This program is described in this treatise. The specific results presented in subsequent chapters attest to the success that can be had when such programs are solidly established and strongly supported. These successes include (1) major reductions in use of insecticides (and acaricides) on some crops (30 to 50 percent and more), with increased profits and corresponding reductions in insecticide loads in the environment and reduced public health hazards; (2) more profitable systems of growing cotton that utilize less energy, less water, and less fertilizer; (3) biological control of a major citrus pest that threatened the whole Florida citrus industry; (4) measures for conserving biological controls in apples and other crops; (5) development of varieties of soybean and cotton offering resistance to a number of major insect pests; and, above all, (6) development of systems of organizing information, research, and advising of farmers on how to make decisions for managing their pest problems so as to optimize, on a long-term basis, returns to themselves in a way that recognizes long-term environmental and social values.

LITERATURE CITED

Baker, K. F., and R. J. Cook. 1974. *Biological Control of Plant Pathogens.* Freeman, San Francisco.

Beck, S. D., and F. G. Maxwell. 1976. Use of plant resistance, in C. B. Huffaker and P. S. Messenger (eds.), *Theory and Practice of Biological Control.* Academic, New York, pp. 615–636.

Birch, M. C. (ed.). 1974. Pheromones. *Frontiers of Biology,* Vol. 32. North-Holland, Netherlands.

Borlaug, N. E. 1965. Wheat, rust and people. *Phytopathology 55*: 1088–1098.

Brown, J. C., and W. E. Jones. 1975. Heavy metal toxicity in plants. I. A crisis in embryo. *Commun. Soil Sci. Plant Anal. 6*: 421–438.

Coons, G. H. 1953. Breeding for resistance to disease, in *Plant Diseases. USDA Yearbook,* Washington, D.C., pp. 174–192.

Day, P. R. 1972. Crop resistance to pests and pathogens, in *Pest Control Strategies for the Future.* National Academy of Sciences, Washington, D.C., pp. 257–271.

DeBach, P. (ed.) 1964. *Biological Control of Insect Pests and Weeds.* Reinhold, New York.

DeBach, P. 1971. The use of imported natural enemies in insect pest management ecology. *Proc. Tall Timbers Conf. Ecol. Anim. Control Habitat Manag. 3*: 211–233.

DeBach, P. 1974. *Biological Control by Natural Enemies.* Cambridge University Press.

Gallun, R. L., K. J. Starks, and W. D. Guthrie. 1975. Plant resistance to insects attacking cereals. *Annu. Rev. Entomol. 20*: 337–357.

Gasser, R. 1966. Use of pesticides in selective manners. *Proc. FAO Symp. Integrated Pest Control 2*: 109–113.

Glass, E. H. (coordinator). 1975. Integrated pest management: Rationale, potential, needs and implementation. E.S.A. Special Publication 75–2. Entomol. Soc. Am., College Park, MD.

Hagen, K. S., R. van den Bosch, and D. L. Dahlsten. 1971. The importance of naturally occurring biological control in the western United States, in C. B. Huffaker (ed.), *Biological Control.* Plenum, New York, pp. 253–293.

Hassell, Michael P. 1976. The dynamics of competition and predation. *Institute of Biology. Studies in Biology No. 72.* Arnold, London.

Horber, E. 1972. Plant resistance to insects. *Agric. Sci. Rev.,* U.S. Dept. Agric. *10.*

Huffaker, C. B. 1970. Biological control and a remodeled pest control technology. *Technol. Rev. 73*(8): 31–37.

Huffaker, C. B., R. F. Luck, and P. S. Messenger. 1976. The ecological basis of biological control. Proc. XV Intern. Congr. Entomol., Washington, D.C., Aug. 1976, pp. 560–586.

Huffaker, C. B., and P. S. Messenger (eds.). 1976. *Theory and Practice of Biological Control.* Academic, New York.

Huffaker, C. B., F. J. Simmonds, and J. E. Laing. 1976. The theoretical and empirical basis of biological control, in C. B. Huffaker and P. S. Messenger (eds.), *Biological Control.* Plenum, New York, pp. 41–78.

Huffaker, C. B., and R. F. Smith. 1976. The principles, strategies and tactics of pest population regulation and control in major crop ecosystems. Progress report, Int. Center for Integr. and Biol. Control, Univ. Calif., Berkeley and Riverside.

Huffaker, C. B., R. L. Rabb, and J. A. Logan. 1977. Some aspects of population dynamics relative to augmentation of natural enemy action, in Ridgway, R. L. and S. D. Vinson (eds.), *Biological Control of Insects by Augmentation of Natural Enemies*. Plenum, New York, pp. 3–38.

Kaae, R. S., J. R. McLaughlin, H. H. Shorey, and L. K. Gaston. 1972. Sex pheromones of Lepidoptera. XXXII. Disruption of intraspecific pheromone communication in various species of Lepidoptera by permeation of the air with Looplure or Hexalure. *Environ. Entomol. 1*:651–653.

Laing, J. E., and J. Hamai. 1976. Biological control of insect pests and weeds by imported parasites, predators and pathogens, in C. B. Huffaker and P. S. Messenger (eds), *Theory and Practice of Biological Control*. Academic, New York, pp. 685–743.

MacPhee, A. W., and C. R. MacLellan. 1971. Cases of naturally-occurring biological control in Canada, in C. B. Huffaker (ed.), *Biological Control*. Plenum, New York, pp. 312–328.

Maxwell, F. G., J. N. Jenkins, and W. L. Parrott. 1972. Resistance of plants to insects. *Advan. Agron. 24*:187–265.

May, R. M. (ed.). 1976. *Theoretical Ecology: Principles and Applications*. Saunders, Philadelphia.

Metcalf, R. L. 1966. Requirements for insecticides for the future. *Proc. FAO Symp. Integrated Pest Control 2*:115–133.

Metcalf, R. L. 1972. Development of selective and biodegradable pesticides, in *Pest Control Strategies for the Future*. Nat. Acad. Sci., Washington, D.C., pp. 137–156.

National Academy of Sciences. 1976. Insect control in the People's Republic of China. CSCPRC No. 2. Natl. Acad. Sci., Washington, D.C.

Newsom, L. D. 1970. The end of an era and future prospects for insect control. *Proc. Tall Timbers Conf. Ecol. Anim. Control Habitat Manage. 2*:117–136.

Newsom, L. D., Ray F. Smith, and W. H. Whitcomb. 1976. Selective pesticides and selective use of pesticides, in C. B. Huffaker and P. S. Messenger (eds.), *Theory and Practice of Biological Control*. Academic, New York, pp. 565–591.

Painter, R. H. 1951. *Insect Resistance in Crop Plants*. Macmillan, New York.

Rabb, R. L. 1971. Naturally-occurring biological control in the eastern United States, with particular reference to tobacco insects, in C. B. Huffaker (ed.), *Biological Control*. Plenum, New York, pp. 294–311.

Ridgway, R. L., and S. B. Vinson. (eds.) 1977. *Biological Control of Insects by Augmentation of Natural Enemies*. Plenum, New York.

Smith, Ray F. 1970. Pesticides—their use and limitations in pest management, in R. L. Rabb and F. E. Guthrie (eds.), *Concepts of Pest Management*. North Carolina State University, Raleigh, pp. 101–113.

Smith, Ray F. 1972. The impact of the green revolution on plant protection in tropical and subtropical areas. *Bull. Entomol. Soc. Am. 10*:7–14.

Smith, Ray F., C. B. Huffaker, P. L. Adkisson, and L. D. Newsom. 1974. Progress achieved in the implementation of integrated control projects in the USA and tropical countries. *EPPO Bull.* 4:221-239.

Stern, V..M., Ray F. Smith, R. van den Bosch, and K. S. Hagen. 1959. The integration of chemical and biological control of the spotted alfalfa aphid. *Hilgardia 29*:81-101.

Taschenberg, E. F., R. T. Carda, and W. L. Roelofs. 1974. Sex pheromones, mass trapping and mating disruption for control of red-banded leaf roller and grape berry moths in vineyards. *Environ. Entomol. 3*:239-242.

Zink, E. W., and J. E. Duffus. 1969. Relationship of turnip mosaic virus susceptibility and downy mildew (*Bremia lactucae*) resistance in lettuce. *J. Am. Soc. Hortic. Sci. 94*:403-407.

2

THE ROLE OF SYSTEMS ANALYSIS IN INTEGRATED PEST MANAGEMENT

Christine A. Shoemaker

Department of Environmental Engineering
Cornell University, Ithaca, New York

INTRODUCTION

The goal of integrated pest management is to make pest control more efficient by coordinating biological, chemical, and cultural means of pest control and by varying control practices with changing conditions in the field. In order to determine the most effective control program for a complex of pests, it is useful to view a crop field as an ecosystem of interacting components.

The major components of the biotic part of an agricultural ecosystem are the crop, its pests, and the beneficial organisms which suppress the pest populations or promote plant growth. The dynamics and interactions among these populations will be strongly influenced by the abiotic environment which includes solar radiation, temperature, humidity, and the availability of water and nutrients in the soil. Moreover, the cultural practices employed in growing the crop will have various interactions with biotic and abiotic factors.

Figure 2.1 is a schematic diagram describing a number of such crop management practices which either directly or indirectly affect pest populations and the damage they inflict. Applications of pesticides are frequently used to control weeds, plant pathogens, and insect pests. Some weeds, nematodes, and insects can be directly suppressed by cultivation or deep plowing. There are also a large number of other crop management practices which when applied to other parts of the ecosystem can have an indirect effect on pest damage. This is especially true of cultural practices such as variation in time of planting or harvesting. Variation in plant spacing can have a primary effect on the crop, but may also indirectly affect a pest by changing its environment. Different kinds of pests may be affected in different ways and degrees.

In assessing the desirability of management practices, it is important to assess their impact on plant growth as well as on pest damage. Such assessments will be influenced by the abiotic environment as well. For example, development of rapidly maturing varieties of cotton has been an important aspect of the developing integrated control programs in Texas since the cotton can be harvested before the cotton boll weevil populations reach their peak densities. However, the same program involving use of an early maturing type of cotton is not effective further east because of the difference in climate.

Some aspects of the abiotic environment such as rainfall and ambient temperature are very difficult to control. However, irrigation, fertilization, and cultivation are routinely used to modify other major aspects of the abiotic environment. Such management practices can enhance the plant's rate of growth and its ability to withstand pest damage. These practices also alter the microenvironment and thereby influence their rates of population growth and mortality. For example, irrigation often causes an increase in

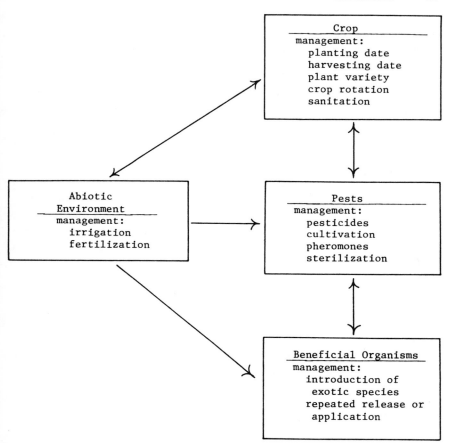

Figure 2.1. Schematic diagram showing the relationships among major components of a crop ecosystem. Management practices listed in each box have a direct effect on the ecosystem component noted. Indirect effects of management practices are noted by arrows between boxes.

humidity, which creates a more favorable climate for some plant pathogens and insect pests.

The advisability of using direct methods of pest control such as pesticides may also depend upon the dynamics of other parts of the ecosystem. Temperature, which affects the rate at which all of the populations are growing and the synchrony of their phases of development with one another, can influence the effectiveness of a pesticide application on beneficial species in preventing yield loss. There may also be an interaction among pest populations. For example, removal of weeds may also remove alternate habitat sites for plant pathogens. Pesticide treatments for one species may also either directly or indirectly affect other pest species or crop growth. Insecticides have destroyed populations of earthworms which are important in maintaining soil porosity. Soil fumigation has been known to retard crop growth by killing soil microorganisms involved in the nitrogen cycle. Outbreaks of insect pest populations have been caused by the damaging effect of insecticides on predators. Insecticide applications can also exacerbate weed problems. For example, the application of insecticides for alfalfa weevil control disrupts insects which feed upon weeds in alfalfa. Modifications in the abiotic environment or in the habitat through changes in cultural practices can also affect beneficial organisms as well as pest species.

Efficient management of a crop production system blends an understanding of the effects of management decisions on biological processes with an analysis of their effects on economic returns. The net return depends upon the size of the harvest, the market price of the crop, and the costs of crop management. The basic object of pest management for individual growers is to increase net economic return, which is not necessarily equivalent to maximizing yields or minimizing pest densities. Pest control actions are economically justified only when the cost of a pest control action is less than the value of the crop damage which is prevented.

In an attempt to develop control policies which are based upon both economic and biological processes and which respond to changing field conditions, entomologists have developed the concept of an economic injury level (Stern et al., 1959). It is the population density P such that

$$\text{cost of pest control} = \text{price} \times \text{yield loss } (P) \tag{1}$$

where yield loss (P) is the amount of crop loss which results from a pest infestation of density P. Equation (1) is used to establish the conditions for which pest control actions are advisable. Growers relying upon use of economic injury levels implement a control practice (usually a pesticide application) only when the density of the pest infestation threatens to exceed this threshold. As a result, the number of pesticide treatments required and the cost of pest control has been substantially reduced for some crops. For example, a recent study of cotton insect control estimated that the implementation of scouting programs, which utilize pest population monitoring by professional insect scouts, would reduce insect control costs by $26 million on 10.6 million acres of cotton throughout the nation (Pimentel et al., 1978).

From Equation (1) it is clear that in order to determine the economic threshold it is necessary to understand the relationship between yield loss and pest density (P). In the past, efforts to establish this relationship have been based upon empirical studies in which fields were artificially infested with varying densities of pests and the resulting yields measured. An example of this approach is a study by Koehler & Rosenthal (1975) which estimated economic thresholds for the Egyptian alfalfa weevil (*Hypera brunneipennis* Boheman).

As originally defined (Stern et al., 1959), the economic threshold depends only upon the density of the pest infestation. However, it is clear that the potential for damage depends upon weather, the vigor and maturity of the crop, the time of pest infestations, the age structure of the pest populations, the size of the beneficial population, and other factors. As a result, the economic threshold should depend upon all the major factors influencing the advisability of a control action.

It is clear why all of these factors have not been considered in the empirical development of economic thresholds. Obtaining statistically significant estimates of a one-dimensional economic threshold is quite difficult and expensive. To establish an economic threshold empirically as a function of several factors would require an extremely large number of field trials— many more than is economically feasible to make.

It is for this reason that there has developed a considerable interest in the use of mathematical models in pest management. The goal of these models is to aid in understanding the way in which the many factors such as weather, population densities, and age structure interact to have an effect on yield and on the success of biological, cultural, and chemical means of pest control. It is hoped that by describing mathematically the way in which pairs or groups of components of the ecosystem respond to one another, that insight can be gained about the behavior of the system as a whole. Such insight can then be used to guide decisions regarding the combinations and timing of pest control practices to be used.

MODELING INSECT POPULATION DYNAMICS—A REVIEW OF EARLY LITERATURE

Attempts have been made for the last half century to use mathematical models to describe various aspects of insect population ecology. Early work in this area includes classic predator–prey and host–parasite models by Lotka (1925), Nicholson & Bailey (1935), Leslie (1945), and Thompson (1939). Most of these models are based upon sets of equations whose solutions can be represented by an algebraic equation.

Unfortunately, it is rare that a closed form solution can be found to a set of equations which incorporate the number of variables, the nonlinearities, and the discontinuities necessary to describe the dynamics of an actual crop system. However, the possibilities for describing the detailed interactions among populations have been greatly enhanced by the development of electronic computers which can numerically calculate solutions to series of equations that are much too complex to yield closed form solutions. Watt (1964) was one of the first to utilize this new possibility to develop a mathematical model to analyze pest management decisions in an actual ecosystem. Watt's model of the spruce budworm, *Choristoneura fumiferana* (Clemens), in New Brunswick, Canada, divided the 6.4 million acre area he was studying into 625 grid squares. He developed a series of equations which attempted to describe the effects of applications of insecticide and a virus and the release of parasites on the spatial distribution of the spruce budworm and the tree damage caused by the infestation. The parameter values in Watt's model were based on data reported in the literature, whenever such data were available. His simulation model calculated the dynamics of the ecosystem as a function of weather and the response of the populations to each other.

Because Watt was not using his model to determine the time within a year that a practice should be implemented, he did not describe the details of the intraseasonal population dynamics of the spruce budworm and related species. However, in field and orchard crops the timing of a control action can be very important. As a result, models currently being developed for management of pests of field crops and fruit trees contain more detailed descriptions of the phenology of pest and related species. Since feeding rates and susceptibility to control are age-specific, it is helpful to use a model to estimate the changes in age distribution as well as in the size of the pest population throughout the growing season.

Age class models of population dynamics have been in existence for over thirty years (e.g., Leslie, 1945), but the incorporation of temperature into such models is much more recent. Hughes and Gilbert (1968) utilized temperature in a model they developed of cabbage aphid populations. The calculations were based on physiologic time, which is a concept first used by Shelford (1927) although the term physiologic time originated with Hughes (1963). This concept has proven quite useful because it simplifies the calculations of development by converting the calculations into a time scale based upon temperature rather than on calendar days. Physiologic time has been widely used in later work (Gutierrez et al., 1976; Ruesink, 1976a; Shoemaker, 1977; Chapters 6, 8) in which it has been more precisely designed as

$$S(x) = \int_0^x R(T(t)) \, dt$$

where

$S(x)$ = physiologic time

x, t = calendar time

R = rate of development as a function of temperature T

STRUCTURE OF DYNAMIC MODELS OF CROP ECOSYSTEMS

Pest control decisions depend not only upon the dynamics of the insect populations but also upon the vigor and maturity of the crop plant population. A plant may be more susceptible to attack during certain stages of development. A vigorous plant may be able to compensate for pest damage which would cause significant yield losses in a less vigorous plant. Weather can influence the periods of time when plants are susceptible to damage and the variations in yields due to cultural pest control practices such as changes in planting or harvesting dates.

As a result, attempts have been made to develop models which describe in detail the effect of the abiotic environment on the dynamics of plant and insect populations and on their interactions with one another. The structure of such models is based upon a qualitative understanding of the factors

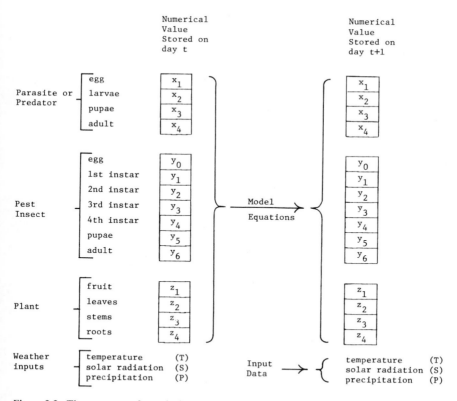

Figure 2.2. The structure of a typical pest management simulation model.

affecting growth, the age-specific feeding behavior of each species, and the effect of pest control methods on pests, natural enemies, and on the crop.

This qualitative knowledge of the structure of the crop ecosystem influences the choice of which aspects of the ecosystem should be described in detail and which can be aggregated. The structure of a typical crop ecosystem model is illustrated in Figure 2.2. In order to predict the effects of weather and pest control measures on yields, it is frequently deemed necessary to examine the dynamics of natural enemy populations, as well as that of the plant and pest populations. Typically each population is divided into a number of categories. Insect populations are categorized into age classes. Plant populations are described by variables which predict the weight of fruit, leaves, stems, and roots. If appropriate, additional categories may be added. For example, insects may be categorized as parasitized or not parasitized, as well as by age.

In the last several years, a number of papers have been written about the use of mathematical models for analyzing pest management systems. An extensive survey of the recent literature in this area has been written by Ruesink (1976b). Therefore, in this chapter we will not attempt to catalog all the studies in this area. Instead we will review and critique in the following sections the alternative mathematical structures used in the analysis of crop ecosystems. Literature references cited are given as examples of certain mathematical approaches and are not intended to represent all the significant work in this area.

MATHEMATICAL MODELS OF POIKILOTHERM POPULATIONS I—FIXED RATE OF DEVELOPMENT

The two columns in Figure 2.2 illustrate the method by which the equations in a simulation model use the population densities and plant biomass from one day (day t) to calculate the population densities and plant biomass in the following day (day $t + 1$). Each of these calculations is based upon equations which describe the growth and death processes affecting the organisms. For insects, the number present each day in each category is the number present the previous day minus those that die or mature out of the class plus those that mature into the class. The most common way of describing this is

$$y_{i+I(T)}(t + 1) = y_i(t)S_i(t) + m_{i+I(T)}(t + 1) \qquad (2)$$

where

$y_i(t) = $ the number of individuals in age class i at time t

$S_i(t) = $ the fraction of individuals in age class i which survive to mature into age class $i + I(T)$

$m_i(t)$ = the number of individuals of age class i which migrate into the eco-system

$I(T)$ = the number of age classes an individual will pass through in one period of time—i.e., from time t to time $t + 1$, given that the temperature during this time is T

The survivorship term S_i is the fraction which survive all causes of death. It can be represented as the product of the fractions surviving starvation, predation, pest control, and extremes of weather. Thus,

$$S_i = \alpha_i(x) \cdot F_i(z) \cdot W_i(T, P) \cdot \mu_i(v) \tag{3}$$

where

$\alpha_i(x)$ = the fraction of age class i which survives predation given that there are x predators searching

$F_i(z)$ = fraction of age class i which obtains adequate nutrition from the host plants to survive

$W_i(T, P)$ = fraction of age class i which survives weather-related causes of mortality given that the temperature pattern is T and precipitation P

$\mu_i(v)$ = fraction of age class i which survives the effects of management decision v

The number of newly oviposited individuals depends upon the rates of oviposition of the members of the reproductive age classes. Thus,

$$y_0(t + 1) = \sum_{i=1}^{n} f_i(t) y_i(t) \tag{4}$$

where $\sum$ means the sum and

$f_i(t)$ = the average number of viable offspring produced in one time period by individuals in age class i

The function f_i may depend upon several variables such as temperature and the availability of oviposition sites.

If time t is expressed in terms of calendar time, the amount of maturation which occurs from time t to time $t + 1$ will vary because the population is poikilotherm. As a result, it is impossible to choose units of t such that $I[T(t)]$ will always be an integer number. To avoid these difficulties, it is much more accurate to express t in terms of physiologic time. Thus, each increment of t represents an equal accumulation of physiologic units. The units of t are chosen so that one unit of t is the age span of one age class. For example, if physiologic development can be related to degree days above a threshold of

$10°C$, t will be in units of degree days above $10°C$. Assume that it takes 200 degree days for an individual to mature from egg to adult. Then if the population is to be modeled by 10 age classes, the physiologic age in terms of degree days will range 20 $DD_{10°C}$ among individuals in each age class. Thus, if A_i is the age in degree days when individuals enter age class i, then

$$A_i = i \cdot 20 \; DD_{10°C} \quad i = 0.19$$

If A_i and t are expressed in units of physiologic time, the basic Equation (2) can be replaced by

$$Y_{i+1}(t + 1) = y_i(t)S_i(t) + m_{i+1}(t + 1) \tag{5}$$

We can now combine Equation (5) with Equation (4) into the vector equation

$$\mathbf{y}(t + 1) = L(t)\mathbf{y}(t) + \mathbf{m}(t + 1) \tag{6}$$

where $\mathbf{y}(t)$ and $\mathbf{m}(t)$ are the vectors $(y_1, \ldots, y_N)$ and $(m_1, \ldots, m_N)$, respectively. The matrix L is defined by

$$L = \begin{bmatrix} f_0 & f_1 & f_2 & \cdots & f_N \\ S_0 & 0 & 0 & \cdots & 0 \\ 0 & S_1 & 0 & \cdots & \\ 0 & 0 & S_2 & \cdots & 0 \\ 0 & 0 & \cdots & S_{N-1} & 0 \end{bmatrix} \tag{7}$$

Readers familiar with the literature in mathematical population ecology will recognize Equation (6) as similar to a Leslie matrix equation with migration. However, the fact that elements of the matrix L may depend upon some of the y_i's can make the mathematical analysis somewhat more difficult; but numerical calculations performed by computer can compute the value of $\mathbf{y}(t)$ even when L is dependent upon t and y.

There are several other ways which the pest population dynamics may be described. In a model by Gutierrez et al. (1976) the population is categorized not by maturity, but rather by time of oviposition. Each day the number of survivors in each class and their physiological maturity is calculated by the following equation:

$$X(j, t) = X(j, t - 1) \cdot S[A(j, t), t] \tag{8}$$

$$A(j, t) = A(j, t - 1) + P[T(t)] \tag{9}$$

where

$X(j, t)$ = number of individuals alive at t which were oviposited at time j

$A(j, t)$ = maturity at time t of individuals oviposted at time j

$P[T(t)]$ = amount of maturation occurring in one time period if the temperature is T at time t

$S(A, t)$ = survival of individuals of maturity A at time t

As in Equation (3), S depends upon temperature, pest control and plant and predator populations.

It is also possible to represent populations by a differential equation in which physiologic age is represented by a continuous variable a. If the only effect on the insect population is an age dependent immortality factor, the rate of change of size of each cohort can be represented by the equation

$$\frac{dN_b}{dt} = -D(t - b)N_b(t) \tag{10}$$

where

$N_b(t)$ = the number of individuals surviving to physiological time t in a cohort oviposited at physiologic time b

$D(a)$ = mortality rate for individuals of age a

Since t and b are expressed in units of physiologic time, the physiologic age is simply computed as the difference between t and b. The solution to Equation (10) is

$$N_b(t) = N_b(b)e^{-K(t-b)} \tag{11}$$

where

$$K(a) = \int_0^a D(a)\, dt$$

The function $N_b(b)$ is the number of eggs laid at time b. This model as well as the Leslie matrix model discussed earlier both assume that all individuals mature at the same rate for a given temperature. Thus, the number of individuals in the ith age class at time t is

$$y_i(t) = \int_{A_i}^{A_{i+1}} N_{t-a}(t - a)e^{-K(a)}da \tag{12}$$

where A_i is the minimum number of accumulated physiologic heat units necessary to enter the ith age class. Shoemaker (1977) (see Chapter 8) used a somewhat more complex form of Equation (12) to describe the interaction between alfalfa weevil, *Hypera postica* (Gyllenhal), and a parasite, *Bathyplectes curculionis* (Thomson).

MATHEMATICAL MODELS OF POIKILOTHERM POPULATIONS II— VARIABLE RATES OF DEVELOPMENT

In the models discussed above, it is assumed that all individuals of the same age will mature at the same rate for a given rate of temperature. However, in fact, many populations display a significant variability in developmental rates. This variability is probably primarily due to genetic differences but may also be caused by environmental heterogenity and other factors. Several attempts have been made to incorporate developmental variability into insect population models. A discrete model similar to Equation (2) has been developed by Ruesink [(1976a) & Chapter 8] to describe the dynamics of alfalfa weevil. The model is based on calendar, not physiologic time, so that $I(T)$ is not an integer. The age classes are so finely divided that under some weather conditions it is possible for individuals to advance more than one age class in one day. As a result, Ruesink apportions individuals between adjacent cells in accordance with the value of $I(T)$. Thus, if $I(T) = 2\frac{2}{3}$, one third of age class i is moved into age class $i + 2$ and two thirds of age class i is moved into age class $i + 3$. The result of this model is that individuals oviposited at the same time will not all reach maturity simultaneously. Thus, developmental variability is incorporated. However, the amount of variability is not quantitatively specified; it is a side-effect of the number of age classes used. The fewer the number of age classes, the greater is the variability. An additional disadvantage of this method is that the amount of variability is increased during cool weather.

A model which attempts to incorporate variability directly was developed by Barr et al. (1973) to describe the dynamics of the cereal leaf beetle. The dynamics of the population are described by a forward Kolmogorov equation of the form:

$$\frac{\partial y(z, t)}{\partial t} = \frac{1}{2} \frac{\partial^2 [v(z, t)y(z, t)]}{\partial z^2} - \frac{\partial [r(z, t)y(z, t)]}{\partial z} - D(z, t)y(z, t) \quad (13)$$

where $y(z, t)$ is the expected number of individuals of maturity z at time t, $D(z, t)$ is the mortality rate, and $r(z, t)$ and $v(z, t)$ are the mean and variance, respectively, of the rate of development.

However, the Barr et al. model and the Ruesink model both treat variability in development as if it were independent of the individual's history. In fact, it is much more reasonable to assume that each individual has a given propensity for speed of development as a function of its genetic characteristics and perhaps of the nutritional history of its mother. Thus, an individual which is a slow developer early in life is much more likely to be a slow developer late in life than an individual which was a fast developer.

Smith and Shoemaker (1978) developed a model which does include this phenomenon. It is based on the assumption that the rate of development is randomly distributed throughout the egg population and that each individual maintains its propensity for fast (or slow) growth throughout its development. The basic equation of this model is

$$N_k(p) = \int_0^p G(k, p_0, p)\beta(p_0)\, dp_0 \tag{14}$$

where

$N_k(p)$ = the number of individuals in age class k at time p

$G(k, p_0, p)$ = the fraction of eggs which are oviposited at time p_0 which are in age class k at time p

$\beta(p_0)$ = the number of eggs oviposited at time p_0

The function G is calculated from the probability distribution of a random variable describing maturity. Details are given in Smith and Shoemaker (1978).

PLANT MODELS

The fundamental equations in models of plant growth describe photosynthate production and the translocation of materials to various parts of the plant. As in Figure 2.2, the plant population is usually divided into fruits, leaves, stems, and roots. Additional categories may also be necessary. For example, Fick (1975) included crown buds because of the significance of alfalfa weevil feeding on this part of the plant upon alfalfa regrowth. The simplest models describe the manufacture of photosynthate and its allocation to various parts of the plant. More detailed models describe the uptake and movement of water and nutrients as well.

The choice of which additional factors need to be included depends very much upon the situation being described. For example, if rate of attack by a pest depends upon the age of the plant part being attacked, it may be more important to describe the age structure of plant parts than in cases where attack rates are independent of age (Gutierrez et al., 1975). In a model of pink bollworm, *Pectinophora gossypiella* (Saunders), in cotton the age of the fruit affects both the rate of attack by pink bollworm larvae and the rate at which larvae mature when they are inside the fruit (Gutierrez et al., 1977).

The effect of the pest insect on the plant is usually described by subtracting the amount of plant material destroyed by insects each day from the appropriate category. For example, the effect of leaf-feeding insects will be to reduce the leaf area and to cause losses due to wounds and healing. The amount of leaf area removed and wound-healing losses can be predicted from an equation which estimates the number of insects present. If d_i is the amount of damage caused by each individual of age i, then the total damage D is commonly estimated from the equation

$$D(t) = [d_0 y_0(t) + d_1 y_1(t) + \cdots + d_N y_N(t)] \tag{15}$$

where $y_i(t)$ is the number of individuals in age class i at time t, and d_i is the damage caused by each individual. By subtracting $D(t)$ from the total leaf area, the instantaneous effect of feeding on photosynthate production is described.

INPUT PARAMETERS AND SENSITIVITY ANALYSIS

The structure of the model chosen to describe the dynamics of a crop ecosystem is based upon a qualitative understanding of the system including a knowledge of which factors are likely to be most important and which thus justify detailed description. Alternative forms of this structure have been discussed in the previous sections.

After developing the qualitative structure of a model, the next step is to quantify the model by specifying the numerical values of the parameters and functions used in the model's equations. This includes the parameter values which specify the functions describing death and development rates as well as weather data, pest control practices, initial population sizes, and emergence and migration rates. Since these values must be specified (e.g., they must be "put into" a model) before numerical results can be calculated, they are often referred to as *input parameters* or *input data*.

An example of an input parameter is the function $\mu_i(v)$, the survival rate from pest control measures. Estimation of $\mu_i(v)$ is usually based upon population counts taken in the field before and after the implementation of control. For insecticides, it may be necessary to take several counts after the treatment in order to determine the rate at which the toxicity of the chemical is decaying. Thus, $\mu_i(v)$ depends upon the number of days that have elapsed since the last insecticide treatments.

Typically, one of the most difficult aspects of crop ecosystem dynamics to measure is the activity of natural enemies. The parameter α_i, the rate of predation or parasitization, is sometimes measured by studies of the behavior of predators in cages. A known number of predators are released into a cage with a known number of prey in an environment approximating their natural habitats. At appropriate time intervals the number of prey attacked is counted. An alternative method is to study the effectiveness of the predator or parasite by comparing the pest population growth in the field in the presence of natural enemies to that which occurs when they are removed by a selective insecticide or a physical barrier such as netting.

It is by varying the values of the input parameters and functions and by observing the effect on the model's results, that the benefits of developing a model are obtained. In this way the model is used to understand the ways in which variations in weather, population densities, and management tactics affect various aspects of the dynamics of the crop ecosystem. By varying the input information, one can obtain estimates of the relative effectiveness of different pest control measures in preventing yield losses. Similarly, the effects of changes in weather or migration rates can be estimated. From the results of such simulations, a function

$$Y(x_0, y_0, z_0, w, v) \tag{16}$$

can be estimated where Y is the yield, w is the weather, v is the set of pest control measures, and x_0, y_0, and z_0 are initial sizes of the populations. In many cases it is also important to estimate the number of pest insects and natural enemies which are left at the end of the season and which influence the size of subsequent generations. These relationships can be denoted by

$$x_0(t + 1) = G_x[x_0(t), y_0(t), z_0(t), v(t)] \tag{17}$$

$$y_0(t + 1) = G_y[x_0(t), y_0(t), z(t), v(t)] \tag{18}$$

$$z_0(t + 1) = G_z[x_0(t), y_0(t), z_0(t), v(t)] \tag{19}$$

Varying the values of input parameters is also a useful technique to determine the sensitivity of the model to different parameter values. Usually there are some parameters such that a small change in their values will cause a significant change in the model's results. When this is the case, the model is said to be *sensitive* to the parameters. Usually there are also some parameters which can change value without significantly affecting the model results. The process of testing the model to see which are the sensitive parameters is called *sensitivity analysis*. It is a very useful procedure because it helps identify which parameters must be measured with a high degree of precision and which do not require such exactness. Given the considerable amount of effort involved in obtaining many of the parameter values, it is very helpful to have some guidance in where best to allocate limited time and effort obtaining additional data to improve the estimates of parameter values.

CALIBRATION AND VALIDATION

The agreement between field observations and model predictions can be improved by modifying the values of some of the model's parameters. Ideally the parameter values are determined independently and the model results are in reasonably close agreement with field observations. Unfortunately, such agreement is rarely obtained. This is not surprising since many of the parameter values are based on laboratory studies conducted under controlled conditions, which differ significantly from field conditions. In addition, there is usually some variance associated with the parameter values measured in these experiments.

Therefore, to increase the predictive ability of a model some of the parameter values may be changed slightly to improve the agreement between model predictions and observed value. This process is called *calibration*. It is a reasonable procedure as long as the modified values of the parameters do not vary too greatly from their original values. The measure of how much is too much depends upon how closely the conditions under which the parameter value was measured agree with field conditions and upon the variance associated with the measurements.

If field data have been used to calibrate a model, it cannot be used again to prove the model's predictive power. If enough parameter values are modified, even a poor model can be made to agree closely with one set of field data. However, a model's predictive power is based upon its ability to simulate the dynamics of actual field populations for a range of conditions including changes in weather, in migration rates, and in pest control practices. Therefore, to test the validity of a model, its results should be compared to field observations made under a variety of different conditions. The field data used for *validation* should be different from that used for calibration. Typically one year's field data are used for calibration and the following years' data are used for validation.

If the model does not appear to provide predictions with a reasonable degree of accuracy, the model may not be valid because one or more fundamental aspects of the ecosystem's dynamics have been omitted. For example, the action of an important class of predators may not have been included. Such errors can be corrected by making changes in the structure of the model. Since a model is originally constructed on the basis of what is *thought* to be important, the discovery of such a structural error can give very useful information about the relative importance of different elements of the ecosystem.

INCORPORATION OF ECONOMIC CONSIDERATIONS

In deciding which pest control methods to use, the expense as well as the effectiveness of the methods must be considered. In order to compare effectiveness with expense, we must convert both into like units. The most convenient units are monetary. In this case, effectiveness is measured in terms of the monetary value of the harvested yield. The total return in 1 year is then price × yield − costs. Costs include pest control costs, fixed costs which are independent of pest control or yield, and finally costs which depend upon the size of the yield. Examples of the latter include harvesting and transportation costs. Thus, the net return in 1 year is

$$R = p \cdot Y(x_0, y_0, z_0, w, v) - C_p(v) - C_y[Y(x_0, y_0, z_0, w, v)] - C_F \quad (20)$$

where

$$p = \text{price}$$
$$Y(x_0, y_0, z_0, w, v) = \text{yield as a function of pest control practices } v \text{ and input parameters } x_0, y_0, z_0, \text{ and } w$$
$$C_p(v) = \text{cost of pest control practices } v$$
$$C_F = \text{fixed costs}$$
$$C_y(Y) = \text{yield-dependent costs}$$

The term C_F can be dropped from Equation (20) because it is a constant and thus does not affect the choice of v which maximizes Equation (20).

In many situations the pest control measures taken in 1 year affect the size of pest populations in the following years as well as in the current year. In such cases, the choice of pest control methods should be based upon long-term returns rather than on the economic returns from a single year. The term over N years can be represented as

$$R_N = \sum_{t=1}^{N} \beta_t [p_t'(Y)Y(x_0^t, y_0^t, z_0^t, w^t, v^t) - C_p(v^t)] \quad (21)$$

where $p' = p - C_Y$, $\sum$ means the sum, and β_t is a discount factor which converts money earned in year t into its value in year 1. In addition, it is necessary to have an equation which relates population size in 1 year to populations, weather, and pest control methods in the previous year. These relationships are given in Equations 17 to 19 and can be calculated by simulation.

In many cases farmers are more concerned about avoiding a disaster than with maximizing average returns. In this case a stochastic model of yield is more appropriate than the deterministic one discussed in the equations above. The objective would be to maximize a weighted sum of the possible outcomes. The weights would be the probability of occurrence multiplied by a measure of the farmer's preference for the outcome. Static forms of such models are discussed in Norton (1976).

USE OF SIMULATION AND OPTIMIZATION MODELS
TO DETERMINE ECONOMICAL MANAGEMENT PRACTICES

In order to determine the pest control program which maximizes returns, we can substitute the pest control program v into the simulation model to calculate the model estimates of yield and the size of overwintering populations (Y, G_x, G_y, and G_z). The values of p' and the cost function $C_p(v)$ can usually be determined directly from information on costs of material and labor and from published market prices. By coupling this information with the simulation models estimates of Y, G_x, G_y, and G_z, the value of the return function [Equation (2)] can be evaluated for each value of v. These values can be examined to determine the v which is most profitable.

The difficulty with this approach is that the simulation model must be calculated for each value of v (the pest control option) in order to estimate the size of the yields and overwintering populations. As mentioned earlier, in many cases there may be thousands of possible combinations of pest control options. For example, determining whether or not to apply an insecticide over 12 time intervals results in $2^{12} = 4096$ different possible insecticide treatment schedules. Given that each simulation may cost several dollars in computer time, a trial-and-error approach is not feasible for examining a large number of pest control options.

In order to avoid this difficulty, attempts have been made to use optimization methods in place of simulation models. Examples of applications of these methods to pest management are discussed in Chapter 8 and in Regev et al. (1976), Taylor (1976), and Shoemaker (1977, 1973). Optimization methods select the best policy without calculating Y, G_x, G_y, and G_z for each possible value of v. These methods are numerical algorithms which are based upon an analysis of the mathematical structure of the model describing the ecosystem and economic returns.

As a result, optimization techniques are most feasible with models which are mathematically tractable. Models with few variables and those with a special form such as linear equations are most likely to have the necessary mathematical characteristics. In some cases, as in Shoemaker (1977), it has been possible to develop special optimization algorithms which are able to handle a nonlinear problem in population management of a fairly high dimension. Unfortunately, it is not possible to develop such algorithms for all problems. Therefore, in order to utilize an optimization method it may be necessary to simplify the model describing the crop ecosystem by reducing the number of variables or by removing some nonlinearities in the equations describing interactions.

Thus, we see that simulation and optimization models each have their advantages and disadvantages. Simulation models can include a large number of variables to give a detailed description of an ecosystem, but are very inefficient at evaluating a large number of management alternatives. Optimization methods, on the other hand, can efficiently evaluate a large number of pest management options but only for systems which can be described by relatively few variables or by a special form of equations.

Unfortunately, most pest management problems require a large number of variables to describe the interactions in the system and also have a large number of possible options. In these cases, the best procedure is to first develop a simulation model which describes the system and which has reasonably good predictive power. The second step is to develop a somewhat simpler version of the simulation model. The simpler version is used in an optimization model. The policies calculated to be optimal can then be substituted into the full-scale simulation model to see if they appear to be effective, economical policies when tested against a more detailed description of the system. The final test, of course, is implementation in the field.

STATIC MODELS TO DETERMINE PEST MANAGEMENT PRACTICES

The models discussed previously have described the dynamics of the ecosystem. It can also be useful to analyze pest management alternatives by examining the ecosystem from a static point of view. Such models have the advantage that because they do not attempt to describe the details of population growth and ecological interactions, they are much simpler and have fewer data requirements.

A static model has been developed by Ruesink (1975) to establish an economic threshold for green cloverworm, *Plathypena scabra* (F.). His approach has been to establish empirical relationships between yield losses and pest population density at each stage in the life cycle of the plant. He then calculates the pest density for which the cost of yield losses equals the cost of an insecticide treatment. This density is the economic threshold. Since the relationship between yield and pest density depends upon the maturity of the soybean crop, the economic threshold also depends upon

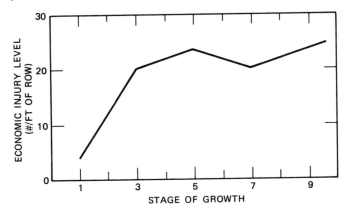

Figure 2.3. Dynamic economic threshold for green cloverworm on soybeans. [From Ruesink, 1975.]

soybean maturity. A graph describing the economic threshold as a function of soybean maturity is given in Figure 2.3.

A static model has also been used by Dover et al., 1979 (see Chapter 9) to analyze the management of the European red mite, a pest of apples, by the predatory mite, *Amblyseius fallacis* (Garman). The goal of this work is to establish what could be described as a two-dimensional economic threshold, which depends upon predator as well as pest density. If the ratio of prey to predator is sufficiently high, insecticide treatments are recommended. Conversely, if the ratio is sufficiently low, insecticide treatments are recommended. If it is between these two levels, a more detailed analysis would be necessary and perhaps additional samples.

A static model has also been used to analyze the integration of cultural, chemical, and biological control of a complex of olive pests (Shoemaker et al., 1979). The pests considered are olive parlatoria scale, *Parlatoria oleae* Colvée, black scale, *Saissetia loeae* Bern., and oleander scale, *Aspidiotus neri* Bouché. It is desirable to analyze the management of the whole complex of pests because in many cases the controls applied against one pest may influence the dynamics of other pests. Insecticides applied for olive parlatoria scale will control oleander scale and black scale as well. Insecticides applied for black scale can upset the biological control of olive parlatoria scale which is provided by two parasites. Frequent pruning, the most important cultural means of control, suppresses black scale, oleander scale, and olive knot infestations. Unfortunately, pruning requires considerable labor and is therefore relatively expensive. It was not clear whether the pest management benefits of frequent pruning outweigh the increase in cost.

The static analysis of the olive pest complex is based upon an equation which estimates the average expenditure for pruning and pest management costs:

$$A = \frac{C_p}{F_p} + \frac{C_{os}}{F_{os}} + \frac{C_{bs}}{F_{bs}} + \frac{C_{is}}{F_{is}} - \frac{C_{bs}}{F_{is}F_{bs}}$$

(22)

where

C_{os}, C_{bs}, C_{is} = cost of pesticide used to control olive parlatoria scale, black scale, and ivy scale, respectively

F_{os}, F_{bs}, F_{is} = frequency of outbreak (i.e., the average number of years between infestations of olive scale, black scale, olive knot, and ivy scale, respectively, which require pesticide treatment)

C_p, F_p = cost and frequency, respectively, of pruning

The last term in Equation (22) is included because a single insecticide application will control both oleander scale and black scale.

Each of the frequencies of pest outbreak depends upon the tree variety, the frequency of pruning, and the presence of biological control. Therefore, the average cost is calculated as a function of each of these three variables. The results indicate that the lowest crop management costs for all varieties were obtained when trees are pruned every 2 years and biological control is relied upon to suppress olive scale. Pesticides are used when outbreaks do occur. It is generally believed that the value of the yield is improved with annual pruning. The improvement in yield necessary to justify annual over biennial pruning is calculated for each variety. For Manzanillo and Sevillano the percentage increase necessary is quite small, about 3 percent. For Mission the necessary increase is significantly larger, about 14 percent. Based on a field study examining the effect of pruning on olive yield, it appears the annual pruning is the most cost-effective option for Sevillano orchards and biennial pruning is best for Mission orchards.

STATISTICAL MODELING

Statistical methods play an important role in pest management modeling. Their most direct use is in calculating the values of the input parameters used in a pest management model. Each of these is based upon data from experiments designed to examine the specific relationship which the parameter is supposed to describe. Examples include the study by Allen (1976) which relates citrus damage to the number of adult days of citrus rust mite, *Phyllocoptruta oleviora* (Ashmead), and a study by Stone and Pedigo (1972) which relates percent yield loss to percent defoliation of soybeans.

Statistical models have also been used to identify the relative importance of different variables. For example, Berryman (1976) used a stepwise multiple-regression method to establish which of nine variables describing lodgepole pine habitat appeared to be most important in determining the productivity of the mountain pine beetle, *Dendroctonus ponderosae* Hopkins. The results indicated that the suitability of the habitat was most closely related to phloem thickness, cortical resin canals, host resistance, and predation by woodpeckers. Tree diameter, which has been reported to be a significant variable in studies of single stands, did not emerge as an important variable in Berryman's study which was based on data collected in a number of different stands.

It is also possible to use population models in conjunction with statistical techniques to obtain estimates of phenomena that are very difficult to measure. An example of this is a study by Smith et al. (1978) which uses parameter identification techniques to estimate rates of oviposition and mortality from age-specific population samples taken throughout the season. Time-varying rates of oviposition are difficult to estimate because it is usually hard to determine the age of the observed eggs. In the case of alfalfa weevil, which was the subject of the Smith et al. paper, it is also difficult to determine the number of eggs since they are hidden inside the alfalfa stems.

In order to generate predictions of the dynamics of populations, it is necessary to estimate the number of eggs being oviposited each day. It is for this reason that a parameter identification method was developed which could utilize population data taken throughout the season to obtain improved estimates of the rate of oviposition. The results of this study were used in an optimization model of alfalfa weevil management (Shoemaker, 1977).

The first step in the method is to convert the samples taken later in the season to their equivalent in terms of oviposition. This relationship can be represented as

$$\theta_i(t_0) = \frac{X(t_i, a_i)}{S(a_i)}$$

where

$$t_0 = t_i - a_i$$

and

t_i = time of sample i

a_i = physiologic maturity of sample i

$S(a_i)$ = fraction of cohort which survives to maturity a_i

$X(t_i, a_i)$ = number of individuals in sample i

t_0 = time of oviposition

$\theta_i(t_0)$ = number of individuals which were oviposited at time t_0 based on sample i

The times t_i and t_0 are in units of physiologic time so that $t_i - a_i$ equals the time of oviposition.

The points $\theta_i(t_0)$ now play the role of observed levels of oviposition. The next step is to find the set of parameters which give the best fit to the scatter of points $\theta_i(t_0)$. To do this a procedure is used to minimize the sum of squares. If the curve used to represent the oviposition curve is linear in the parameters, a fairly simple routine can be used. If the curve is nonlinear in the parameters, a nonlinear optimization routine must be utilized. Results for several years and different fields are presented in Smith et al. (1978) and Smith (1978). The analysis was based upon data collected by R. G. Helgesen (unpublished).

CONCLUSION

In conclusion, we see that there are a number of different types of models, developed to answer different questions. Mathematical models of crop eco-systems have been used to understand relationships between populations, to predict the effect of weather on population densities, and to analyze manage-ment alternatives. The power of models in addressing these issues is that they can be used to increase the usefulness of information gathered about inter-actions between individual components of a crop ecosystem in order to understand the dynamics of the ecosystem as a whole. In all cases, there is a possibility of a discrepancy between the estimate provided by a model and the observed data. However, the question is not whether or not a model gives an exact answer, but rather whether the gain in understanding, predictive ability, and management efficiency exceeds that which would have been obtained without using a model. It is felt that the pest management models described in this book illustrate that modeling can, in fact, be a valuable tool in pest management research.

ACKNOWLEDGMENTS

This chapter was written while the author was on leave at the Division of Biological Control, University of California, Berkeley. A. P. Gutierrez, C. B. Huffaker, and G. E. Smith made many helpful suggestions for improve-ments in the manuscript.

LITERATURE CITED

Allen, J. C. 1976. A model for predicting citrus rust mite damage on Valencia orange fruit. *Environ. Entomol.* 5:1083–1088.

Barr, R. O., P. C. Cota, S. H. Gage, D. L. Haynes, A. N. Kharkar, H. E. Koenig, K. Y. Lee, W. E. Ruesink, and R. L. Tummala. 1973. Ecologically and economically compatible pest control. *Mem. Ecol. Soc. Aust.* 1:241–264.

Berryman, Alan A. 1976. Theoretical explanation of mountain pine beetle dynamics in lodgepole pine forests. *Environ. Entomol.* 5:1225–1233.

Berryman, A. A. 1974. Dynamics of bark beetle populations, toward a general productivity model. *Environ. Entomol.* 3:579–85.

Dover, M. J., B. A. Croft, S. M. Welch, and R. L. Tummala. 1979. Biological control of *Panonychus ulmi* (Acarina: Tetranychidae) by *Amblyseius fallacis* (Acarina: Phytoseiidae) on apple: a prey-predator model. *Environ. Entomol.* 8:282–292.

Fick, G. W. 1975. ALSIM I (Level I) User's manual. Cornell University, Department of Agronomy. Mimeo 75 20.

Gutierrez, A. P., J. B. Christensen, C. M. Merritt, W. B. Loew, C. G. Summers, and W. R. Cothran. 1976. Alfalfa and the Egyptian alfalfa weevil (Coleoptera: Curculionidae). *Can. Entomol.* 108:635–648.

Gutierrez, A. P., G. D. Butler, Y. Wang, and D. Westphal. 1977. A model for pink bollworm in Arizona and California. *Can. Entomol.* 109:1457–1468.

Gutierrez, A. P., L. A. Falcon, W. Loew, P. A. Leipzig, and R. van den Bosch. 1975. An analysis of cotton production in California: A model for Acala cotton and the effects of defoliators on its yields. *Environ. Entomol.* 4:125–136.

Hughes, R. D. 1963. Population dynamics of the cabbage aphid, *Brevicoryne brassicae* (L.). *J. Anim. Ecol.* 32:393–424.

Hughes, R. D., and N. Gilbert. 1968. A model of an aphid population—a general statement. *J. Anim. Ecol.* 37:553–563.

Koehler, C. S., and S. S. Rosenthal. 1975. Economic injury levels of the Egyptian alfalfa weevil or the alfalfa weevil. *J. Econ. Entomol.* 68:71–75.

Leslie, P. H. 1945. On the use of matrices in certain population mathematics. *Biometrika* 33:183–212.

Lotka, A. J. 1925. *Elements of Physical Biology.* Williams & Wilkins, Baltimore.

Nicholson, A. J., and V. A. Bailey. 1935. The balance of animal populations. *Part I. Proc. Zool. Soc. London*:551–598.

Norton, G. A. 1976. Analysis of decision making in crop protection. *Agro-Ecosystems* 3:27–44.

Pielou, E. C. 1969. *An Introduction to Mathematical Ecology.* Wiley, New York.

Pimentel, D., C. A. Shoemaker, E. LaDue, R. Rovinsky, and N. Russell. In press. Alternatives for reducing insecticides on cotton and corn: Economic and environmental impact. Environmental Protection Agency (1978).

Regev, U., A. P. Gutierrez, and G. Feder. 1976. Pests as a common property resource: A case study of alfalfa weevil control. *Am. J. Agric. Econ.* 58:188–196.

Ruesink, W. G. 1975. Analysis and modeling in pest management, in R. L. Metcalf and W. H. Luckman (eds.), *Introduction to Insect Pest Management*, Wiley, New York, pp. 353–378.

Ruesink, W. G. 1976a. Modeling of pest populations in the alfalfa ecosystem with special reference to the alfalfa weevil, in R. L. Tummala, D. L. Haynes and B. A. Croft (eds), *Modeling for Pest Management*, Michigan State University Press, East Lansing, pp. 80–89.

Ruesink, W. G. 1976b. Status of the systems approach to pest management. *Annu. Rev. Entomol. 21*:27–44.

Shelford, V. E. 1927. An experimental investigation of the relations of the codling moth to weather and climate. *Bull. Ill. Nat. Hist. Survey 16*:311–440.

Shoemaker, C. A. 1977. Pest management models of crop ecosystems, in C. A. Hall and J. Day (eds.), *Ecosystem Modeling in Theory and Practice*. Wiley-Interscience, New York, pp. 546–574.

Shoemaker, C. A. 1973. Optimization of agricultural pest management III: Results and extensions of a model. *Math. Biol. 18*:1–22.

Shoemaker, C. A., C. B. Huffaker, and C. E. Kennett. 1979. A systems approach to the management of a complex of olive pests. *Environ. Entomol. 8*:182–189.

Smith, G. E. 1978. Variable development models of poikilotherm populations: parameter estimation, error analysis, and optimal control. Ph.D. thesis, Department of Biological Sciences, Cornell University.

Smith, G. E., and C. A. Shoemaker. 1978. Analysis of the dynamics of piokilotherm populations with randomly distributed rates of development. Dept. of Environ. Eng. Tech. Rep., Cornell University.

Smith, G. E., C. A. Shoemaker, and R. E. Helgesen. 1978. Stepwise linear estimation of insect oviposition and mortality. Dept. of Environ. Eng. Tech. Rep., Cornell University.

Stern, V. M., R. F. Smith, R. van den Bosch, and K. S. Hagen. 1959. The integrated control concept. *Hilgardia 29*:81–101.

Stone, J. D., and L. P. Pedigo. 1972. Development and economic-injury level of the green cloverworm on soybean in Iowa. *J. Econ. Entomol. 65*:197–201.

Taylor, C. R. 1976. Determining optimal sterile male release strategies. *Environ. Entomol. 5*:87–95.

Thompson, W. R. 1939. Biological control and the theories of the interactions of populations. *Parasitology 31*:299–388.

Watt, K. E. F. 1964. The use of mathematics and computers to determine optimal strategy and tactics for a given insect pest control problem. *Can. Entomol. 96*:202–220.

3

GENERAL ACCOMPLISHMENTS TOWARD BETTER PEST CONTROL IN SOYBEAN

L. D. Newsom

Department of Entomology
Louisiana State University, Baton Rouge, Louisiana

M. Kogan

Department of Agricultural Entomology
University of Illinois, Urbana, Illinois

F. D. Miner

Department of Entomology
University of Arkansas, Fayetteville, Arkansas

R. L. Rabb

Department of Entomology
North Carolina State University, Raleigh, North Carolina

S. G. Turnipseed

Department of Entomology and Zoology
Clemson University, Clemson, South Carolina

W. H. Whitcomb

Department of Entomology and Nematology
University of Florida, Gainesville, Florida

INTRODUCTION

Soybean has rapidly developed into one of the main sources of food for man and feed for his livestock. The accelerating demand for oil and protein has resulted in an increase in acreage planted to soybean in the United States from an average of 18,045,000 during the period 1950 to 1959 to 52,460,000 acres in 1974. This amounted to a 2.91-fold increase. It almost doubled from 27,340,000 in 1961 to the 52,460,000 acres in 1974. Soybean produced for export on this rapidly expanding acreage has contributed substantially toward maintaining a favorable balance of trade for the United States.

Although expansion of acreage planted to soybean has increased significantly since 1960, in all areas where the crop is grown commercially the most dramatic increase has been in southern United States (Table 3.1). The acreage planted to soybean in 11 southern states increased from an average of

Table 3.1. Acreage Devoted to Production of Soybean for Grain in Midwestern and Southern United States During the Period 1950 to 1976

State	1950–9	1961	1971	1974	1976[a]
Twelve Midwestern States (1000 acres)					
Ohio	1,206	1,722	2,634	3,190	2,880
Indiana	1,968	2,753	3,377	3,910	3,280
Illinois	4,318	5,570	7,150	8,470	7,560
Michigan	169	285	500	630	555
Wisconsin	73	109	128	217	155
Minnesota	1,940	2,341	2,780	4,040	2,900
Iowa	2,190	3,445	5,500	7,110	6,560
Missouri	1,782	2,674	3,605	4,350	4,300
North Dakota	108	199	208	179	147
South Dakota	152	125	232	393	277
Nebraska	131	277	609	1,190	1,030
Kansas	397	703	871	1,030	880
Subtotal	14,434	20,203	27,594	34,709	30,524
Eleven Southern States (1000 acres)					
Virginia	213	378	353	430	395
North Carolina	347	609	990	1,420	1,150
South Carolina	218	604	1,070	1,250	1,190
Georgia	50	80	600	1,010	950
Florida	24	36	207	279	265
Kentucky	136	201	705	1,170	1,050
Tennessee	240	463	1,219	1,520	1,840
Alabama	100	146	655	1,020	1,160
Arkansas	1,231	2,612	4,300	4,300	4,320
Mississippi	588	1,044	2,632	2,525	3,400
Louisiana	98	197	1,644	1,760	1,950
Subtotal	3,245	6,370	14,375	16,684	17,670
Total U.S.	18,045	27,340	42,790	52,460	45,401

[a]Preliminary estimates of November 10, 1976.

Source: Crop Production Annual Summaries, Crop Reporting Board, Statistical Reporting Service, U.S. Department of Agriculture, Washington, D.C.

3,245,000 acres during 1950 to 1959 to 17,670,000 in 1976, a 5.45-fold increase. Comparative figures for the 12 states of the Midwest where soybean has been produced traditionally are 14,434,000 acres for 1950 to 1959 and 30,524,000 acres in 1976, a 2.11-fold increase. Any additional expansion of soybean acreage in the United States must occur predominantly in the South because of the availability of agricultural land in this area.

Unfortunately, from the standpoint of insect pest problems of soybean, the South is an area of much higher pest hazard than the Midwest. Problems caused by insect pests have been relatively insignificant in the latter area. In the Midwest, only the green cloverworm, *Plathypena scabra* (F.), is considered to be a major pest. Even it occurs in populations too low to require treatment except sporadically and on a very small percentage of the total acreage. The extent of problems caused by soybean insect pests in the Midwest has been accurately described by Ignoffo et al. (1976) when they stated that ". . . in most instances soybean growers in Missouri and probably throughout the Midwest do not need to do anything " about soybean insects. It has been this lack of serious insect pest problems in areas of traditional soybean production that has been mainly responsible for the relative lack of entomological research on soybean pests. As recently as 1969 there were only eight entomologists working full-time on the crop.

The situation with insects is entirely different in the South, particularly in the states bordering on the Gulf of Mexico and the South Atlantic. Expansion of soybean production in this area has exposed the crop to a high degree of hazard from attack by a large and diverse pest species complex, several members of which are capable of becoming key pests. Soybean, in this area, offers unique opportunities for the development of effective, economical, environmentally safe, and relatively stable pest management systems.

1. No key pests of the crop, species causing serious perennial problems, have developed yet.
2. There has as yet been comparatively little to no degradation of soybean ecosystems in the area from previous use of pesticides.
3. Many of the soybean growers either are, or have been, cotton producers and thus are acutely aware of the necessity for developing pest management systems for soybean that will avoid the catastrophic problems that have developed in cotton insect control from excessive use of insecticides (Chapter 5).

It is imperative that systems be developed for control of soybean pests that avoid the pitfalls and problems attendant upon unilateral systems based preponderantly upon excessive use of conventional chemical pesticides. Much of the expanded acreage of soybean is in recently cleared forested areas comprising prime habitat for upland game animals, waterfowl, and fish. These areas are especially sensitive to pollution by pesticides. In addition, the per acre returns from soybean production are too low to allow for extensive use of conventional insecticides. Thus, integrated pest management programs provide the only reasonable approach to control of soybean insect pests.

Recognition of the critical need for avoiding past mistakes was a key feature in planning research for the Soybean Subproject of the IBP sponsored, NSF/EPA supported Integrated Pest Management Project (IPM) which this book is about. The overall objective, as stated in other chapters, has been the development of effective, economical, environmentally acceptable, and relatively stable integrated pest management systems for control of soybean insect pests.

ASSESSMENT OF THE PROBLEM AND PLAN OF ATTACK

Research on soybean insect pests had received little support prior to 1970. By this time soybean production had expanded into many areas where the crop had not been produced before, and few of its pests had attracted much attention from researchers. Only four species, the velvetbean caterpillar, *Anticarsia gemmatalis* Hubner, bean leaf beetle, *Cerotoma trifurcata* (Forster), and Mexican bean beetle, *Epilachna varivestis* Mulsant in the South and green cloverworm, *Plathypena scabra* (F.) in the Midwest had received significant attention. Relatively little was known about the biology of the other eight species now known to be major pests and more than two dozen additional species that attack soybean with sufficient intensity and regularity over relatively large areas to be considered minor or potential pests (Table 3.2). In addition to the demonstrated or potential importance of these species as direct pests, there are several others known to be vectors of virus diseases of soybean.

Thus, soybean in the United States is attacked directly by a complex of 12 major species and more than twice that many minor pests, plus at least 15 species that are vectors of three important virus diseases. All parts of the plant are attacked directly. The most versatile of these as a pest is the bean leaf beetle, *C. trifurcata*, that attacks both foliage and fruit as an adult, nodules and roots as a larva, and is the major vector of bean pod mottle mosaic virus (BPMV) as an adult. At the other end of the spectrum the platystomatid fly, *Rivellia quadrifasciata* (Macquart) limits its attack on soybean to nodules by the larvae.

At the time the IPM project was initiated the pest status of the great majority of the insect pests that attack soybean was poorly, or not at all, understood, realistic economic injury thresholds were virtually nonexistent,

Table 3.2. Insects and Spider Mites Known to Attack Soybean in the United States, and Damaging Stage

Foliage Feeders

Major Pests

Anticarsia gemmatalis Hubner	Larva
Plathypena scabra (Fabricius)	Larva
Pseudoplusia includens (Walker)	Larva
Epilachna varivestis Mulsant	Adult and larva
Cerotoma trifurcata (Forster)	Adult
Diabrotica balteata LeConte	Adult

Minor Pests

Spodoptera exigua (Hubner)	Larva
Spodoptera frugiperda (J. E. Smith)	Larva
Spodoptera ornithogalli (Guenee)	Larva
Estigmene acrea (Drury)	Larva
Trichoplusia ni (Hubner)	Larva
Colias eurytheme Boisduval	Larva
Urbanus proteus (Linnaeus)	Larva
Diabrotica undecimpunctata howardi (Barber)	Adult
Colaspis brunnea (Fabricius)	Adult
Popillia japonica Newman	Adult
Melanopus spp.	Adult and nymph
Sericothrips variabilis (Beach)	Adult and larva
Frankliniella fusca (Hinds)	Adult and larva
Trialeurodes abutilonea (Haldeman)	Adult and larva
Empoasca fabae (Harris)	Adult and nymph
Tetranychus urticae Koch	Adult, larva, and nymph

Stem Feeders

Minor Pests

Spissistilus festinus (Say)	Adult and nymph
Dectes texanus texanus (LeConte)	Larva
Elasmopalpus lignosellus (Zeller)	Larva

Roots and Nodules

Major Pests

Cerotoma trifurcata (Forster)	Larva
Diabrotica balteata LeConte	Larva
Rivellia quadrifasciata (Macquart)	Larva

56

Table 3.2. (*continued*)

Roots and Nodules (continued)	
Minor Pests	
Diabrotica undecimpunctata howardi Barber	Larva
Colaspis brunnea (Fabricius)	Larva
Phyllophaga spp.	Larva
Agriotes spp.	Larva
Conoderus spp.	Larva

Blossoms and Pods	
Major Pests	
Heliothis zea (Boddie)	Larva
Cerotoma trifurcata (Forster)	Adult
Nezara viridula (Linnaeus)	Adult and nymph
Acrosternum hilare (Say)	Adult and nymph
Euschistus servus (Say)	Adult and nymph
Euschistus tristigmus (Say)	Adult and nymph
Minor Pests	
Heliothis virescens (Fabricius)	Larva
Diabrotica balteata LeConte	Adult
Frankliniella tritici (Fitch)	Adult and larva
Diabrotica undecimpunctata howardi (Barber)	Adult
Euschistus spp.	Adult and larva

few reliable sampling methods were available, and insecticides recommended for control were usually those recommended for cotton insect control, and they were applied at the same relatively high rates required for control of cotton pests. In effect, researchers in the Soybean Subproject were faced with a relatively new crop in many areas that was attacked by a large and highly diverse complex of species, few of which previously had been the object of significant research. Fortunately, little insecticide has been applied to soybean. Therefore, the situation provided an excellent opportunity "for starting anew" with a new crop and new pests and, as well, a pressing challenge to develop integrated pest management systems for a crop of more than 50 million acres.

Based on this assessment of the problems posed by soybean insect pests, it appeared that the most immediate needs were to accomplish the following:

1. Define the pest status of the complex of species known to attack soybean.
2. Develop reliable economic injury thresholds for the species determined to be major pests.
3. Develop accurate sampling methods for the various species.
4. Discover more selective insecticides and ways of using broad spectrum chemicals more selectively in order quickly to reduce substantially, or eliminate entirely, the common practice of applying heavy rates of mixtures of organochlorine and organophosphorus insecticides to soybean.

Ecological studies of pests and their natural control agents among entomophagous insects and entomopathogenic organisms were recognized as being more fundamental and long range in importance. It was decided to conduct these latter studies, plus the assessment of various tactics that might be used in developing integrated pest management systems, concurrently with the four efforts listed above. Research on the Soybean Subproject has been carried out along these lines.

Accomplishments toward achieving the objectives of the Soybean Subproject research effort have been many and highly gratifying. There have also been some disappointing failures. Some of the most significant of both categories are discussed below.

ECOLOGICAL STUDIES OF PESTS, ENTOMOPHAGOUS INSECTS, AND ENTOMOPATHOGENIC ORGANISMS

Because soybean is a relatively new crop in areas at high hazard from insect pests and because insect pests are of relatively minor importance in areas of traditional soybean production, comparatively little was known about the ecology of most of the important pests as recently as 1973 when research began in the Subproject. Also, as a new crop is introduced into an area, and planted on huge acreages, it is often colonized by species not previously associated with the crop, and by some not considered to be pests. Turnipseed & Kogan (1976) pointed out the patterns of colonization of recently established soybean ecosystems by three main components of the native fauna: (1) Species complexes with wide host ranges such as the corn earworm, stink bugs, cucumber beetles, and grasshoppers that readily accept soybean as a new host; (2) species with a comparatively narrow host range adapted to both wild and cultivated species of the Leguminose (e.g., bean leaf beetle,

Mexican bean beetle, and green cloverworm) that also readily colonize recently established soybean ecosystems; and (3) species with a narrow host range that slowly adapt to a new host, an example being the stem borer, *Dectes texanus texanus* (Le Conte). This species has been associated with common cocklebur, *Xanthium pennsylvanicum* Wallroth, an important, widely distributed weed species. The association was close enough for this cerambycid at one time to have been considered for introduction and colonization into Australia for control of *Xanthium*. However, *D. texanus* has accepted soybean readily and appears capable of becoming a major pest in some areas.

It is not unexpected for differences to occur in the extent to which soybean, produced in widely separated geographically and climatically different areas, is colonized by native fauna. However, unexpected differences may occur in ecosystems that are separated by only a few miles and are generally the same climatically. Wuensche (1976) reported substantial differences in seasonal populations of bean leaf beetle, velvetbean caterpillar, and soybean looper in three closely adjacent soybean systems (Table 3.3). Knowledge of such differences is essential in developing integrated pest management systems.

Detailed ecological knowledge of pests and potential pests is the base upon which successful integrated pest management systems are constructed. Therefore, major effort was devoted to the accumulation of an adequate data base as rapidly as possible, concentrating effort on those species where the need was most urgent. Advantage was taken of previously accumulated data on

Table 3.3. Comparative Populations on Soybean of Six Insect Pests in Three Louisiana Soybean Ecosystems During the Growing Seasons of 1969 and 1975—Average Number per Acre per Weekly Sample During the Growing Season (Wuensche, 1977)

Species	Soybean-rice-pasture ecosystem		Cleared hardwood swamp ecosystem		Soybean-cotton-corn sugarcane ecosystem	
	1969	1975	1969	1975	1969	1975
Cerotoma trifurcata	12	15	5,658	2,362	485	427
Anticarsia gemmatalis	15,473	4,323	10,078	7,938	78	2,037
Pseudoplusia includens	253	1,000	13,473	986	15,635	1,346
Plathypena scabra	623	2,468	903	1,917	1,336	2,097
Nezara viridula	1,150	7,555	258	205	590	1,856
Euschistus servus	585	1,493	150	698	100	767

the various species wherever it was available including information from other crops for such polyphagous species as the corn earworm and southern green stink bug, for example.

Velvetbean Caterpillar

The velvetbean caterpillar, *Anticarsia gemmatalis* Hubner attracted attention of American entomologists in the early 1900s as a pest of various leguminous crops. It is considered to be the major pest of soybean in Florida and often develops highly destructive populations, if left uncontrolled, as far as 150 miles inland in other Gulf coastal states (Greene, 1976). Substantial amounts of sound data on the biology of this species were available from early studies (Watson, 1916). Lack of information on the source of the immigrants that annually invade the Gulf Coast and South Atlantic states was a major weakness in available data.

Because of its major importance in Florida, researchers there have concentrated their efforts on the velvetbean caterpillar. A major contribution of their ecological studies was elucidation of its overwintering areas in Florida. They were able to establish that it overwinters successfully at latitudes south of approximately 28 degrees (W. H. Whitcomb, personal communication). Four species of leguminous hosts, *Vigna luteola* (Jacquin) Bentham, *Phaseolus lathyroides* L., *Dolichos lablab* L., and *Pueraria lobata* (Willdenew) Ohwi, appear to be of major importance as overwintering hosts. All of these are introduced species. This fact raises the interesting question of whether the velvetbean caterpillar was capable of overwintering in Florida prior to their introduction.

It appears that temperature limits its ability to overwinter successfully north of 28 degrees latitude because *Vigna luteola*, *Pueraria lobata*, and alfalfa remain green and survive in good condition in areas of north central Florida, yet the velvetbean caterpillar does not overwinter there even though these are favorable hosts. Because the southern limits of the other Gulf Coast states, except Texas, are north of 28 degrees, it is not surprising that the species does not overwinter in those states or the other South Atlantic states.

The probability exists that much of the area infested by the velvetbean caterpillar each year is by immigrating moths from Central America, South America, and Mexico, flying directly across the Gulf of Mexico. Velvetbean caterpillar larvae often appear in South Louisiana as soon as they are found in North Florida. This phenomenon suggests that the overwintering population in Florida contributes little, if any, to populations that develop in the other Gulf Coast states. There is relatively little difference in distance from the central Gulf Coast to the north coast of Yucatan and much of northeastern

Mexico and to areas of Florida where it is capable of overwintering. This is not so for the South Atlantic states. The Florida population may contribute significantly to the velvetbean caterpillar population in that area.

Soybean Looper

The soybean looper, *Pseudoplusia includens* (Walker), has many similarities to the velvetbean caterpillar. It is not capable of overwintering in the soybean producing states except in the southern part of Florida and the lower Rio Grande Valley of Texas. Thus, it too is an annual immigrant into the Gulf Coast and South Atlantic Coast States, probably from the same areas of South and Central America and Mexico that are believed to be the sources for immigrating velvetbean caterpillars. Its life history is similar to that of the velvetbean caterpillar. Unlike the latter, it has a wide host range among cultivated and wild species in several plant families other than the Leguminosae. Also, unlike the velvetbean caterpillar, the areas in which its populations develop to damaging proportions are almost exclusively limited to those in which cotton is a component of the ecosystem. Jensen et al. (1974) discovered the reason for this relationship. They reported that adults provided water alone, produced few eggs, none of which hatched. Those provided a source of carbohydrate in the form of honey or nectar from cotton blossoms laid the normal complement of eggs, a high percentage of which hatched. Thus, it appears that the soybean looper requires a source of carbohydrate as an adult in order to develop large enough populations to reach economic injury levels in soybean ecosystems.

Wuensche's (1976) research substantiated Burleigh's (1972) earlier report that populations of the soybean looper never reached economic injury levels in the rice–pasture–soybean or cleared hardwood swamp–soybean ecosystem of South Louisiana but reached outbreak proportions each year in nearby cotton–corn–sugarcane ecosystems. Wuensche was able to take advantage of a fortuitous change in crop production in the latter area in 1975. Because of uncertainties about markets for cotton and effectiveness of controls for tobacco budworm, *Heliothis virescens* (F.) growers in the cotton–soybean–sugarcane–corn ecosystem, where soybean looper populations had been under study for several years, chose not to plant cotton in 1975, but to plant soybean instead. Prior to 1975, outbreaks of the soybean looper had occurred in some fields in the area every year for at least 10 years. During 1975, populations of the pest failed to reach economic injury levels in a single field in the area, yet the usual outbreak numbers occurred in a similar area less than 10 miles away. During 1976 the economics of cotton production was more favorable, the usual acreage devoted to cotton production in the area was planted to the crop again, and the usual outbreaks of soybean looper occurred on soybean. These field data confirm laboratory data that showed the necessity for a source of carbohydrate in order for adult soybean loopers to produce enough eggs to give rise to economically damaging infestations. This need is apparently provided by nectar from cotton in areas where the two crops are grown in rotation on a substantial amount of acreage. In areas where a source of abundant carbohydrate is not available, the soybean looper does not appear to be a threat to soybean with the rare exception of areas where insecticides have been misused (Greene et al., 1974).

One of the major objectives of research on the Soybean Project is to develop pest management systems that require use of insecticides in such relatively small amounts and low frequencies of application that insecticide-resistant populations will not develop among soybean pests. At the outset, however, it was recognized that there were two sources of potential difficulty in attaining and maintaining such an objective, situations over which neither entomologists nor soybean producers had control. Both situations involve species that infest other crops or soybean in other areas, that are treated intensively with insecticides. The corn earworm provides an example of a species that is treated heavily on cotton in the United States and has already developed high levels of resistance, in some areas, to some of the insecticides recommended for its control on soybean. The velvetbean caterpillar provides an example of an immigrant species that is exposed to comparatively heavy insecticide pressure in some areas of Latin America. The soybean looper fits into both categories.

Late in the season during 1976, an extensive outbreak of soybean looper occurred in Georgia which was not controlled satisfactorily with methomyl, the chemical relied upon almost exclusively for control of this pest (J. W. Todd, personal communication). Information provided on the circumstances involved in the outbreak suggested that resistance to methomyl might be involved. Pupae were collected from the Georgia populations and sent to Louisiana State University to establish a laboratory culture for toxicological studies to determine if this was the case. Comparison of response of the Georgia population to that of a laboratory population and to the response of field populations from Louisiana and Florida in previous years, showed the existence of comparatively high levels of resistance to methomyl. The Georgia population was much less susceptible to methomyl at the LD_{90} than the L.S.U. laboratory, the 1970 Louisiana, and the 1970 Florida populations, respectively (Table 3.4). This raised the question of how resistance in the Georgia population could have come about. It was learned later that similar problems had occurred in South Carolina and north Florida.

Insecticide pressure from treatment of soybean had been too little, affected too small a percentage of the total population of this highly mobile species, and had been applied over too short a time span to select resistant populations. .

Table 3.4.   Comparative Susceptibility of Populations of Soybean Looper from Various Sources to Methomyl Applied Topically to Third Instar Larvae Reared in the Laboratory on a Standard Synthetic Diet, 1976

Source of Population	LD_{50}	LD_{90}	Slope of ldp line
LSU laboratory culture, 1976	0.0061	0.45	0.69
Krotz Springs, LA population collected and tested during 1970	0.031	0.49	1.06
Florida population collected and tested during 1970	0.041	0.37	1.30
Georgia population collected and tested during 1976	0.14	24.90	0.57

The logical places to look for the answer to the question are those areas where the species is resistant. The soybean looper overwinters in South Florida where some of its hosts, tomato for example, may receive as many as 40 applications of insecticide per year. However, there is a large chrysanthemum-producing area on the west coast of South Florida at Fort Myers which receives as many as 100 applications of methomyl per year for control of soybean looper. Furthermore, resistance to methomyl in the population attacking chrysanthemum had been suspected 2 years earlier, tested by scientists of the company producing the insecticide, and found to have developed substantial levels of resistance to methomyl at that time (Sidney Poe, personal communication).

Evidence of the relationship of the South Florida population to the one in Georgia that was found to be resistant to methomyl is circumstantial. But the fact that it was later reported from North Florida and South Carolina, but not from anywhere else, strongly suggests the probability that the resistant population originated from South Florida. If subsequent research proves this to be so, an example will have been provided of how pest control practices in one state, or even one nation, may have an important impact on the development of integrated pest management systems in another.

Stink Bugs

A complex species of stink bugs attacks soybeans. It consists primarily of the southern green stink bug, *Nezara viridula* (L.), green stink bug, *Acrosternum hilare* (Say), brown stink bug, *Euschistus servus* (Say), dusky stink bug, *E.*

tristigmus (Say), and onespot stink bug, *E. variolarius* (Palisot de Beauvois). The first four of these often occur in the same field at the same time in many areas of the Gulf Coast and South Atlantic states. A substantial amount of data on the biology and ecology of the stink bugs was available prior to the initiation of research on the Subproject. The southern green stink bug especially had attracted worldwide attention as an important pest of a number of crops, including soybean. A substantial body of useful data has been developed for the species in several foreign countries but it has received relatively little attention in the United States. Japanese workers have published extensively on their very thorough and excellent ecological studies of the species on rise in Japan. It became a serious pest of rice in Japan only after the practice of early planting was adopted thereby advancing the time of heading and providing the second generation with an abundance of a favorable host which had been lacking previously (Kiritani 1963; Kiritani et al., 1963; Kiritani & Sasaba, 1969). A similar change in pest status of the species has taken place in the United States where soybean has provided huge quantities of host material favorable for the third and fourth generations.

The stink bugs were considered to be relatively minor pests of soybean in the United States prior to expansion of the crop into lower latitudes favorable for species such as the southern green stink bug. Stink bugs rapidly colonized this new crop and have exploited it very successfully. Large populations develop on soybean each year and stink bugs have become the most important pests of soybean in many areas of the Southern states.

In spite of the availability of a relatively large amount of data on *N. viridula*, there was a conspicuous need for more information on its population dynamics. The greatest apparent need was for a better understanding of its spatial and temporal distribution with special reference to soybean. Research in this area was given a major priority. Substantial progress has been made toward obtaining the necessary data on this pest, both for developing an effective pest management program for it and modeling its population on soybean (see Chapter 4).

In Lousiana *N. viridula* overwinters as diapausing adults, the female unmated. Diapause is induced by decreasing photoperiod and the fifth nymphal instar is the sensitive stage. Data in Table 3.5 show results of a laboratory study conducted at 70°F, light regimes of 11 and 13 hours, and various combinations of exposures of third, fourth and fifth nymphal instars (Pitts, 1977).

Dissections of adults collected from soybean show that diapause begins to be expressed in September. By October a high percentage of the population has undergone diapause. Diapause in *N. viridula*, as in most species undergoing adult diapause, is characterized by hypertrophy of the fat body and atrophy of the gonads. In addition, a substantial number of diapausing *N.*

Table 3.5.   Incidence of Diapause After Reciprocal Transfers of Southern Green Stink Bug Nymphs from Diapause Inducing to Diapause Inhibiting Photoperiod Regimes—Temperatures Held at 70°F During Day and 65°F at Night—Baton Rouge, Louisiana, 1976

Photoperiod Regimes				
Third Instar	Fourth Instar	Fifth Instar		
Light:Dark	Light:Dark	Light:Dark	Total No. Exposed	Percent Diapause
13:11	11:13	11:13	108	40.74
13:11	13:11	11:13	69	37.68
11:13	13:11	11:13	97	42.27
11:13	11:13	11:13	67	47.76
11:13	13:11	13:11	97	6.19
13:11	11:13	13:11	82	2.44
11:13	11:13	13:11	77	6.29
13:11	13:11	13:11	95	4.21

viridula exhibit various degrees of color change (i.e., from green to reddish). Usually the reddening appears at the base of the wings and as large dorsal areas of the abdomen extending in a roughly triangular pattern from the base on each side.

Many of the adults overwinter under the bark of trees, in plant litter of forested areas, at the base of low-growing perennial grasses, in "Spanish moss," and in shrubby evergreens such as arborvitae and ornamental junipers. However, large numbers also overwinter on winter annual vegetable crops such as turnip, mustard, and beet. The overwintering bugs become active during periods of mild weather in winter and feed readily on various types of vegetation. Emergence from winter quarters depends upon temperature. Mating was observed during late February 1975 after a protracted period of relatively mild temperatures. Ability to overwinter successfully in a mild winter such as 1975 to 1976 and in a severe winter such as 1976 to 1977 was compared in outdoor cages. Data in Table 3.6 (Pitts, 1977) show that the provision of broad leaf mustard as a source of food during winter substantially increased survival of the population in 1975 to 1976. Survival during the unusually severe winter of 1976 to 1977 was relatively low and was not increased by provision of mustard. " Virtually no *N. viridula* survived last winter " in South Carolina (S. G. Turnipseed, personal communication).

Table 3.6.   Winter Mortality of *Nezara viridula* in Outdoor Cages at Baton Rouge, Louisiana—Survival to March 1 (Pitts 1977)

Winter	Total No. Tested	No. per Cage	Plant Litter Only		Plant Litter Plus Mustard	
			Number	Percent	Number	Percent
1976–1977	1980	165	297	30.00	517	52.22
1976–1977	1920	160	18	1.88	19	1.98

Adults emerging from overwintering quarters feed on a wide variety of hosts, including wild and cultivated species of Cruciferae, Chenopodiaceae, and Leguminosae. These plants serve for both feeding and breeding hosts and some, such as mustard, turnip, beet, and red clover, often produce large numbers of first generation southern green stink bug adults. The second generation develops on the same hosts when they are available. However, corn, wheat, and oats, especially corn, are becoming increasingly important for the development of second generation *N. viridula*. The stink bug also utilizes soybean as a second generation host. Early planted soybeans are highly attractive to first generation adults as well as to survivors of the overwintering generation. Large numbers of adults invade soybean fields and oviposit on the plants while they are still in the early stages of vegetative growth and development. However, the vegetative soybean is such a poor host that *N. viridula* can barely maintain its population on soybean prior to the time pods are set and begin to fill. Data in Table 3.7 show the comparative rate of population increase in field cage experiments on both fruiting and vegetative soybeans and snap beans. Considering a sex ratio of 1 : 1 and that one female will produce about 200 eggs, the population on vegetative soybean little more than maintained itself during one generation while that on

Table 3.7.   Effect of Food Quality on Populations of *Nezara viridula*—Baton Rouge, Louisiana, 1975

Food Source	Stage of Growth	No. Eggs Placed on Plants	No. Hatched	No. Adults Produced
Soybean	Vegetative	216	168	5
Soybean	Reproductive	249	187	99
Snap bean	Vegetative	195	154	7
Snap bean	Reproductive	240	170	66

plants with developing pods increased about 20-fold. In this experiment, predation and parasitism were prevented. However, field data of the sort presented in Table 3.8 support data obtained in the cage tests discussed above. Data from both experiments show the same trend in population, little more than maintenance of the populations during the roughly 2-month period of vegetative growth and development and virtually explosive increase of the third generation after pods have set and seeds have begun to develop. Thus, the southern green stink bug must have hosts that are reproducing in order to develop damaging population levels.

Table 3.8. Comparative Rates of Population Increases of *Nezara viridula* on Red Clover and on Vegetative and Reproductive States of Soybean Varieties of Different Maturity Groups—Baton Rouge, Louisiana, 1975–1976

Crop	Variety and Maturity Group	Stage of Development	Time Period	Population Increase
Red clover		Reproductive	May 7–June 12	34-fold
Soybean	McNair 400 (IV)	Vegetative	June 16–July 22	2-fold
		Reproductive	July 22–August 14	20-fold
Soybean	Dare (V)	Vegetative	July 14–July 27	4-fold
		Reproductive	July 27–August 24	49-fold
Soybean	Davis (VI)	Vegetative	June 16–July 27	2-fold
		Reproductive	July 27–August 26	27-fold
Soybean	Bragg (VII)	Vegetative	June 15–July 27	2-fold
		Reproductive	July 27–September 30	94-fold

From the beginning of pod set and seed development until adults enter overwintering quarters, a high percentage of *N. viridula* populations utilizes soybean as a feeding and breeding host. Cowpea appears to be preferred over soybean and where it is produced on substantial acreage it supports high populations of stink bugs. Some species of wild legumes, *Cassia* and *Sesbania* for example, may support relatively small populations.

Other species of stink bugs are similar to *N. viridula* in temporal and spatial distribution. They exhibit major differences in their choice of alternate hosts but all colonize soybean and exploit the crop effectively. Combined, *Acrosternum hilare* and *Euschistus* spp. may comprise one-third or more of the total stink bug population in some soybean fields in some areas. Their damage is sufficiently similar to that of *N. viridula* to justify considering all of them as equal in establishing economic injury thresholds (Table 3.9).

Table 3.9.   Comparative Damage Potential to Soybean of Four Species of Stink Bugs in Field Cage Studies—1976[a]

Species		Heavy	Percent Damage Mod.	Light	None	Rating[b]
N. viridula	Adult	11.9	2.5	10.6	74.5	0.52
	5th instar	12.7	4.3	13.8	69.3	0.60
A. hilare	Adult	9.4	2.6	8.0	80.0	0.41
	5th instar	10.3	3.4	10.9	75.4	0.53
E. servus	Adult	9.3	1.9	9.7	79.2	0.41
	5th instar	7.5	2.6	11.8	78.2	0.40
E. tristigmus	Adult	8.5	1.8	8.0	81.6	0.37
	5th instar	6.2	1.2	5.3	87.2	0.27

[a]5 individuals per foot of row, allowed to feed 7 days.
[b]A rating of 1.0 indicates severe damage.

Morphologic changes that occur in the reproductive system of female *N. viridula* provide an effective means of estimating age distribution and predicting trends in spatial and temporal ebb and flow of stink bug populations. Five distinct stages of development ranging from undeveloped ovaries with undifferentiated egg chambers and no oocytes visible to completed oogenesis are readily distinguished by dissection of the females.

Knowing the preoviposition period, the average number of egg masses produced, the average interval between deposition of egg masses, and the average postoviposition period, reasonably accurate estimates can be made of the age distribution of females in a population. The weakest point in the method is inability to judge whether more than one egg mass has been deposited. No technique has been developed thus far to determine whether one or several masses have been produced.

The male reproductive system of *N. viridula* also undergoes similar changes but they are not as clearly distinct as those for females.

Bean Leaf Beetle

The bean leaf beetle, *Cerotoma trifurcata* (Forster), is by far the most versatile of the insect pests of soybean in the United States. Adults attack foliage, flowers, and pods, and transmit bean pod mottle mosaic virus. Larvae attack roots and nodules and drastically reduce the capacity of infested plants to fix atmospheric nitrogen (Eastman, 1976). It is also one of the most widely distributed soybean insect pests native to the United States, occurring from the Gulf of Mexico into Canada and from the Atlantic Ocean westward to Arizona and South Dakota (Waldbauer & Kogan, 1976). It attracted attention as early as 1915 because of its influence on the nitrogen-fixing capacity of cowpea (McConnell 1915; Leonard & Turner, 1918).

The host range of the bean leaf beetle appears to be restricted to the Leguminosae. It is known to feed readily on species of *Amphicarpa*, *Apios*, *Centrosema*, *Desmodium*, *Lespedeza*, and *Strophostyles*, all native to the United States. In addition, it has readily adapted to introduced species of legumes including cowpea, soybean, and numerous species of *Phaseolus*. Its association with soybean demonstrates how readily a native pest species may adapt to an introduced host plant. Because it is a major pest of soybean, it is surprising that it received so little attention during the 40-year period from 1930 (Eddy & Nettles, 1930) until the initiation of the IPM Project. Recognizing the obvious need for the biological and ecological data required for developing integrated pest management systems for the bean leaf beetle, Illinois researchers have devoted a major effort to study of the species. Results of these excellent studies have been summarized recently (Waldbauer & Kogan, 1976).

Other Major Pest Species

Comparatively, large amounts of ecological data are available from previous research for the corn earworm, *Heliothis zea*, and the Mexican bean beetle, *Epilachna·varivestis*. For example, the corn earworm is a major polyphagous pest of many crop species. Consequently, a vast amount of data is available on this species. Researchers in the Subproject were especially fortunate to have available a large amount of excellent, directly applicable data emanating from a recent major research effort devoted to studies of the population dynamics and modeling of *H. zea* in North Carolina (Stinner et al., 1974; Johnson et al., 1975). No such extensive and immediately applicable masses of data are available for Mexican bean beetle or bean leaf beetle. Nevertheless, there are substantial amounts available. Therefore, there was much less need for major new research efforts to be devoted to these species in the early stages of development of the Soybean Subproject.

PEST STATUS

Soybean, like many crop species, can tolerate substantial amounts of apparent injury by insect pests with little or no adverse effect on yield or quality. Its increasingly demonstrated ability to compensate for injury, rapid expansion of the crop into areas of relatively high hazard from pests, and increased per acre value of the crop are elements that have made necessary a reassessment of the status of established and potential pest species. Rapidly increasing costs of insect control and the necessity for developing control strategies based on sounder ecological principles than those of previous years also contributed to the necessity for reassessment.

Species such as the corn earworm, *Heliothis zea*, velvetbean caterpillar, *Anticarsia gemmatalis*, green cloverworm, *Plathypena scabra*, and Mexican bean beetle, *Epilachna varivestis*, have long been known to be capable of inflicting intolerable levels of damage to soybeans. But, other pests and potential pests have been studied relatively little and information on their pest status prior to initiation of the IPM Project has been either weak or completely lacking. Lack of data on pest status has been especially serious in recently established soybean ecosystems, such as formerly forested areas, cleared and drained hardwood swamps, and areas where soybean culture has displaced other crops. In many such ecosystems the arthropod fauna is still undergoing drastic changes and it may be many years before substantial levels of faunal stability are attained.

Data in Table 3.10 summarize the changes in pest status that have been discovered since research was begun on this project. Such changes have an important impact on the development of integrated pest management systems. Unfortunately, it is impossible, in most cases, to determine whether the species have changed in some way or, on the other hand, whether more thorough study has simply resulted in a better understanding of their status.

Table 3.10.   Changes Discovered During the Period 1973–1976 in Pest Status of Insect Species Associated with Soybean

Cerotoma trifurcata	Discovered to be economically important as a pod and nodule feeder as well as foliage feeder.
Euschistus servus and *E. tristigmus*	Proved to be similar in damage potential to *N. viridula* and *A. hilare* when compared as individuals at comparable stages of growth and development.
Nezara viridula	Found to be associated with various species of pathogenic fungi and bacteria in addition to the yeast spot organism *Nematospora coryli*.
Heliothis virescens	Found to attack soybean with sufficiently increasing frequency and intensity to suggest possible rapid adaptation to this new host.
Spissistilus festinus	Found to be associated with southern blight, *Sclerotium rolfsii*, infection of soybean.
Rivellia quadrifasciata	Discovered during 1975 to be an important pest of soybean; occurs widely and attack nodules heavily.

It had not been known previously that the bean leaf beetle causes serious economic damage to the pods and significant damage to nodules as well as to foliage. *Euschistus servus* and *E. tristigmus* were found to be sufficiently damaging to require including them equally with *N. viridula* and *A. hilare* in establishing economic injury levels. Previously unknown associations of both *N. viridula* and *Spissistilus festinus* (Say) with important pathogenic organisms makes necessary more intensive study of the relationship of these species to dissemination of pathogens under field conditions.

It is difficult to explain how the damage by a species that occurs throughout the major soybean producing areas of the United States and destroys more than 40 percent of the nodules in some fields during some years had remained undetected. Yet this is the case of the platystomatid fly, *Rivellia quadrifasciata*, whose status was discovered in 1975 to 1976 by a persistent, observant graduate student (Eastman, 1976). Prior to this time, it had not even been considered a potential pest of soybeans.

In the examples listed above, it is probable that a change in pest "status" is simply recognition of previously unrecognized status. This does not appear to be the case with the tobacco budworm, *Heliothis virescens*, however. Here, a definite and rapid change in adaptation of this pest to a new host appears to be taking place. The rapidly increasing numbers of this species being collected during sampling for the corn earworm could hardly have gone unnoticed. The possibility exists that the tobacco budworm may be developing into a pest of soybean in a manner analogous to its development on cotton. Its change on that crop from a status of being virtually unknown to a key pest that now poses greater hazard to cotton production than any other pest species in the United States and Mexico occurred in a relatively short period of time. Indications that it may be adapting to soybean as a new host give cause for much concern.

ECONOMIC INJURY LEVELS

The economic injury level is appropriately defined as the level of pest infestation required to produce damage that if prevented would at least offset the cost of controls imposed to suppress the population.

The fundamental importance of developing accurate economic injury levels to the development of effective integrated pest management systems is widely recognized. Much progress has been made in this area, beginning with "experienced" subjective guesses on percentage of injury, leading on to accurate, quantitatively expressed densities of pests of a particularly damaging stage per unit of the crop.

Personnel involved in the Subproject were unanimous in the conviction that establishment of accurate economic injury thresholds for the major pests of soybean should have one of the highest priorities for initial study. They also agreed that although the ultimate goal of research on economic injury levels was their use in the development of dynamic models for both plant and insect populations, the greatest immediate need of such studies was for developing empirical models or general guidelines to aid implementation of integrated pest management programs at the grower level.

Kogan (1976) pointed out that a major weakness of empirically established static models is that they assume pest populations will continue to increase beyond the economic injury level. This assumption is true for most pest species in most situations for most of the time. But these populations do not follow such a trend for a continuing significant percentage of the time. Therefore, use of static models may result in overtreatment. Despite this obvious weakness, such empirically determined economic injury levels must suffice until appropriate dynamic models of insect populations are developed. Before such models can be developed to the levels of accuracy required for predicting population trends at the farm level, much more accurate weather forecasts than are presently possible must become available.

A critical lack of information on food intake and damage potential for the various developmental stages of major soybean insects has severely handicapped the development of accurate economic injury levels. Research in the project was concentrated in this area and the necessary data have been obtained for velvetbean caterpillar (Reid, 1975; Strayer, 1973), bean leaf beetle (Kogan, 1976), soybean looper (Kogan & Cope, 1974), Mexican bean beetle (D. Lau, W. Campbell, & G. Carlson, unpublished data), corn earworm (Turnipseed, 1973), and southern green stink bug (J. W. Thomas & R. M. McPherson, unpublished data).

Prior to the development of data on the amount of injury one individual of a pest species is capable of causing during its development, economic injury levels had been based on estimated percentage injury by visual inspection and related to the effects on yield as recorded for mechanically simulated injury. Economic injury levels derived from such data have critical weaknesses. For example, in the case of mechanical removal of foliage to simulate defoliation by lepidopterous larvae, many experiments have involved excising leaflets or removing portions of leaves with a paper cutter or some similar device, to the extent required to obtain a desired level of defoliation. In most of these experiments all of the defoliation has been done on one date instead of being spread over a period of time corresponding to that caused by the insect injury for which simulation is being attempted. Such a procedure fails to take into account the plant's ability to repair wounded tissue and compensate for injury. Thus, mechanically simulated injury of the kind used in a great majority of experiments may significantly underestimate injury by defoliating insects. The need for further refinement of economic injury thresholds is an area of high priority for additional research.

However, refinements continue to be made in economic damage thresholds as data accumulate to show that soybean has the ability to compensate for substantial amounts of injury. Research in Arkansas suggests that the economic damage threshold for *Heliothis zea* on soybean at flowering and early pod set stages of development may be increased from 3/foot of row to 5 to 8/foot of row, with no loss of yield or adverse effects on quality. Work there further indicated that soybeans, prior to the beginning of flowering, can tolerate as much as 50 percent destruction of the plants by the three-cornered alfalfa hopper without loss of yield (Mueller & Dumas, 1975) under the recommended planting rate of 12 seeds/foot. Illinois workers have concluded that damage caused by the unprecedented outbreak of the soybean thrips, *Sericothrips variabilis* (Beach), on seedling soybean during 1975 had no effect on yields and that such infestations of soybean do not require treatment. They also raised the economic injury levels for the green cloverworm to 12/foot of row, as opposed to 6/foot of row as adopted up to 1975.

Two important problems have developed recently that present new dimensions of both soybean insect sampling procedures and establishment of accurate economic injury levels. The first involves a complex of nodule feeding species. Larvae of the bean leaf beetle, banded cucumber beetle, and the platystomatid fly, *Rivellia quadrifasciata*, all feed on nodules of soybean. Research during 1976 showed that larvae of the bean leaf beetle and *Rivellia* are capable of destroying 40 to 50 percent of the nodules under field conditions if present at moderate population densities. The banded cucumber beetle, also damaging, is not this destructive. Damage to nodules causes a substantial reduction in the nitrogen-fixing capacity of the plant. No adequate techniques are available for sampling larval populations of these three nodule-destroying pests. Neither is there information available on how much damage to the nodules, with consequent reduction in nitrogen-fixing capacity, the soybean can tolerate without yield loss.

Also, it has been found in greenhouse tests that defoliation of soybean by the soybean looper results in reduction of nitrogen-fixing capacity by the nodules of affected plants. The reduction in nitrogen-fixing capacity appears to be roughly equal to the percentage of defoliation. Again, there are no data on the effects of reduction in nitrogen-fixing capacity of the nodules on yield. However, defoliation by the soybean looper and the velvetbean caterpillar usually occurs under field conditions during the pod filling stage when the requirements for nitrogen are considered to be most critical.

This complexity of the problem involving a reduction in nitrogen fixation caused by several insect species that attack the nodules directly and by those that destroy the foliage, requires reevaluation of economic injury thresholds. Not only must we consider the effect of reduced nitrogen fixation on yields but the economic importance of a reduction in nitrogen fixation on the amount of nitrogen produced by legumes that are attacked by any pest requires investigation.

Despite the obvious weaknesses in the economic injury thresholds as presently indicated for major soybean insect pests and the need for additional research, use of these thresholds in pest management systems at the grower level is necessary and effective. A large volume of data from field experiments provide assurance that the changes made in economic injury levels used for soybean insect control recommendations during the last 4 years (Table 3.11) have maintained yield and quality at maximum levels while reducing the costs of control, adverse effects on nontarget organisms (including important natural control agents), environmental pollution, and the possibility of selecting insecticide-resistant pest populations. They also provide conclusive evidence that a high percentage of the total soybean crop requires *no treatment* for control of insect pests. Not more than an average of 20 percent of the total acreage requires treatment in areas at greatest hazard from insect

Table 3.11.   Changes in Economic Injury Thresholds for Some Major Insect Pests of Soybean During the 4-Year Period 1973–1976

Species	Economic Injury Thresholds	
Velvetbean caterpillar[a]	40% defoliation	8 larvae larger than $\frac{1}{2}$ inch in length per row-foot
Soybean looper[a]	40% defoliation	8 larvae larger than $\frac{1}{2}$ inch in length per row-foot
Green cloverworm[a,b]	40% defoliation	8 larvae larger than $\frac{1}{2}$ inch in length per row-foot
Corn earworm[c]	1 larva per 3 row-feet	3 larvae per row-foot
Stink bugs[d]	1 stink bug per 3 row-feet	1 bug larger than $\frac{1}{4}$ inch diameter (3rd and later nymphal instars and adults) per row-foot

[a]May consist of either species alone or combinations of all three.
[b]Recent data obtained at the University of Illinois suggest that the number should be increased to 12 per row-foot.
[c]Additional data obtained at the University of Arkansas suggest that further increase to 5–8 larvae per row-foot may be appropropriate.
[d]May consist of *Nezara viridula, Acrosternum hilare, Euschistus servus, E. tristigmus* alone or in any combination.

infestation, the Gulf Coast and the South Atlantic states. Even this percentage needs only an average of one application per acre in years when such treatment is required.

One of the most disappointing facets of research in the Subproject has been failure to make any substantial progress on development of economic injury levels for complexes of pests. The economic injury levels established thus far have been for individual species or limited complexes of a few species that cause very similar injury to soybean—lepidopterous defoliators or stink bug pod feeders, for example. In most areas, soybean is attacked simultaneously by a large number rather than a single pest species. In many areas foliage may be simultaneously attacked by six or more species, stems by two, roots and nodules by three, and pods by five, all at the same time. Problems posed in development of economic injury thresholds by the bewildering intricacies of interrelationships between such a large complex of species have proved to be enormously difficult. Their solution will require use of far more sophisticated models than any developed thus far.

ELEMENTS AND TACTICS OF POPULATION REGULATION AND CONTROL

Host Plant Resistance

Subproject personnel are convinced that use of host plant resistance has greater potential for development as the key component of a successful integrated pest management system than any other tactic or strategy. It is highly compatible with all other control tactics and is considered more likely to contribute stability to pest management systems in an annual row crop situation than any other available tactic. The traditional approach of screening as much material as possible to identify resistant soybean genotypes followed by selection, crossing, and backcrossing to transfer desirable characters to agronomically desirable varieties has been followed by researchers in the Subproject.

Vigorous research efforts in the area of host plant resistance have been underway by personnel at most of the cooperating institutions and at cooperating laboratories of the U.S. Department of Agriculture. Among the thousands of plant introductions (PIs), breeding lines, and varieties that have been screened, several hundreds have been discovered to possess measurable levels of resistance to one of more of the major insect and disease pests. A few have shown enough promise to justify transfer of the resistant characters to agronomically acceptable varieties. Genotypes have been identified that possess resistance to one or more of the following major insect pests of soybean: velvetbean caterpillar, soybean looper, Mexican bean beetle, bean leaf beetle, corn earworm, and southern green stink bug.

Especially good progress has been made in breeding for resistance to the Mexican bean beetle. For example, 216 advanced breeding lines and 600 F_3 lines have been developed in South Carolina that have high levels of resistance to this pest and moderate to low levels of resistance to soybean looper (Fig. 3.1). This material was also evaluated during 1976 for resistance to velvetbean caterpillar. Although most lines were susceptible, some possessed levels of resistance sufficiently high to be selected for further study.

Several advanced lines and the PIs 229358, 227687, and 171451 that possess good levels of resistance to Mexican bean beetle were also evaluated for possible resistance to the southern green stink bug in South Carolina during 1976. Useful levels of resistance were identified. Nymphs reared on pods of the PIs 229358, 227687, and 171451 and the breeding line ED 73-308 exhibited significantly higher mortality, slower rates of development, and lower weights than those reared on the commercial variety "Bragg." Extensive evaluations made in North Carolina of PIs and established breeding lines for resistance to a complex of pests showed some to possess high levels of resistance to the

Figure 3.1. Resistance to the Mexican bean beetle, *Epilachna varivestis*, in soybeans. Susceptible variety "Bragg" on left; resistant line with P.I. 229358 parentage on right. [Photograph courtesy Sam G. Turnipseed, Clemson University.]

Mexican bean beetle and low to moderate levels to the soybean stem borer and corn earworm. Hatchett et al. (1976) found resistance to *H. zea* in PI 227687 sufficiently high to cause 100-percent mortality of larvae fed on its foliage. Feeding on the foliage of two other PIs resulted in about 90-percent mortality of larvae (Table 3.12).

Table 3.12. Comparative Development and Mortality of *Heliothis zea* Reared on Leaves of Eight Soybean Cultivars— Stoneville, Mississippi, 1976 (Hatchett et al., 1976)

Cultivar	Mean Weight of 16-Day-Old Larvae (mg)[a]	Mean Weight of Pupae (mg)[a]	Percent Mortality[a]
Forrest	315.1	227.8	5.0
Bragg	190.7	233.0	43.3
Davis	292.0	231.6	18.4
Lee 68	263.2	244.5	33.3
Bragg	201.3	196.3	56.7
PI 171451	117.4	170.3	88.3
PI 229358	136.6	167.1	90.0
PI 227687	70.5		100.0

[a]Averages of two experiments.

Studies have been done to determine possible interactions of insect-resistant soybean germ plasm with control tactics or agents. Thus far there appears to be little, if any, difference in infection of corn earworm or soybean looper by the fungus, *Nomuraea*, when these species develop on resistant as contrasted to susceptible lines. The amount of insecticides required to control field populations of *H. zea* appears to be lower on the resistant line ED 73-371 than on the susceptible "Bragg" variety (Table 3.13). In small plot field experiments similar rates of application of methyl parathion or methomyl produced higher levels of control of corn earworm on plants possessing resistant than on those possessing susceptible germ plasm. In the case of *Bacillus thuringiensis*, formulations of the bacterium gave a much higher degree of control of *H. zea* on ED 73-371 than on the commercial variety "Bragg."

Although host plant resistance offers more promise of contributing effectiveness and stability to pest management systems for soybean than any other tactic of pest population regulation, progress toward developing acceptably resistant commercial varieties has been disappointingly slow.

Table 3.13.   Influence of Varietal Resistance on Effectiveness of Three Insecticides for Control of *Heliothis zea* Larvae—Blackville, South Carolina, 1976 (Turnipseed, 1973)

	Percent Control of *H. zea* Larvae with Recommended Rates of Application of Insecticide		
Cultivar	Methyl parathion	Methomyl	*B. thuringiensis*
Bragg (susceptible)	27.0[a]	45.5 a	4.4 b
ED 73-371	52.0 a	63.60 a	56.0 a

[a]Means in each column followed by the same letter are not significantly different.

Yet, levels of resistance have been transferred to several lines that are sufficiently high to eliminate any need for insecticide to control Mexican bean beetle, for example (Table 3.14). Some of these lines are fully as good as the best commercial varieties in all characteristics except for the most important of all—yield. In addition to possessing resistance to Mexican bean beetle, bean leaf beetle, corn earworm, and soybean looper, protein and oil content of seed is as high and of as good quality, plants are as erect and as resistant to seed "shattering," and resistance to major nematode pests and plant pathogens is as good in some lines as in the best commercial varieties. None, possessing satisfactory levels of resistance plus all other desirable agronomic characteristics except yielding ability, has consistently yielded as as well as the major commercial varieties. Yields of the better lines that are in advanced stages of testing usually fall short of the performance of commercial varieties by about two bushels per acre. At current prices for soybean the difference amounts to about $16 per acre. The Mexican bean beetle can

Table 3.14.   Effect of Varietal Resistance on Populations of the Mexican Bean Beetle—Average Number of Larvae per Foot of Row on Dates Indicated —Blackville, South Carolina, 1974

Cultivar	8/9	8/21	9/4	9/18
Bragg (susceptible)[a]	3.27	4.73	6.17	2.92
ED 73-371 (resistant)[a]	0.46	0.07	0.58	0.28

[a]Planted in semi-isolated fields of approximately 0.6 hectare each.

be controlled effectively with carbaryl at rates of use sufficiently low as to cause minimum adverse environmental effects, and at a cost of about $4 per acre. Therefore, growers will not be willing to plant a resistant variety unless it yields about as well without treatment as susceptible varieties with treatment.

Natural Control Agents

Insect pests of soybean are attacked by a complex of natural enemy species even larger and more diverse than the complex of pests of the crop (Dietz et al., 1976). The role of predators, parasites, nematodes, protozoa, fungi, bacteria, and viruses that comprise these natural enemy complexes in regulating populations of insect pests is not adequately understood. There is much less understanding of the interactions among the various components of this natural enemy complex. What are the effects of an epizootic of one of the entomopathogens of velvetbean caterpillar or soybean looper on polyphagous predators of these pests, on narrowly specific parasites, on other entomopathogens? Such questions have barely reached the point of being asked, much less answered. However, ample data have been obtained to show that natural enemies regulate, at subeconomic levels, the populations of soybean insect pests on a vast majority of the total United States acreage every year.

Despite the effectiveness of natural enemies and the indispensable role they play in control of soybean pests, populations of corn earworm, Mexican bean beetle, bean leaf beetle, velvetbean caterpillar, soybean looper, stink bugs, and other lesser pests, often reach levels that exceed, at least temporarily, the capacity of the natural control agents to control them in a small percentage of the acreage during most years. A major challenge to researchers on the Subproject, and to entomologists generally, is to develop the methodology required to understand more fully and to manipulate populations of natural control agents in a way that will reduce the extent and severity of such outbreaks, or to prevent them completely. It must be emphasized however, that in row crops subjected to annual environmental perturbations and periods of absence of hosts, automatic continuity of the locally present natural enemies is not wholly dependable. However, Rabb et al. (1976) have described various manipulations that may partially offset the adverse effects of such annual disturbances.

Polyphagous Predators

The contribution of general predators to control of insect pests has been seriously underestimated by entomologists. The often spectacular resurgence of pest species, and elevation of secondary and potential pests to key pest status following disruption of natural enemy populations by use of broad spectrum synthetic insecticides suggested to a few entomologists that the role of general predators needed reexamination. Information that has been generated recently has brought about a generally better appreciation of their importance. Much of the previous lack of understanding of the role they play in regulating pest populations probably stems from studies that have concentrated on individual species of the complex, or the emphasis that has been placed on tree crop situations wherein various entomophagous parasitoids have had a key role in regulating certain pest species at very low densities. Failure to show numerical response to host prey density is characteristic of polyphagous predators. This characteristic has often caused them to be dismissed as being of little consequence, but Murdoch and Marks (1973) stress the significant role that density-related prey-switching may present. Additionally, study of predator complexes has proved to be enormously difficult.

As research on soybean insects expanded and accelerated under the impetus of the NSF/EPA, IPM Project, the importance of the complex of polyphagous predator species associated with the various pests became more obvious. Buschman et al. (1977) have undertaken an extensive study of the role of predaceous arthropods in soybean fields of Florida. They have documented the occurrence of more than 1000 predaceous species that are associated with the soybean crop, less than 100 of which prey directly on major pests. Impressive amounts of data of the sort reported in their studies on the ants of soybean fields and predators of velvetbean caterpillar eggs (Whitcomb et al., 1972 have been accumulated. They have found that 11 species of ants prey extensively on soybean pests, 20 species of predators prey on eggs of the velvetbean caterpillar, and 13 on the eggs of soybean looper. They found that the members of a 14-species complex in Florida soybean fields averaged 5.5 and 3.6 per foot of row during July and August, respectively. Shepard et al. (1974) reported similar data from South Carolina. Similar but less extensive studies in other states have given comparable results. For example, eggs and nymphs of the southern green stink bug are preyed on by a 21-species complex in Louisiana (Pieter Stam, unpublished).

Thus, it may be argued that polyphagous predators are probably the most effective natural control agents of soybean insect pests. These native enemies have been criticized as not being sufficiently host-specific to be effective in regulating pest populations. However, it is this very lack of specificity that allows them to be effective in annual row-crop ecosystems that are characterized by a high degree of instability. Only the "generalists" are likely to be able to adapt to the hazards of the annual disruptions involved in preparing land for planting, cultivation, and harvesting operations of a crop such as soybeans. The relative instability of agricultural ecosystems subjected to such annual perturbations pose almost insurmountable hazards to the "specialists" that have performed so well for control of pests of many perennial crops.

Parasites

A complex of several species of humenopterous and dipterous parasites attack the major insect pests of soybean (Deitz et al., 1976). It appears that in many cases they have less impact on the populations of most species than the predators as a complex, but there are two important exceptions. Research in Arkansas indicates that the green cloverworm is often controlled effectively by one parasite, *Apanteles marginiventris*, which attacks the first stage larvae. Parasitization as high as 92 percent has been observed there.

The annual release of the Mexican bean beetle parasite, *Pediobus foveolatus* (Crawford), shows great potential as a tactic for integration into soybean pest management systems in areas where it cannot overwinter. In a pilot test in North Carolina, trap crops of early planted snap beans were used to attract and concentrate overwintered adults of the Mexican bean beetle, and *Pediobus* was released into these small, heavily infested trap crop areas. The parasite built up rapidly on first-generation beetle larvae, dispersed into adjacent fields as the season progressed, and inflicted very heavy mortality on the Mexican bean beetle population.

In Florida, *Pediobus* is performing exceptionally for control of *A. varivestis* (F. W. Maxwell, personal communication).

Ooencyrtus submetallicus, an egg parasite of stink bugs, was found to be equal or superior to *Trissolcus basalis* as a parasite of southern green stink bug eggs in the laboratory. However, in small-plot field experiments in Louisiana its performance has been extremely poor when released, singly or jointly with *T. basalis*. It appears to offer little promise as an egg parasite of the stink bug complex.

A complex of dipterous parasites consisting of *Trichopoda pennipes*, *Cylindromyia euchenor* (Walker), *Euthera tentatrix* Loew, *Gymnosoma fuliginosum* Robinson-Desvoidy and two as yet unidentified species attack adults of the stink bug complex. The average percentage of parasitism for the growing season in Louisiana has been as follows: For *Euschistus tristigmus*, 1.75; *Acrosternum hilare*, 2.39; *Euschistus servus*, 2.53; *Podisus maculiventris*, 5.26; and *Nezara viridula*, 32.0. Parasitism by *Trichopoda pennipes* accounts for nearly all of the parasitism of *N. viridula*. In general, it may be concluded that these parasites are relatively ineffective although, in the case of *N. viridula*, the percentage of parasitism appears to be high enough to exert a substantial regulatory effect on populations of this pest. However, parasitization of adult female *N. viridula* by *T. pennipes* does not prevent a high percentage of the normal number of eggs being deposited.

Microbial Pathogens

Insect pests of soybean are attacked by a large number of native and introduced pathogens, some of which play important roles in regulating populations of major pests. Researchers of the Soybean Subproject have concentrated efforts on two species of fungi and three viruses—*Nomuraea rileyi* and *Entomophthora gammae* and the nuclear polyhedrosis viruses (NPV) of soybean looper, alfalfa looper, and velvet bean caterpillar.

Nomuraea rileyi. This naturally occurring entomogenous fungus has a wide host range. It has been particularly effective on the larval lepidopterous pest complex of soybean—green cloverworm, corn earworm, velvetbean caterpillar, and soybean looper. Because of its wide spectrum of activity, *N. rileyi* has received more attention than any of the other soybean insect pathogens.

N. rileyi, like other entomopathogens of soybean insect pests, has a major weakness. Naturally occurring epizootics often develop so late in the season that substantial damage has occurred before the pest population is significantly affected. The possibility of initiating epizootics earlier in the season than occurs naturally has been the objective of a substantial amount of research. Promising results have been obtained in a few instances but the degree of success can best be described as generally disappointing. However, researchers in North Carolina modified the most often used technique of applying fungi to insect populations—spraying aqueous suspensions of spores to the host plants—and obtained highly promising results (Sprenkel & Brooks, 1975). They cultured *N. rileyi* in larvae of the tobacco budworm, *Heliothis virescens*, cut the dead diseased larvae into 3-mm fragments, and distributed them over field plots. The results of these experiments indicated that the fragments of cadavers distributed in the field provided for the continuing release of inoculum for a period of about 3 weeks. Epizootics initiated by this technique occurred earlier and were more intense than those that occurred naturally or by conventional application of spore suspensions. The results suggest that a carefully timed application of fragments of larvae infected by *N. rileyi* could exert useful levels of control of host populations throughout the period of highest hazard to soybean from lepidopterous larvae (Table 3.15).

The effect of certain cultural management practices on incidence of infection by *N. rileyi* in key lepidopterous pests was studied in North Carolina (R. K. Sprenkel, W. M. Brooks & J. W. Van Duyn, unpublished). They found that the epizootic potential of *N. rileyi* was significantly enhanced in early planted soybeans as contrasted with late-planted plots (Table 3.16). Seeding rate proved to be the least significant factor. Although differences among plots of varying row widths were not statistically significant, a definite trend

Table 3.15. Effect of Treatment of Soybean Plots with Fragments of Cadavers of Corn Earworm on Incidence of Infection of *N. rileyi* in populations of *Heliothis zea* and *Plathypena scabra*, North Carolina, 1974— No. Larvae Infected by *N. rileyi* per 15 Feet of Row on the Indicated Number of Days After Application of Fragments of Cadavers of Larvae Killed by the Fungus (Sprenkel & Brooks, 1975)

Species		14	21	28	35	42
P. scabra	Control	9.6	9.3	14.0	12.8	4.8
	Treated	8.6	4.9	9.5	5.0	1.4
H. zea	Control	0	0	2.6	5.3	1.2
	Treated	0	0	0.7	2.6	0.1
Number killed	Control	0	0	0.3	1.1	5.0
by *N. rileyi*	Treated	0.2	0.6	3.7	3.7	6.1

Table 3.16. Average Incidence of Infection by *N. rileyi* in Noctuid Larvae in Plots of Various Planting Date, Row Widths and Seeding Rates at Peak Incidence of Infection—September 20, 1973, Plymouth, North Carolina

Treatment[a]	Average Number *N. rileyi* 1/1000 A[b]	Percent Infection
E-30-6	6.5 a,b	69.4
E-30-12	8.5 a	54.6
E-48-12	5.5 a,b	47.3
E-48-6	4.1 b,c	34.1
L-30-12	1.3 c,d	12.5
L-48-6	0.5 c,d	11.9
L-48-12	0.6 c,d	8.5
L-30-6	9 c,d	0

[a]E = early planted, June 5; L = late planted, July 11; 30 = 30-inch rows; 48 = 48-inch rows; 6 = 6 seeds/row ft.; 12 = 12 seeds/row ft.
[b]Means followed by a common letter do not differ significantly ($p = 0.05$).

toward higher seasonal incidences of the infection was apparent in plots with narrow rows. These data suggest that planting soybeans early and in narrow rows tend to increase the epizootic potential of *N. rileyi*.

Microenvironmental monitoring studies in Florida defined the conditions that are favorable for the initiation of epizootics of *N. rileyi*. Results from three seasons of these studies support the following assumptions:

1. Dry, windy conditions promote conidial dispersal but high densities of conidia are also dependent upon host population densities and infection rates.
2. Dry, windy conditions retard germination and infection but promote infection if followed by humid conditions, if excess free water is not present.
3. Fungus ontogeny within the host, up to and including death of the host, is independent of weather conditions.
4. Conidiophores form independently of fluctuations in relative humidity so long as the host cadaver is not subjected to rapid dessication following death.
5. Development of conidiophores and conidia although arrested for a week or longer by adverse weather conditions, may resume development when conditions again become favorable.
6. The minimum relative humidity for conidia production is 70 percent. Conidia production increases with increase in relative humidity.
7. Rain and vegetative wetting promote conidiophore formation and conidiogenesis.
8. Conidia are washed off cadavers by extended periods of light rain, brief periods of heavy rain, or long periods of vegetative wetting by heavy dew formation.
9. Alternation of wet and dry conditions is necessary for spread of infection.
10. The presence of excess free water during the peak of an epizootic has little net effect on the course of the epizootic.
11. Excess free water in the early stages of an epizootic (infection rate less than about 10 percent) may retard the spread of infection if it follows conidia formation but precedes conidia dispersal.
12. Alternation of short periods of vegetative wetting with longer periods of dry, windy conditions favor the increase and spread of infection.

Observations on the overwintering survival of *N. rileyi* were also made by Sprenkel and Brooks (1977). They demonstrated the possibility that infectious conidia can survive through the winter in North Carolina. A more likely mode of overwintering of the fungus, however, is in the sclerotial form of hyphal bodies in larval cadavers, on the surface or buried in the soil.

These studies have contributed significantly to a better understanding of the biology and epidemiologic relationships of *N. rileyi* in lepidopterous pest populations affecting soybean. However, there has been little information developed that will lead soon to effective and economical techniques for increasing the impact of naturally occurring epizootics on pest populations. Except for the general recommendation that soybeans should be planted as early as possible, the possibility of a direct application of the fungus under field conditions is still in the experimental phases of development and unlikely to be available as a control tactic in the near future.

Entomophthora gammae. Unlike *Nomuraea rileyi*, the entomogenous fungus *E. gammae* has a narrow host range. It affects only the soybean looper among the complex of lepidopterous larvae that defoliate soybean. However, it can cause spectacularly effective control of this pest, often virtually wiping out huge populations within a period of a few days (Brousseau, 1975). But the annually occurring epizootics in many areas where soybean looper is a pest of soybeans often come too late to prevent severe defoliation. Like *N. rileyi*, its effects are often too little and too late relative to the occurrence of severe damage.

Environmental factors affecting conidial sporulation and germination of *E. gammae* were described by Newman and Carner (1975). They reported a relatively narrow range of favorable temperature and humidity for these processes. Conidia cannot survive the extended periods of low humidity and high temperature that are characteristic of August conditions in most areas where soybean looper is an important soybean pest. Furthermore, the resting stage of the fungus, the azygospores, is exceedingly refractory to all techniques known to stimulate germination. Thus, there is apparently even less possibility for effective manipulation of *E. gammae* for stimulating early development of epizootics than is the case for *N. rileyi*.

Nuclear Polyhedrosis Viruses. The soybean looper NPV, velvetbean caterpillar NPV, and alfalfa looper NPV all appear to have good potential for control of soybean pests. Arkansas researchers have assumed major responsibility for research on these agents and have developed a large amount of excellent basic and applied data. The soybean looper virus was obtained from Guatemala and has been tested extensively in Arkansas and Louisiana. It has demonstrated good potential for practical control of soybean looper in laboratory, small-plot, and large-scale field trials. At rates of foliage application of 100 larval equivalents per acre (100 L.E.) it has given results comparable to those provided by conventional chemical insecticides recommended for control of the soybean looper (W. C. Yearian, unpublished data).

This virus has an advantage over the chemical insecticides in never requiring more than one application per season for control of the looper.

It is so narrowly specific that it affects no other species. Also, timing of applications requires much less precision than that for chemical insecticides. Good control of the soybean looper may also be obtained by applying the virus to the soil at planting time. It also persists in the soil, in low titers, from one season to the next (Kuo, 1975). Even at these low levels of occurrence in the soil, in some cases the virus applied one year has persisted through the winter and has built up to sufficient levels the next year to give substantial control of looper populations without reapplication (Table 3.17).

Table 3.17. Persistence of *Pseudoplusia includens* NPV in Soybean Fields from One Season to the Next—Alexandria, Louisiana, 1975

| | No. Soybean Loopers/100 Sweeps, 1975 | | Percent Control |
Date	Treated with NPV 1974	Untreated 1974	Without Additional Treatment, 1975
7/29	0.4	0.5	20
8/8	3.3	7.0	53
8/15	11.9	35.0	66
8/22	73.7	173.3	57
8/27	132.0	256.8	49
8/29	127.0	330.0	62
8/31	38.0	63.5	40
9/3	44.0	72.0	39
9/10	0.0	0.0	

A multiple embedded nuclear polyhedrosis virus of the velvetbean caterpillar collected in Brazil was tested in Florida and South Carolina. It shows much promise for control of this pest. In small-scale field tests in South Carolina during 1976 it gave more than 70-percent control at rates of application as low as 20 larval equivalents per acre. However, carbaryl applied at the rate of 0.24 lbs. per acre in the same experiment gave 99-percent control (Table 3.18).

The alfalfa looper virus is an especially interesting pathogen. It has an unusually broad spectrum of activity. In small-scale field tests in Arkansas it gave control of soybean looper amounting to about three-fourths that obtained with the soybean looper NPV, about equal to that of the *Heliothis* NPV, and about equal to that of a commercial formulation of *Bacillus thuringiensis* (W. C. Yearian & S. Young, unpublished data).

Data obtained thus far clearly indicate a good potential for use of these pathogens in integrated pest management systems for control of soybean

Table 3.18.   Comparative Effectiveness of a Nuclear Polyhedrosis Virus and Carbaryl for Control of Velvetbean Caterpillar—Blackville, South Carolina, 1976

Treatment	Rate/acre	Mean No. Larvae Alive/ft. of Row 12 Days Post-Treatment	Percent Control
Virus	5 LE[b]	4.13 d[a]	39
Virus	10 LE	3.56 cd	48
Virus	20 LE	1.84 bc	73
Virus	40 LE	1.44 b	79
Carbaryl	0.25 lb.	0.06 a	99

[a]Values followed by the same letter are not significantly different.
[b]LE = larval equivalent.

pests. They more readily lend themselves to manipulation to achieve immediate results than do other natural control agents. However, their status from the standpoint of registration for use is so uncertain that little additional progress is likely until questions pertaining to the safety of their use are resolved. One of the most serious obstacles toward their eventual use on a commercial scale is the lack of a satisfactory methodology for large-scale production, especially for the viruses. Research on the methodology of commercial production is expensive. Industry is unlikely to become involved in this sort of research as long as the future of microbial pathogens for insect control remains so uncertain. There is a critical need for breaking out of the seeming cul-de-sac in development of microbial pathogens for insect control, especially so for soybean pests. Some problems that have been encountered recently in research on microbial pathogens illustrate the imperative need for interdisciplinary cooperation in pest management research. Plant pathology research on soybean diseases clearly shows that control of a complex of fungal pathogens that infect soybean by application of benomyl significantly increases yields and improves quality. In some areas, particularly with relatively high rainfall, yield increases consistently amount to 3 to 5 bushels per acre. In areas where climatic conditions are less favorable for fungus development, increases in yields resulting from applications of fungicides have been inconsistent and much smaller.

Unfortunately, the methodology for determining when soybean fields require treatment has not been developed, so fungicidal applications are made "automatically," based on stage of plant development. Even more unfortunately, benomyl has an adverse effect on populations of entomogenous fungi. Its impact on *Nomuraea* and *Entomophthora* is often sufficient to release lepidopterous larvae from the regulating effects of these natural control agents. Insect increases following applications of benomyl have often been dramatic and frequently have reached levels that require treatment. Applications of benomyl alone, or in combination with insecticides, may delay development of *N. rileyi* for about 3 weeks and result in significant decreases in yield of treated plots. This phenomenon poses a potentially serious adverse effect on development and use of integrated pest management systems for soybean. Solutions to the problem are being sought in closely coordinated, cooperative research involving both entomologists and plant pathologists.

Cultural and Physical Methods

Small-acreage trap crops of early-maturing soybean varieties effectively attract and hold populations of the southern green stink bug. Such trap crops can also be used for control of the bean leaf beetle and the important virus disease of soybean, bean pod mottle virus, for which it is the only important vector. The trap crop principle has been tested in grower-scale field trials for the third successive year and has proved to be highly satisfactory. It is an effective, economical, ecologically acceptable tactic that is highly compatible with other components of pest management systems and with general farming practices (Newsom et al., 1975; see also "Plant Disease/Vector Relationships," below).

Choosing varieties of appropriate maturity groups and using planting dates that help to obtain maturity at the most desirable time, as far as is consistent with good agronomic practices, is proving to be an effective tactic for control of some major pests. Early maturity of the crop, mid-to-late August, avoids damage by the velvetbean caterpillar, corn earworm, and soybean looper in most areas.

One of the most interesting and effective methods for avoiding damage by *H. zea* is to reduce space between rows to promote early "closing in" of the plant canopy before blossoming begins. Corn earworm moths are not attracted to soybean as an oviposition site except during the period of blossoming. Fields in which the canopy has closed before blossoming are not satisfactory for oviposition. Therefore, damaging infestations do not occur in such fields.

Early "closing in" of the canopy has another important effect. It provides an environment that is highly favorable for development of *Nomuraea* epizootics. Reducing the distance between rows and between the plants within the row results in significant, early increases in the populations of this useful pathogen, (see also, "Microbial Pathogens," above).

The effectiveness of plowing under stubble for control of the cerambycid stem-borer, *D. texanus*, has been demonstrated for 2 successive years in North Carolina. In stubble buried 2 or 6 inches deep, mortality of larvae was 33 and 54 percent, respectively. Fewer adults were able to emerge from the 6-inch depth. Soil types also affected adult emergence. Burying stubble in dry, clay soils almost eliminated emergence of adults of this pest.

Selective Use of Insecticides

Some of the most immediately applicable research on the Soybean Subproject has involved the selectivity principle in use of insecticides. Data in Table 3.19 show the changes that have been made in the kinds and amounts of insecticides recommended by Cooperative Extension Service personnel for control of soybean pests. Reductions in rates of application for most of the materials used on soybean were made possible by an intensive cooperative effort to establish minimum dosage rates required for adequate control of soybean pests (Turnipseed et al., 1974).

Table 3.19. Change in Dosage Rate of Insecticides Recommended for Control of Soybean Pests During the Period 1973–1977

Pest Species	Insecticides	Recommended Rate, lbs. a. i./ Acre	
		1973	1977
Bean leaf beetle	Carbaryl	1.5	0.50
	Methyl parathion	0.5	0.25
	Azinphosmethyl	0.25	0.125
Velvetbean caterpillar and green cloverworm	Carbaryl	1.5	0.5
	Methyl parathion	0.5	0.25
	Bacillus thuringiensis		0.25–0.5
Corn earworm	Carbaryl	2.0	0.5–1.0
	Methyl parathion	1.0	1.0
Soybean looper	Methomyl		0.45
	Orthene		0.75
	Bacillus thuringiensis		0.5
Stink bugs	Carbaryl	1.5	
	Methyl parathion	0.5	0.25–0.5

Reduction in rates of application alone provides a measure of selectivity favoring some species of natural control agents. However, much greater selectivity is provided by substituting low rates of application of carbamate and organophosphorus insecticides for the much heavier rates of mixtures of organochlorine and organophosphorous chemicals that had been used previously.

The kinds of truly narrow-spectrum insecticides that are needed for control of soybean insect pests have not been discovered. *Bacillus thuringiensis* has a desirably narrow spectrum of activity and is effective for control of lepidopterous defoliators such as the velvetbean caterpillar and soybean looper. Unfortunately, in many areas where these pests are a problem their infestations often coincide with infestations of stink bugs that are not controlled by *B. thuringiensis*. In such situations growers are economically compelled to use an insecticide that controls all three pests. Therefore, selectivity has to be achieved by reduction in dosage rates and in basing recommendations on more realistic economic injury thresholds. Excellent research progress has been made along these lines.

Proper timing of insecticide applications is another way of achieving selectivity. Research in this Project at many locations and under varied conditions has demonstrated the critical importance of the principle of proper timing. One of the most dramatic examples is provided by data collected in North Carolina in 1976 during a heavy outbreak of *H. zea* on drouth stressed, late planted soybeans. An application of methyl parathion made just prior to invasion of the fields by *H. zea* eliminated the predator population and allowed *H. zea* larval populations to develop to such high levels that the crop was practically destroyed. Yields were reduced to only 6 bushels per acre by this treatment as compared to 18 bushels per acre in untreated plots, and 30 bushels per acre in plots treated with properly timed applications of carbaryl, methomyl, or methyl parathion (Table 3.20). Not only is proper timing critical for successful control, but it may often be used to help conserve populations of natutal control agents.

Table 3.20.   Effect of Timing Insecticide Applications on Control of *H. zea* and Yield of Soybean in Large Scale Field Trials—North Carolina, 1976

Time of Application	Yield Bu./Acre
Immediately prior to invasion of fields by *H. zea*	6
When *H. zea* larval populations reached economic injury levels	30
Control	18

Plant Disease/Vector Relationships

Substantial progress has been made toward a better understanding of the association between insects and the viral, fungal, and bacterial diseases of soybean. The number of species of aphids proved capable of transmitting soybean mosaic virus (SMV) has been increased to more than 12. However, the cowpea aphid, *Aphis craccivora* Koch, is the major vector in field transmission of the disease in most areas.

Relationships of the chrysomelid vectors of bean pod mottle mosaic virus (BPMV) have been clarified. Four species, bean leaf beetle, grape colaspis, banded cucumber beetle, and spotted cucumber beetle, have all been proved capable of transmitting BPMV. However, only the bean leaf beetle is of any significance as a vector of this important virus disease. The other three species are inefficient vectors even in laboratory experiments.

Bean pod mottle mosaic virus may be controlled very effectively and economically by use of trap plots to control its bean leaf beetle vector (see above). Planting small acreage trap plots of soybean about 2 weeks earlier than the main planting and controlling with insecticide the bean leaf beetles attracted to the trap crop before they move into the main planting is effective and inexpensive. By choosing an early-maturing variety, the bean leaf beetle trap crop also serves as a trap crop for stink bugs later in the season, thus a two-fold purpose is served.

Tobacco ringspot virus of soybeans (TRSV), "bud blight," continues to exhibit some puzzling aspects with regard to its epidemiology. It is well known that TRSV epidemics occur every few years in some areas (Ford & Goodman, 1976). Characteristically, the disease develops first along field margins. Eventually, it may spread throughout affected fields. In some cases the spread is dramatically rapid. Proven vectors are grasshoppers, *Melanoplus* spp., onion thrips, *Thrips tabaci* L., spider mites, *Tetranychus* spp. and nematodes, *Xiphinema americanum* Cobb. The characteristic appearance of the disease first in the field margins may be explained by assuming any of the proven vectors, except *Xiphinema*, to be responsible for movement of the virus from surrounding reservoirs of weeds or other crop plants known to be hosts. But the further, rapid spread of the virus cannot be satisfactorily explained with present knowledge. The adult onion thrips is sufficiently mobile to fit the pattern of a vector capable of spreading the disease in epidemic proportions throughout a field if it were not for the fact that only the nymphs acquire and transmit the virus. Substantially more research is needed on this problem.

A disease of unknown etiology is occurring with increasing frequency in many soybean-producing areas. Some of its symptoms resemble those of tobacco ringspot virus but they more closely resemble those of a disease of similar unknown etiology that affects soybean in the Cauca Valley of Colombia (Granada, 1976). Affected plants become indeterminate in growth, often continuing to grow and fruit until killed by frost. The disease agent is readily transmissible through seed and by grafting. It is not mechanically transmissible and has not been transmitted in laboratory tests using aphids, chrysomelid beetles, spider mites, thrips, or whiteflies. Plant pathologists cooperating in the study have thus far been unable to demonstrate the presence of any causative agent in diseased tissues by use of light and electron microscopy or serologic techniques.

The association of the southern green stink bug with "yeast spot" has been studied intensively (Daugherty, 1967). An association with a large number of additional organisms including both fungi and bacteria has been reported recently (Ragsdale, 1977). Surface-sterilized-field-collected southern green stink bugs allowed to feed on soybean infusion agar (SIA) under aseptic conditions readily transmitted species of seven genera of bacteria. After incubation of the SIA plates on which they had been allowed to feed, those bugs that had thus transmitted bacteria were dissected under asceptic conditions, and the labium, stylets, salivary receptacle, foregut, midgut, and hindgut removed. Each organ was placed in brain heart infusion broth (BHI) and incubated 24 hours. From the tubes that showed growth, BHI agar plates were streaked with an inoculating loop and incubated at 30°C for 24 hours. The results were as follows:

Bacterium	Organs from which Bacteria Were Isolated
Brevibacterium	Salivary receptacle
Cellulomonas	Labium, foregut, midgut
Corynebacterium	Labium, stylets, salivary receptacle, foregut, midgut, hindgut
Flavobacterium	Salivary receptacle
Micrococcus	Stylets
Pseudomonas	Foregut, midgut, hindgut
Xanthomonas	Hindgut

Among these bacteria are three genera, *Corynebacterium*, *Pseudomonas*, and *Xanthomonas* that include species responsible for important bacterial diseases of soybean in many areas throughout the world. The fact that they occur generally in many of the internal organs of the southern green stink bug is a significant finding. It is also significant that these bacteria possess extra enzymes that are capable of digesting lipids, carbohydrates, and proteins of soybean. Determining the biological significance of these findings will require additional research.

Monitoring Populations and Sampling

Improvements made during the last 4 years in techniques of monitoring pest populations and in refinements of economic damage thresholds have led to substantial reductions in costs of pest control to the soybean grower. By taking into account information that has been accumulated on the biology and behavior of insect pest species and varietal differences in rates of growth and development in soybean, accurate predictions can be made of the time during which various varieties will be susceptible to attack by the various insects in specific areas. Through partitioning the crop by maturity group, planting dates, and row widths, the portion of the acreage susceptible to a particular pest at a specific time can be predicted and monitoring efforts can be concentrated and restricted at these focal points in space and time. Thus the number of times a field must be inspected during the season can be drastically reduced. This approach to monitoring pest populations is most effective for the corn earworm, Mexican bean beetle, southern green stink bug, soybean looper, and velvetbean caterpillar, five of the most important pests of soybean. By using it, one inspection date of any field is all that is required in the treatment decision process for the corn earworm in North Carolina and population assessment (monitoring) of the stink bug complex, soybean looper, and velvetbean caterpillar in Louisiana need not be initiated prior to August 15.

Refinements in sampling methods for soybean pests have also been made (Jensen & Rudd, 1977). A procedure has been developed for converting populations of some of the major soybean pests to numbers obtained by the " drop cloth " method of sampling and vice versa. The following are conversions of the numbers of five major pest species, collected by sampling with a 15-inch-diameter sweep net, to numbers per foot of row.

Species	Economic Injury Thresholds, Numbers/ Foot of Row	Numbers/sweep of a 15-Inch Diameter Net
Nezara viridula	1	0.37 ± 0.004
Pseudoplusia includens	8	1.54 ± 0.09
Anticarsia gemmatalis	8	3.1 ± 0.4
Plathypena scabra	8	3.3 ± 1.5
Heliothis zea	3	0.38 ± 0.24

Sampling populations with a sweep net has two major advantages. It can be used effectively in fields in which the plants have fallen down or become so intertwined from row to row that the drop cloth cannot be used effectively, and more area can be sampled in an equal amount of time.

CONCLUDING STATEMENT

Grower-scale integrated pest management systems have been tested during the past two seasons in both Louisiana and North Carolina. The following tactics have been successfully incorporated into the system in North Carolina:

1. Using economic damage thresholds for defoliation and numbers of pod feeding insects as the basis for insecticide use decisions
2. Scouting at regular intervals to assess pest populations and damage
3. Protecting and enhancing arthropod natural enemies of insect pests
4. Selecting early maturing varieties to reduce populations of lepidopterous larvae, especially the corn earworm
5. Selecting planting dates to render some varieties unattractive to the corn earworm through canopy enhancement and asynchronization of flowering with peak moth flight
6. Planting all double-crop soybeans (late-planted soybeans following wheat) in narrow rows to promote rapid canopy development for corn earworm management
7. Using minimum rates of selective insecticides when pest populations or damage exceeds economic threshold levels.

The system developed for control of soybean pests in Louisiana incorporates the following tactics:

1. Using economic damage thresholds as the basis for all insecticide use decisions
2. Scouting fields at regular intervals, during periods of growth and development in which the crop is at risk only, to assess damage and monitor populations of both pest species and their natural enemies
3. Taking maximum advantage of the regulatory effects on pest populations of biological control agents, including complexes of polyphagous predators, egg, larval, and adult parasites and microbial pathogens
4. Using trap crops of small acreages of early planted, early maturing Group V soybean varieties for control of the bean leaf beetle and bean pod mottle mosaic virus for which it is the only important vector, and for control of stink bugs.
5. Using minimum rates of application of the most selective insecticides when economic damage levels are exceeded.

The principles outlined above have been incorporated into Cooperative State Extension Service recommendations in Louisiana for general use by growers during 1977. They are being accepted enthusiastically and rapidly.

Adoption of these systems has allowed the amount of insecticide on a per acre basis required for the control of soybean insect pests to be reduced by one-half to one-third that used prior to the beginning of research on the

Soybean Subproject. It has also resulted in substituting less persistent and more selective carbamate and organophosphorus insecticides for the more persistent and environmentally hazardous organochlorine compounds. With these systems less than 20 percent of the approximately 20 million acres in the United States that are at greatest risk from insect attack are required to be treated an average of not more than once with insecticide during any one season. Much less treatment is required for the remainder of the acreage planted to soybean.

As the sixth year of research on the Soybean Subproject begins, it is clear that its major objective has been achieved. Integrated pest management systems for the soybean ecosystem have been developed that are effective, economical, ecologically acceptable, and appear to have the quality of maintaining stability that is so desirable. The systems are far from being as sophisticated as they will evolve to be eventually. In many ways they are as yet relatively crude and simple. Nonetheless, they are so soundly based ecologically that there is every reason to believe that they will afford the level of control required to prevent loss from insect pests and at the same time, avoid the critical problems that have characterized so much of insect control in the United States during the last quarter century. Furthermore, the systems possess the dynamic character that favors continuing addition of new tactics and refinements required to adapt to the rapidly changing problems of pest control for the more than 50 million acre soybean crop—an acreage that continues to expand.

LITERATURE CITED

Brousseau, D. E. 1975. A field study of the seasonal history of *Entomophthora gammae* Weiser and its relationship to the soybean looper, *Pseudoplusia includens* (Walker). Master's thesis, Louisiana State University.

Burleigh, J. G. 1972. Population dynamics and biotic controls of the soybean looper in Louisiana. *Environ. Entomol. 1*:290–294.

Buschman, L. L., W. H. Whitcomb, R. C. Hemenway, D. L. Mays, Nguyen Ru, N. C. Leppla, and B. J. Smittle. 1977. Predators of velvetbean caterpillar eggs in Florida soybeans. *Environ. Entomol. 6*:403–407.

Daugherty, D. M. 1967. Pentatomidae as vectors of yeast spot disease of soybean. *J. Econ. Entomol. 60*:147–152.

Dietz, L. L., R. L. Rabb, J. W. Van Duyn, W. M. Brooks, J. R. Bradley, Jr., and R. E. Stinner. 1976. A guide to the identification and biology of soybean arthropods in North Carolina. *N. C. Agric. Exp. Stn. Tech. Bull.* 238.

Eastman, C. E. 1976. Infestation of root nodules of soybean by larvae of the bean leaf beetle, *Cerotoma trifurcata* (Forster) and the platystomatid fly *Rivellia quadrifasciata* (Macquart). Doctoral dissertation, Louisiana State University.

Eddy, C. O., and W. C. Nettles. 1930. The bean leaf beetle. *S. C. Agric. Exp. Stn. Bull.* 265.

Ford, R. E., and R. M. Goodman. 1976. Epidemiology of soybean viruses, in L. D. Hill (ed.), *World Soybean Research.* Interstate Printers and Publishers, Danville, Ill., pp. 501–512.

Granada, G. A. 1976. A leafhopper transmitted disease of soybean in the Cauca Valley, Colombia. *Am. Phytopath. Soc. Proc. Abstr.* Vol. 3.

Greene, G. L. 1976. Pest management of the velvetbean caterpillar in a soybean ecosystem, in L. D. Hill (ed.), *World Soybean Research.* Interstate Printers and Publishers, Danville, Ill., pp. 601–610.

Greene, G. L., W. H. Whitcomb, and R. Baker. 1974. Minimum rates of insecticide on soybean: *Geocoris* and *Nabis* populations following treatment. *Fla. Entomol.* 57: 114.

Hatchett, J. H., Beland, G. L., and E. E. Hartwig. 1976. Leaf-feeding resistance to bollworm and tobacco budworm in three soybean plant introductions. *Crop Sci. 16:* 277–280.

Ignoffo, C. M., N. L. Marston, B. Puttler, D. L. Hostetter, G. D. Thomas, K. D. Biever, and W. A. Dickinson. 1976. Natural biotic agents controlling insect pests of Missouri soybeans, in L. D. Hill (ed.), *World Soybean Research.* Interstate Printers and Publishers, Danville, Ill., pp. 561–578.

Isely, D. 1930. The biology of the bean leaf beetle. *Ark. Agric. Exp. Stn. Bull.* 248.

Jensen, R. L., L. D. Newsom, and J. Gibbens. 1974. The soybean looper: Effects of adult nutrition on oviposition, mating frequency and longevity. *J. Econ. Entomol.* 67:467–470.

Jensen, R. L., and W. G. Rudd. 1977. Sweep-net and ground cloth sampling for insects in soybeans. *J. Econ. Entomol.* 70:301–304.

Johnson, M. W., R. E. Stinner, and R. L. Rabb. 1975. Ovipositional response of *Heliothis zea* (Boddie) to its major hosts in North Carolina. *J. Econ. Entomol. 4:* 291–297.

Kiritani, D. 1963. Oviposition habit and effect of parental age upon the post-embryonic development in the southern green stink bug *Nezara viridula. Jpn J. Zool.* 13:88–96.

Kiritani, K., N. Hokyo, and J. Yukawa. 1963. Co-existence of the two related stink bugs *Nezara viridula* and *N. antennata* under natural conditions. *Res. Popul. Ecol.* 5:11–22.

Kiritani, K., and T. Sasaba. 1969. The differences in bio- and ecological characteristics between neighboring populations in the southern green stink bug, *Nezara viridula* L. *Jpn J. Ecol. 19:*177–184.

Kogan, M. 1976. Evaluation of economic injury levels of soybean insect pests, in L. D. Hill (ed.), *World Soybean Research.* Interstate Printers and Publishers, Danville, Ill., pp. 515–533.

Kogan, M. and D. Cope. 1974. Feeding and nutrition of insects associated with soybeans. 3. Food intake, utilization, and growth in the soybean looper, *Pseudoplusia includens. Ann. Entomol. Soc. Am.* 67:65–72.

Kuo, Shau-yu. 1975. Persistence of the nuclear polyhedrosis virus of soybean looper *Pseudoplusia includens* (Walker) (Lepidoptera: Noctuidae) in the soybean ecosystem. Master's thesis, Louisiana State University.

Leonard, L. T., and C. F. Turner. 1918. Influence of *Cerotoma trifurcata* on the nitrogen-gathering functions of the cowpea. *J. Am. Soc. Agron. 10*:256–261.

McConnell, W. R. 1915. A unique type of insect injury. *J. Econ. Entomol. 8*:261–266.

Mueller, A. J., and B. A. Dumas. 1975. Effects of stem girdling by the three cornered alfalfa hopper on soybean yields. *J. Econ. Entomol. 68*:511–512.

Murdoch, W. W., and J. R. Marks. 1973. Predation by coccinellid beetles: Experiments on Switching. *Ecology* 54:160–167.

Newman, G. G., and G. R. Carner. 1975. Environmental factors affecting conidial sporulation and germination of *Entomophthora gammae*. *Environ. Entomol. 4*:615–618.

Newsom, L. D., R. L. Jensen, D. C. Herzog, and J. W. Thomas. 1975. A pest management system for soybeans. *La. Agric. 18*(4):10–11.

Pitts, J. 1977. Photoperiodic induction of diapause in the southern green stink bug, *Nezara viridula* (L.). Master's thesis, Louisiana State University.

Rabb, R. L., R. E. Stinner, and Robert van den Bosch. 1976. Conservation and augmentation of natural enemies, in C. B. Huffaker and P. S. Messenger (eds.), *Theory and Practice of Biological Control.* Academic, New York, pp. 233–254.

Ragsdale, D. W. 1977. Isolation and identification of bacteria and fungi transmitted during feeding from various organs of *Nezara viridula* (L). Master's thesis, Louisiana State University.

Reid, J. C. 1975. Larval development and consumption of foliage by the velvet bean caterpillar, *Anticarsia gemmatalis* Hubner (Lepidoptera: Noctuidae) in the laboratory. Doctoral dissertation, University of Florida.

Shepard, M. G., R. Carner, and S. G. Turnipseed. 1974. Seasonal abundance of predaceous arthropods in soybeans. *Environ. Entomol. 3*:985–988.

Sprenkel, R. K., and W. M. Brooks. 1975. Artificial dissemination of *Nomuraea rileyi*, an entomogenous fungus of lepidopterous pests of soybeans. *J. Econ. Entomol. 68*: 847–851.

Sprenkel, R. K., and W. M. Brooks. 1977. Winter survival of the entomogenous fungus *Nomuraea rileyi* in North Carolina. *J. Invertebr. Pathol. 29*:262–266.

Stinner, R. E., R. L. Rabb, and J. R. Bradley. 1974. Population dynamics of *Heliothis zea* (Boddie) and *H. virescens* (F.) in North Carolina: A simulation model. *Environ. Entomol. 3*:163–168.

Strayer, J. R. 1973. Economic threshold studies and sequential sampling for management of the velvetbean caterpillar, *Anticarsia gemmatalis* Hubner on soybeans. Doctoral dissertation, Clemson University.

Turnipseed, S. G. 1973. Insects, in B. E. Caldwell (ed.), *Soybeans: Improvement, Production and Uses.* Am. Soc. Agron., Madison, Wis., pp. 545–572.

Turnipseed, S. G., J. W. Todd, G. L. Greene, and M. H. Bass. 1974. Minimum rates of insecticides on soybeans: Mexican bean beetle, green cloverworm, corn earworm and velvetbean caterpillar. *J. Econ. Entomol. 67*:287–291.

Turnipseed, S. G., and M. Kogan. 1976. Soybean entomology. *Annu. Rev. Entomol. 21*: 247–282.

Waldbauer, G. P., and M. Kogan. 1976. Bean leaf beetle: Phenological relationship with soybean in Illinois. *Environ. Entomol. 5*:35–44.

Watson, J. R. 1916. Life-history of the velvetbean caterpillar (*Anticarsia gemmatilis* Hubner). *J. Econ. Entomol. 9*:521–528.

Whitcomb, W. H., H. A. Denmark, A. P. Bhatkar, and G. L. Greene. 1972. Preliminary study of ants in soybean fields. *Fla. Entomol. 55*:130–142.

Wuensche, A. L. 1976. Relative abundance of seven pest species and three predacious genera in three soybean ecosystems. Master's thesis, Louisiana State University.

4

THE SYSTEMS APPROACH TO RESEARCH AND DECISION MAKING FOR SOYBEAN PEST CONTROL

W. G. Rudd

Departments of Computer Science and Entomology
Louisiana State University, Baton Rouge, Louisiana

W. G. Ruesink

Illinois Natural History Survey, Urbana, Illinois

L. D. Newsom
D. C. Herzog
R. L. Jensen

Department of Entomology
Louisiana State University, Baton Rouge, Louisiana

N. F. Marsolan

Department of Chemical Engineering
Louisiana State University, Baton Rouge, Louisiana

Research supported in part by the Graduate Research Council of Louisiana State University.

Present address of D. C. Herzog is: AREC, Quincy, Florida; and of N. F. Marsolan is: Department of Chemical Engineering, Louisiana Technological University, Ruston, Louisiana.

INTRODUCTION

This chapter furnishes a second aspect of the progress of the IBP-sponsored, NSF-EPA supported National Integrated Pest Management project on soybean insects. The other aspect was covered in Chapter 3.

The so-called systems approach applies several tools—computer-based modeling and simulation, mathematics, statistics and optimization—to the solution of complex problems in which there may be several strongly interacting components. A typical target problem for the systems approach is represented by the analysis of the soybean crop ecosystem with the objective (one among many possible objectives) of controlling soybean insect pests in such a way as to maximize grower profits while holding harmful environmental effects at a minimum. To date the primary emphasis in this work has been construction of mathematical and computer-based simulation models to use as aids in designing and evaluating pest management strategies and tactics (Chapter 1). Thus our objective has been to develop (sub) models that show typical behavior in population dynamics, soybean yield, and response to external influences. Experiments and/or optimization techniques with the models are carried out to help choose the best tactics and strategies from sets of possible alternatives.

In soybean pest management, no real effort has yet been made to use these models as part of an actual pest management system. The models already have the necessary predictive capability for such use, but funds for the development of a delivery system are still being sought.

Comparatively little is known about the biology and ecology of the insect components of the soybean system. Therefore, much of the emphasis of modeling research has been in the development of prototype model systems. These prototype structural outlines require a minimum of biological data, which is fortunate, since few are available; generally, these models incorporate most of the quantitative information that is known about the species involved.

Since the models are " up to date " with current research results, they supply guidelines for further field and laboratory experimental work. At any given time, the modeling process is awaiting vital new pieces of information in order to progress further. The models must deal with the real world. This means that the modeling effort forces experimental work into channels which will provide the data of most use in describing the dynamics of or predicting insect populations, soybean growth and yields, and other parameters that are needed in the use of the model for " experiments " with pest management strategies and tactics.

Two factors—the relative scarcity of information about the soybean system and the need for simple models to help guide research efforts as discussed above—dictate that we follow a " top-down " approach in modeling the soybean crop system. In the top-down approach, the guideline is that the simplest model that explains and/or predicts those phenomena that are to be explained and/or predicted is the preferred model. In modeling the behavior of a system process—the population dynamics of an insect species or soybean plant growth, for example—one starts with a crude model and adds mathematical descriptions of basic biological information to the model until the desired level of accuracy of the model is obtained. The decision about whether or not the model " works " is based upon comparison of the model results with experimental data.

For example, in the insect population dynamics models described below, the starting model consists of a set of computer programs that predict insect populations from developmental (life-table) data and counts of initial populations. The output from the starting model rarely agrees quantitatively with measured field populations. One must incorporate other information, such as effects of food quality or dispersal behavior, before the model results begin to agree with field data.

Unlike many other groups working on similar problems, we have not yet found it necessary to include in our models the effects of weather on the processes being modeled. The model results agree well with field measurements without considering the effects of temperature, solar radiation, rainfall, and other such factors on developmental and growth rates. Louisiana weather during the soybean growing season varies little; mean daily temperatures change less than 5°F from mid-May through mid-September. Furthermore, it is apparent that food quality effects, dispersal, predation, the effects of disease epizootics, and other similar processes have a much larger effect on population dynamics than do climactic variations. While this approach helps keep the models as simple as possible, which is one of our objectives, it does at present prevent us from modeling such processes as overwintering population dynamics.

The remainder of this chapter should be considered an interim progress report on the development and use of models in the soybean crop system. The work was carried out under the auspices of the NSF-EPA National Integrated Pest Management project indicated above.

All of the topics discussed below are still the focus of active research. We begin by describing a powerful descriptive technique we have developed to aid communications among research workers from various disciplines involved in research problems such as this one. We then consider our insect population dynamics models and a preliminary model of stinkbug dispersal. We describe a soybean plant growth model and present results of a preliminary model-based economic injury level study. Finally, we report the results of an application of optimal control techniques in soybean ecosystem management modeling.

HIERARCHICAL DESCRIPTION OF SOYBEAN ECOSYSTEMS

A prerequisite to work on ecosystem management problems is a comprehensive description of the system under study. Investigators from all disciplines involved must have a basic understanding of how the various system components interact and must be able to communicate with other investigators. They must know what the most important processes in the system are and how these processes influence other processes in the system.

There are several system block diagramming techniques that have been developed just for such applications. A typical example is DYNAMO (Forrester, 1961), in which one links together reservoirs of state variables and controls flows from one reservoir to another by sets of valves, the degrees of opening of which are determined by the rates of the various processes. For our purposes, DYNAMO and other similar techniques are less than sufficient. We have developed a descriptive technique that emphasizes the hierarchical structure of an ecosystem and simplifies the task of establishing and documenting the relationships between the various dynamic processes in the system. The technique proves to particularly effective as an aid in crossing communications boundaries between disciplines.

The technique we use is an adaptation of HIPO—Hierarchy, Input, Process, Output—which has recently been developed for use in documenting large computer program systems (e.g., Katzan, 1977).

The technique is based upon the fundamental relationships between systems states and processes. At any given instant, an ecosystem within prescribed spatial boundaries is in a well-defined *state*. The state of a system is determined by the current values of its *state variables*, all of which are in principle measurable. Some examples of state variables for a crop ecosystem are the weights of various plant components, the numbers of insects of various species and developmental stages, and the concentrations of various soil components.

In theory, one would have to measure a large number of values of state variables in order to determine the state of a system exactly. In practice, we at first select a few quantities which appear to give an overall indication of the state of the system. These are quantities chosen to give, as nearly as possible, the most information about the underlying processes in the system.

A *process* is an event or group of events that moves the system or a subsystem from one state to another. In other words, processes provide dynamic behavior of the system, moving it from state to state as time progresses. Some of the major processes that form the dynamics of an ecosystem are photosynthesis, translocation, insect development and mortality, and evaporation, for example.

The hierarchical structure of an ecosystem becomes evident when we consider the relationships between processes. In general, a given process consists of several subprocesses, each of which can in turn be divided into more fundamental processes, and so on. For example, some processes that lead to insect population dynamics are dispersal, development, and mortality. Mortality, in turn, consists of such subprocesses as parasitization, predation, and death due to insecticide application. Figure 4.1 shows an example of the small portion of the soybean crop ecosystem with which we have been concerned up to now. A more complete set of diagrams is available from the senior author.

Each process has an associated set of inputs and outputs and may or may not be influenced by extraneous factors which might modify how a process occurs or its rate of occurrence without actually being used up or created by the process. For example, some of the inputs for the photosynthetic process are light, CO_2, and water, while an output is carbohydrate.

The primary factor that makes the use of our descriptive system attractive in modeling efforts is the close analogy between processes, (sub) models, and

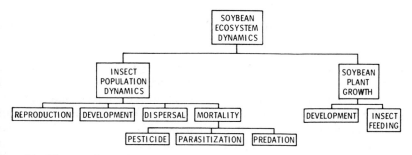

Figure 4.1. Diagram of a portion of the hierarchy of processes in the soybean crop ecosystem.

computer programs. When one describes a process by giving its inputs, outputs, and a set of rules, equations, or an algorithm for calculating the products of the process from its inputs, one is in fact defining a model of the process and designing a computer subroutine to implement the model.

POPULATION DYNAMICS

There are several insect population dynamics models that have been used in soybean insect pest population dynamics studies. Menke (1973) was the first to apply modeling techniques to soybean pest management in his stochastic population dynamics model of the velvetbean caterpillar, *Anticarsia gemmatalis* Hubner.

Waddill et al. (1976) describe a model for Mexican bean beetle, *Epilachna varivestis* Mulsant, populations using a " box car " technique similar to that described in Tummala et al. (1975). Marsolan and Rudd (1975) describe a partial differential equation (Forrester) model for Southern green stinkbug, *Nezara viridula* (L). Stinner et al. (1974) model populations and dispersal of *Heliothis* spp. in North Carolina (see also Chapters 2, 6, 8, 9, 11).

The most intensive model-based studies of the soybean crop system began with Rudd (1975). The general approach is based upon two fundamental ideas. First, while insect species may differ in their appearance, behavior, fecundity, dispersal, patterns, and many other aspects, the processes involved in their development and reproduction are, at a high level in the process hierarchy, structurally and mathematically independent of species. Individuals enter a development stage, spend a variable amount of time there, and then progress to the next stage. Similarly, the second fundamental idea is actually a philosophy: the simplest model that works (in the sense of providing accurate explanations and/or predictions) is the preferred model.

The mathematical and computational details of our model are described in Rudd (1975). Expressed as a system of ordinary differential equations, the fundamental model equations are

$$\frac{dx(t)}{dt} = A \cdot Q(t) - M(t) \tag{1}$$

where $x(t) = [x_1, (t), \ldots x_n(t)]$ and the vectors $Q(t)$ and $M(t)$ are defined in an analogous fashion. Here $x_i(t)$ is the number of individuals in stage i at time t. The variable $Q_i(t)$ is the flux or rate of movement *from* stage i, while $M_i(t)$ is the mortality rate for stage i. The A matrix describes the direction of movement of individuals from stage to stage; a positive A_{ij} indicates that individuals are entering stage i from stage j while a negative A_{ij} means that individuals are leaving stage i.

In the computer solution of the discretized version of Equation (1), we first use the past histories of entries and exits into and from each stage to compute the $Q_i(t)$. We then compute the mortalities $M_i(t)$ and use a numerical procedure to integrate the equations for the next time step. While we have not yet done so, it would be a reasonably simple procedure to convert the temporal units in the equations represented in Equation (1) to physiologic units, thereby incorporating the effects of temperature variations into the models.

There is no restriction in the model as to what a " stage " is, other than that it be possible to determine a function that describes how individuals progress through the stage in time. Equation (1) has been used to compute population numbers of several species simultaneously, a single species whose individuals move about among several fields, and an insect pest and its fungal pathogen. The plant growth model described below could be formulated in the above terms, but we have not yet done so.

A large number of validation studies have been done using the model. Data required include initial populations, developmental data, and mortality rates. We use measured field populations for the initial values and

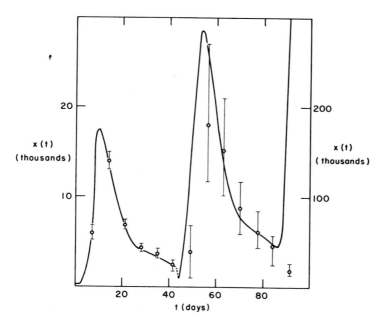

Figure 4.2. Model results (this article) (——————) and stochastic model results (circles and error bars) vs. time (days) from Menke (1973) for third instar and higher velvetbean caterpillar, *A. gemmatalis*, larvae.

developmental data from the literature or local laboratories. Since field data on mortality rates are not available for the insect and locations we have studied, we must adjust the model mortality rates until a good fit with a typical set of field survey data is obtained. Thus the modeling process provides an indirect way to determine actual field mortality rates. As mentioned above, it is sometimes necessary to include food quality effects and/or dispersal behavior in order to obtain reasonable results.

Figure 4.2 shows a comparison of the model results with the output from Menke (1973) for velvetbean caterpillar, *A. gemmatalis*. The results of both models show good agreement with field data (Menke & Green, 1976). Developmental data for these studies are from Watson (1916).

In Figure 4.3 we show model results and typical field data for the soybean looper, *Pseudoplusia includens* (Walker). The field data are from Burleigh (1972). The developmental data for the model were taken from Mitchell (1967). The soybean fields from which the data were taken are in an area in which a significant amount of cotton is grown. The rapid rise in population numbers starting about day 60 is due to an increase in oviposition rate by a factor of 60. Jensen et al. (1974) showed that an increase of this magnitude occurs when the adult females have a supply of nectar available, as they do

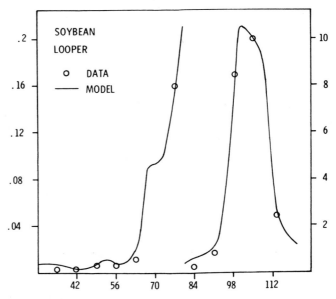

Figure 4.3. Model results (————) and field data (circles) from Burleigh (1972) vs. time (days) for soybean looper, *P. includens* (third and higher instar larvae).

when cotton begins to bloom on about July 1. Without knowing the actual factor by which the oviposition rate increases, modelers determined that the factor must be approximately 60.

Most of the population modeling efforts to date have focused on the southern green stinkbug, *Nezara viridula* (L.). These insects have been found in clover in early May and their population monitored over large tracts of land throughout the growing season in pilot pest management program studies. Figure 4.4 compares the model output with early season populations in clover in Louisiana for second through fifth instar and adult southern green stinkbugs. Developmental data used are from Kiritani et al. (1963) and Kiritani (1964). This match between data and model is used to align the model time with calendar time. Figure 4.5 compares mid-season model adult populations and real data from a field in which an early-maturing variety (Dare) was planted. This population match between model and data has been found to be typical over a period of 4 years for *N. viridula* in early-maturing varieties.

In order to successfully model *N. viridula* populations throughout the growing season, it is necessary to model the movement of these highly mobile

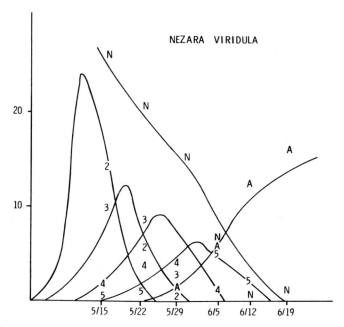

Figure 4.4. Model results (——————) and field data (2–5: second-fifth instar, A: adults, N: total nymphs) vs. time (days) for *N. viridula* in clover early in the growing season.

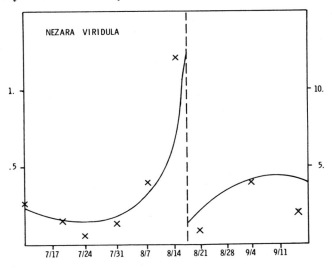

Figure 4.5. Model results (————————) and adult populations (×) for *N. viridula* adults vs. time (days), mid-season in an early-maturing variety.

insects from field to field. Our initial approach to this problem is to run the model for three fields simultaneously. The fields differ in the parameter values used in the plant growth model described below. One field represents an early season alternate host, the clover discussed above in connection with Figure 4.4. The other fields might represent, for example, an early (Dare) and an intermediate (Lee) maturing variety. Mortality rates in these fields depend strongly upon the developmental stage of the crops, as determined by Thomas (unpublished data). To model the insect dispersal and settling behavior, we use an approach similar to that developed by Stinner et al. (1974). We assume that adult stinkbugs choose to move to and settle in various fields in proportion to the relative attractiveness of the fields. A potential destination field's attractiveness depends upon several factors, including its distance from the insect source, its area, and, most importantly, the degree of development of the soybean crop in the field.

In Figure 4.6 we show early season populations of *N. viridula* in soybeans. The model curves are the results of several runs of the dispersal and settling model in each of which we used a different set of parameters to represent actual fields in which early- and late-maturing varieties were planted.

Figure 4.7 shows the full-season data and model results for a typical field planted with Dare soybeans. The model agrees well with data until the end of August, at which time the field populations show a large increase about 2

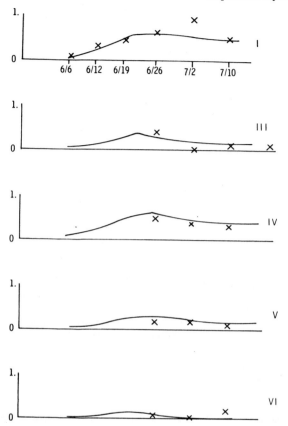

Figure 4.6. Early-season model results (————) and adult populations (×) of *N. viridula* vs. time (days) in several soybean fields.

weeks before the model shows a similar increase. A similar phenomenon appears when comparing data and model for the fields planted with Lee soybeans. Examination of population and dissection data from corn, cowpeas, and other hosts for the southern green stinkbug indicates that these hosts produce another cohort of stinkbugs about 2 weeks younger than those that come from the clover. This earlier cohort is responsible for the earlier increase in population. Currently, work is underway to model simultaneously 11 of the fields in the pilot project—three alternate host fields and eight soybean fields.

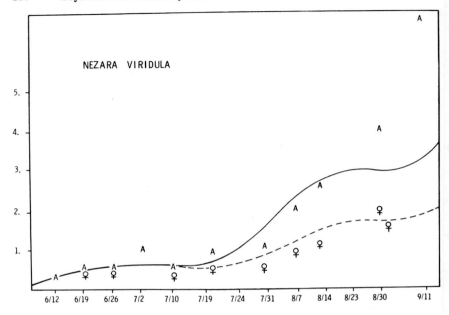

Figure 4.7. Full-season model adults (————————), model females (– – – – – –), adult data (A), and adult female data (♀) for *N. viridula* vs. time (days) in an early-maturing variety.

PLANT GROWTH MODEL

In order to do model experiments with various pest and crop management strategies and tactics it is necessary to have a model of the plant growth that does the following:

1. Predicts yield with no insect damage for several varieties under typical growing conditions
2. Responds correctly to differences in planting dates, row spacing, and other cultural variables
3. Responds correctly to simulated damage by insects

We have developed a first-generation plant growth model which partially satisfies these requirements. The model is based upon the soybean dry matter increase versus leaf area index data in Shibles and Weber (1965). A fit to their data yields

$$R(LAI) = \begin{cases} (5.4 + 47.8LAI - 6.23LAI^2)PTSMAX & \text{if} \quad LAI < 3.835 \\ (1 - e^{-0.914LAI})PTSMAX & \text{if} \quad LAI > 3.835 \end{cases} \quad (2)$$

where LAI is the leaf area index (ratio of leaf area to ground area), $PTSMAX$ is the maximum closed canopy rate of dry matter production, and R is the rate, in units of weight per unit area per unit time, of dry matter production.

Figure 4.8 shows the data from which the design and parameters for the original model were based. The data are for the Dare variety grown in Louisiana in 1975. $PTSMAX$ is simply the slope of a regression line through the total above-ground dry weight data (T) after canopy closure, which occurred on about day 47 for the data shown.

We define the following symbols:

t_E = time of emergence

t_p = time of first visible pods

t_s = time at which dry matter production effectively ceases

T = total above-ground dry weight = $P + S + L$

P = dry weight of hulls and beans

L = dry weight of leaves

S = dry weight of stems

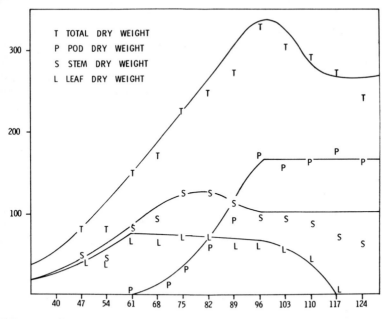

Figure 4.8. Model (——————) and experimental soybean dry weights versus time (days after planting). T = total dry weight, S = stem dry weight, L = leaf dry weight (not including petioles), P = pod dry weight.

After emergence and before the onset of fruiting, all above-ground dry matter goes to leaves and stems. Thus we have

$$\frac{dT}{dt} = R(LAI)$$

$$\frac{dL}{dt} = f\frac{dT}{dt}$$

$$\frac{dS}{dt} = (1 - f)\frac{dT}{dt} \qquad \text{if} \quad t_E < t < t_p \qquad (3)$$

$$\frac{dP}{dt} = 0$$

where t is the time and f, a constant, is the ratio of leaf dry weight to total dry weight before t_p.

In the absence of severe defoliation, we assume that all new dry matter produced between first podset and the termination of growth goes to the pods eventually. However, it is evident that during the early stages of pod formation and pod fill the pods are incapable of absorbing all the dry matter produced. We suggest that the reason for the "hump" in the stem-weight is that the excess dry matter produced during this period is stored temporarily in the stems. Temporary storage in the leaves probably occurs, but the effect appears to be negligible. We therefore postulate that during this period pods can accept no more dry matter than a constant C times their current weight. Thus, letting K be the rate of senescence of leaves, we have the following equations:

$$\frac{dT}{dt} = R(LAI)$$

$$\frac{dP}{dt} = CP$$

$$\frac{dS}{dt} = \frac{dT}{dt} - CP \qquad \text{if} \quad t_p < t < t_S \qquad (4)$$

$$\frac{dL}{dt} = -K$$

While the pods are small, dS/dt is positive but decreases as the pods grow. When the pods become large enough to absorb dry matter faster than the plant produces it, the stems yield stored dry matter back to the pods. In the solution to the model we do not let the stem weight decrease to a level below the stem weight at time t_p.

In the model, dry matter production and movement of materials stops at t_s. Leaves senesce and drop off and some dry matter is lost from the stems while the pod weight remains constant. After t_s, insects can have no effect upon yield other than by reducing the pod weight by feeding on the fruit. We therefore place little emphasis on the details of the model after $t = t_s$.

Figure 4.8 also shows the model results for leaf, pod, stem, and total dry weights. Figure 4.9 compares the computed LAI with the LAI measured using an optical integrating device.

We have found it possible to model other varieties simply by changing the four model parameters t_p, t_s, $PTSMAX$, and f. Model experiments show good agreement with the 3 years of fruit and total dry matter data for the intermediate Lee variety taken in North Carolina by Henderson and Kamprath (1970). Similar agreement has been obtained from partial sets of data taken here for the late-maturing Bragg variety.

Briefly, while the plant model should not yet be considered to be fully validated, the agreement between the model and data over a range of varieties and climates is most encouraging.

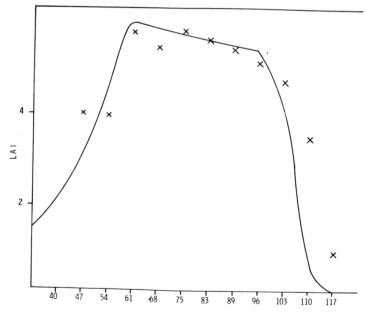

Figure 4.9. Model (——————) and experimental (×) leaf area index vs. time (days after planting).

It is important that the model respond properly to simulated defoliation by insects. Figure 4.10a shows yield reduction *versus* mechanical defoliation as a function of plant developmental stage at several levels of defoliation (Kalton et al., 1949). Without modification the model described above seriously over estimates the effect of defoliation at stage 5 (just before first pods are visible). This is evidently because the unmodified model has no way to "compensate" for defoliation while the plant evidently does. Given the current lack of information on just how the plant compensates for defoliation, we introduce the postulate that, if defoliation is severe enough to cause an eventual significant reduction in yield, a portion of the excess photosynthate destined for temporary storage in the stems goes to produce extra leaves instead. This can happen only during early podfill, since before t_p there is effectively no temporary storage. And, after the pods grow larger, there is no excess to go either to temporary storage in stems or to replace lost leaf area.

For the model we implement compensation as discussed above by replacing Equations (4) with the following if $LAI < LMAX$ ($LMAX$ is arbitrarily chosen to be 4):

$$\frac{dT}{dt} = R(LAI)$$

$$\frac{dT}{dt} = C \cdot P$$

$$\frac{dL}{dt} = k\,\frac{dT}{dt} - \frac{dP}{dt} - K \qquad t_p < t < t_s \qquad (5)$$

$$\frac{dS}{dt} = (1 - k)\left(\frac{dT}{dt}\right) - \left(\frac{dP}{dt}\right)$$

where k is the fraction of the excess production that goes to leaves. We currently use $k = 0.5$ if there is an excess—that is, $dT/dt - dP/dt > 0$, and $k = 0$ otherwise. Figure 4.10b shows the model results for a simulated duplication of the experiment of Kalton et al. (1949). There is reasonable agreement between the model and data; the largest difference is that the model does not compensate enough for extreme levels of defoliation. The differences between model and data at intermediate and low levels of defoliation may be attributed to differences in variety and leaf growth habits between the indeterminate variety used by Kalton et al. and the determinate variety used in the model.

The agreement between simulated mechanical depodding and experimental results (Thomas et al., 1974) is shown in Figure 4.11. No changes in the basic

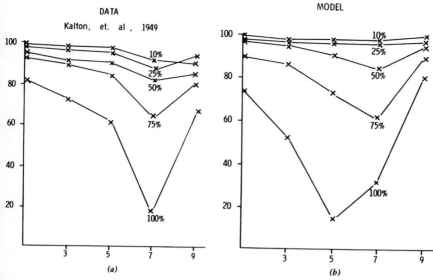

Figure 4.10. Yield (10% maximum) versus plant developmental stage at several levels of defoliation. *a*: Data of Kalton et al. (1946). *b*: Model results.

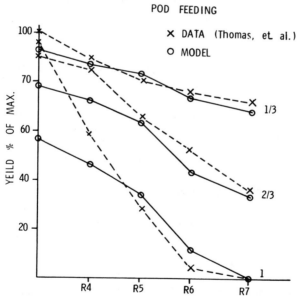

Figure 4.11. Comparison of model (○) and data (×) (Thomas et al., 1974) yield (% of maximum) versus plant developmental stage at several levels of mechanical depodding.

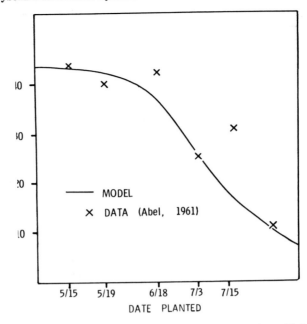

Figure 4.12. Yield versus planting date for Lee soybeans. Data (×) from Abel (1961). Model results (−).

model were necessary to obtain this level of agreement, indicating that, with the possible exception of the case of heavy damage early in podfall, the plant does nothing special to compensate for depodding.

Finally we show in Figure 4.12 a comparison of model and experimental yield for Lee soybeans (Abel, 1961). To obtain the model results we simply adjusted the model parameters f and $PTSMAX$ to the proper yield on May 15. We then adjusted t_E, t_p, and t_s as given in Abel (1961) to reflect the changes in these times due to variations in planting date. The model cannot predict the decrease in yield that Abel found when the soybeans were planted before May 15. It appears that differences in leaf area at podset can explain most of the observed dependence of soybean yield on planting date.

ECONOMIC INJURY LEVELS

An immediate application of modeling work lies in the area of computing how large a population of insect pests a crop can tolerate without a significant reduction in yield. Treatment policies could then be based upon economic injury levels (EILs) computed in this manner.

Ruesink (1975) developed a static model for the calculation of EILs of soybean defoliators. This model is based upon a quadratic fit of the yield loss versus mechanical defoliation data of Kalton et al. (1949). Given the amount of defoliation per individual insect, the market price of soybeans, the cost of insecticide application, the leaf area in various growth stages of the plant, and the expected yield, one can use Ruesink's model to compute the insect population at which it becomes profitable to apply insecticide at each stage of plant development for which the defoliation yield data are available. Kogan (1976) uses this model in a thorough study of economic injury levels of several defoliating pests of soybeans. Kogan also computes EILs that depend upon insect developmental stage.

We have begun a study in which we use dynamic models for EIL computation. By coupling an insect pest population dynamics model to a plant growth model and varying the insect population, we can examine the relationship between insect populations and economic yield.

Experiments with the plant model indicate the dominant role leaf area plays in the growth of normal and of damaged soybeans. We conclude that economic injury levels and treatment thresholds should depend upon the

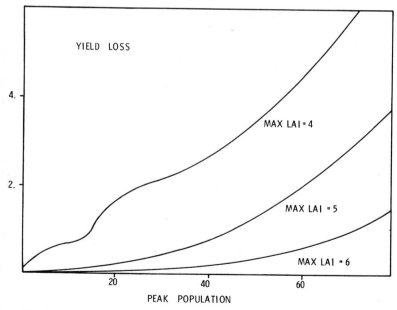

Figure 4.13. Yield reduction as a function of peak soybean looper population for various maximum leaf areas.

leaf area of the crop, which in turn depends among other factors, upon the variety of soybean planted, and the planting date.

Figure 4.13 shows the results of a study in which the basic model for looper population dynamics was coupled to the soybean plant growth model for a late-maturing variety. Defoliation rates are from Kogan and Cope (1974). The primary peak in late August and early September shown in Figure 4.3 has been found to be the typical population behavior for looper populations in soybean fields in the vicinity of cotton fields in Louisiana. The peak varies in height from year to year but its position in time does not change more than a week in either direction.

In Figure 4.13 we have plotted relative yield reduction versus the population at the early September peak. *MAX LAI* is the leaf area index at the beginning of podset in the absence of defoliating insects. The vertical scale has been left unlabeled purposely because this model study has yet to be validated.

The model study clearly indicates that economic injury levels and treatment thresholds depend strongly upon the leaf area of the plant. In fact, these results suggest that it might be a serious mistake to plant varieties which have low leaf areas in regions where defoliating insects threaten soybean crops.

OPTIMAL CONTROL

As a final example of the use of models in pest management, we present a discussion of our example of the application of optimal control techniques in pest management. Mathematical details and a thorough discussion of the results of this study are presented in Marsolan and Rudd (1975, 1976) and in Marsolan et al. (1977). Here we present a brief summary of the method and results.

The object of the optimal control methodology is to combine mathematical minimization or maximization techniques with a model of the processes involved to determine that control strategy which optimizes a cost function. In our studies we minimized the combined cost, in dollars, of pesticide application and yield loss due to insect activities.

The basis of the technique is the insect population and plant growth models. These were reformulated as a coupled set of partial differential equations (equivalent to the ordinary differential equation models described earlier) so that the established methods of variational calculus could be applied to find the optimal control policy.

The most complex system studied in the optimal control project consisted of the podfeeding southern green stinkbug *N. viridula*, the defoliating velvetbean caterpillar, *A. gemmatalis*, the soybean plant (intermediate-maturing variety) and a predator–parasite complex. The latter represents a lumping together of the effects of all of the predators and parasites of the velvetbean caterpillar (VBC). We did this because it appears that there is no dominant predator or parasite species responsible for most of the natural control of VBC; natural control of VBC by insects is due to a complex of controlling species. The population dynamics of the predator–parasite complex is modeled using the logistic equation. The predator–parasite complex is assumed to be much more sensitive to insecticide application than to population levels of either of its host species. Feeding rates for *N. viridula* are from Thomas and Newsom (unpublished data) while other developmental data and parameters are from sources previously referenced (e.g., Watson, 1916; Kiritani et al., 1963).

Figure 4.14 shows typical results of optimal control studies using this model. Plotted is the optimal daily fraction of pests killed by insecticide versus time after the first appearance of pods in the plant. Little is known about how well the predator–parasite complex recovers from application of pesticide. We therefore show two possible cases—the solid line, in which the complex recovers its original strength in about 2 weeks, and the dashed line, in which the complex never recovers.

It is certainly not feasible to apply insecticide continuously as Figure 4.14 suggests. Our interpretation of the results is that peaks in the optimal control

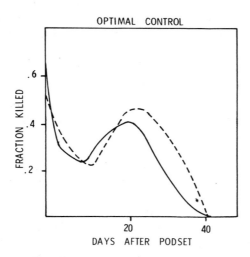

Figure 4.14. Optimal control policy results, fraction of insects killed vs. days after first pods visible. (——————) predator–parasite complex recovers in about 14 days. (– – – – –) predator–parasite complex does not recover.

policy represents times at which insecticide should be applied. This interpretation has been verified by computing, via a pattern-search technique, the optimal dates for discrete application of insecticide. The dates for application are the same as the dates at which the peaks in the continuous policy occur.

The control strategy suggested in Figure 4.14 for soybeans is exactly contrary to current recommended policies. The use of an economic treatment threshold usually means that the need for control action tends to increase as time progresses, or as populations increase. The optimal control policy suggests heavy treatments early and decreasing treatments as time progresses. The reasons for this behavior is twofold. First, the model is " smart " enough to know that there is no sense in applying insecticide late in the season, since defoliation can no longer reduce yield and impending harvest of this intermediate variety means that the cumulative effect of the remaining pod feeders decreases as time progresses. Second, the model knows that, if one is going to pay the price to apply an insecticide, the *total* damage by insects is minimized by applying the material as early as possible.

CONCLUSIONS

We have presented a brief survey of the current state of the science of developing and applying soybean ecosystem models for insect pest management. We have demonstrated the feasibility of our approach and we have indicated how these models can be used in the development of pest management programs.

The work remaining to be done far outweighs that accomplished to date. Some of the projects now underway include efforts to

1. Extend and generalize the stinkbug population dynamics and dispersal and settling model so that we can use it to analyze alternative geometrical arrangements of trap crops and the timing of pesticide application in trap crops
2. Carry out further validation tests of the plant growth model
3. Incorporate a dispersal model and settling into the optimal control system
4. Incorporate a fungal pathogen submodel into our defoliation population dynamics models so that the effects of fungicide application on insect populations can be studied
5. Determine how our models can be used to solve the "inverse" problem: given field survey data, how can we run the model "backwards" to obtain information on field development and mortality.

In many cases further advances await experimental determination of the nature of, and parameters in, the fundamental processes. For example, most EILs and treatment thresholds for soybean defoliators are based on mechanical defoliation data, which are assumed to mimic the effect of insect defoliation. There is no evidence that this is or is not the case. Much more experimental data are needed to determine the species that are the major natural predators and parasites of soybean insect pests and to determine their effectiveness, developmental patterns, and dispersal behavior.

As more knowledge is developed, this new information can be quickly added to the appropriate models to increase their accuracy and validity. For example, the effect of food quality on *N. viridula* mortality rates is already in the stinkbug models. As major predators and/or parasites are identified and their effects quantified, we can incorporate these data into the model. In this way the models serve as a stimulus for new experimental work, while the results of the experimental work are the basis upon which the models are built.

LITERATURE CITED

Abel, G. H., Jr. 1961. Response of soybeans to dates of planting in the Imperial Valley of California. *Agron. J. 53*:95–98.

Burleigh, J. G. 1972. Population dynamics and biotic controls of the soybean looper in Louisiana. *Environ. Entomol. 1*: 290–294.

Forrester, J. W. 1961. *Industrial Dynamics*. M.I.T. Press, Cambridge, Mass., pp. 369–381.

Henderson, J. B., and E. J. Kamprath. 1970. Nutrient and dry matter accumulation by soybeans. *N. C. Agric. Exp. Stn. Tech. Bull.* 197.

Jensen, R. L., L. D. Newsom, and J. Gibbens, 1974. The soybean looper: Effects of adult nutrition on oviposition, mating frequency, and longevity. *J. Econ. Entomol. 67*: 467–470.

Kalton, R. R., C. R. Weber, and J. C. Eldredge. 1949. The effect of injury simulating hail damage to soybeans. *Iowa Agric. Exp. Stn. Res. Bull.* 359.

Katzan, H., Jr., 1977. *Systems Design and Documentation: An Introduction to the HIPO Method*. Van Nostrand Reinhold, New York.

Kiritani, K., N. Hokyo, and K. Kimura. 1963. Survival rate and reproductivity of the adult southern green stinkbug, *Nezara viridula*, in the cage. *Jpn. J. Appl. Entomol. Zool. 7*:113–118.

Kiritani, K. 1964. The effect of colony size upon the survival of larvae of the southern green stink bug, *Nezara viridula*. *Jpn. J. Appl. Entomol. Zool. 8*:45–49.

Kogan, M. and D. Cope. 1974. Feeding and nutrition of insects associated with soybeans. 3. Food intake, utilization, and growth in the soybean looper, *Pseudoplusia includens*. *Ann. Entomol. Soc. Am. 67*:66–72.

Kogan, M. 1965. Evaluation of economic injury levels for soybean insect pests. *World Soybean Research*, September 1976: 515–533.

Marsolan, N. F., and W. G. Rudd. 1975. Continuous optimal control applied to insect pest management. *Proc. 6th Annu. Pittsburgh Conf. on Modeling and Simulation*, pp. 831–835.

Marsolan, N. F., and W. G. Rudd. 1976. Modeling and optimal control of insect pest populations. *Math. Biosc. 30*: 231–244.

Marsolan, N. F., W. G. Rudd, D. C. Herzog, R. L. Jensen, and L. D. Newsom. 1977. Optimal pesticide control strategies for soybean insect pests. *J. Theor. Biol.* (submitted)

Menke, W. W. 1973. A computer simulation model: The velvetbean caterpillar in the soybean agroecosystem. *Fla. Entomol. 56*: 92–101.

Menke, W. W., and G. L. Green. 1976. Experimental validation of a pest management model. *Fla. Entomol. 59*: 135–142.

Mitchell, E. R. 1967. Life history of *Pseudoplusia includens* (Walker) (Lepidoptera: Noctuidae). *J. Ga. Entomol. Soc. 2*: 53–57.

Rudd, W. G. 1975. Population modeling for pest management studies. *Math. Biosci. 26*: 283–302.

Ruesink, W. G. 1975. Analysis and modeling in pest management, in R. L. Metcalf and W. H. Luckman (eds.), *Introduction to Insect Pest Management*. Wiley-Interscience, New York, pp. 353–376.

Shibles, R. M., and C. R. Weber. 1965. Leaf area, solar radiation interception and dry matter production in soybeans. *Crop. Sci. 5*: 575–577.

Stinner, R. E., R. L. Rabb, and J. R. Bradley, 1974. Population dynamics of *Heliothis zea* (Boddie) and *H. virescens* (F.) in North Carolina: A simulation model. *Environ. Entomol. 3*: 163–168.

Thomas, G. D., C. M. Ignoffo, K. D. Biener, and D. B. Smith. 1974. Influence of defoliation and depodding on yield of soybeans. *J. Econ. Entomol. 67*: 683–685.

Tummala, R. L., W. G. Ruesink, and D. L. Haynes. 1975. A discrete component approach to the management of the cereal leaf beetle ecosystem. *Environ. Entomol. 4*: 175–186.

Waddill, V. H., B. M. Shepard, J. R. Lambert, G. R. Carner, and D. N. Baker. 1976. A computer simulation model for the population of Mexican bean beetles on soybeans. *S.C. Agric. Exp. Stn. Bull. 590*.

Watson, J. R. 1916. Life history of the velvetbean caterpillar. *J. Econ. Entomol. 9*: 521–528.

5

GENERAL ACCOMPLISHMENTS TOWARD BETTER INSECT CONTROL IN COTTON

J. R. Phillips

Department of Entomology
University of Arkansas, Fayetteville, Arkansas

A. P. Gutierrez

Division of Biological Control
University of California, Berkeley, California

P. L. Adkisson

Department of Entomology
Texas A & M University, College Station, Texas

INTRODUCTION

Cotton is the world's most important fiber crop, and is grown on more than 74 million acres (30 million hectares) of land in some 80 countries. In the United States it is grown from coast to coast in the southern states and northward as far as Illinois. While cotton is grown primarily for its fiber, the seeds are highly valued for their oil and protein content for use in human food and livestock feeds. The oil is also used for various industrial purposes.

The cotton plant is naturally drought resistant and has the ability to produce lint and seed under extreme environmental adversities. Because it is well adapted to hot, dry, tropical and subtropical areas of the world it is grown as a dryland crop in areas of low rainfall. Cotton may be profitably grown in areas where alternative crops such as soybeans or corn (maizes) may not do as well because of their high demand for water and lack of ability to compensate for drought and other environmental adversities. Cotton produces its highest yields when grown in desert areas under full irrigation.

In all areas of the world where cotton is produced, there are major insect pest problems. The present technology for control of these pests in most areas depends on the heavy use of costly insecticides. This preventive strategy of insect control in cotton, based upon insecticides, evolved soon after World War II. The very use of these chemicals has, by and large, been self-defeating, because the insecticides have often created problems more serious than those for which they were originally applied (see below). These problems include target pest resurgence, secondary pest outbreaks, development of insecticide-resistant pest strains, poisoning of farm workers and insecticide applicators, toxic residues on adjacent crops, adverse impact on nontarget organisms, and pesticide biomagnification in ecological food chains.

The development and spread of pesticide resistant strains of the pest, and the other problems associated with heavy applications of pesticides marked the onset of what has been termed the "crisis phase" of cotton production (Doutt & Smith, 1971). This pattern has emerged not only in the United States but in other cotton growing areas of the world. It has left bankruptcy and ecological ruin in its wake. Losses to insects in the United States alone are estimated to exceed half a billion dollars annually with some $150 million spent on insecticides to control them (Huffaker & Croft, 1976). Cotton has been estimated to absorb about 45 percent of the total insecticides applied to field crops, this despite the fact that about 50 percent of the crop is not treated in most years. The excessive use of pesticides has contributed significantly to the present antipesticide syndrome in the United States (Bottrell & Adkisson, 1977).

The NSF/EPA supported IBP/IPM project has helped develop an orderly approach to agricultural research which promises to make significant contributions toward the solution of cotton pest control problems. The research undertaken during the project was multidisciplinary, and may be arbitrarily divided into three categories: (1) host plant resistance and biological control, (2) basic and applied research on the factors affecting the growth and development of the cotton crop, and (3) basic research to elucidate the physiologic, behavioral, and other processes regulating the dynamics of insect pest populations. The research results have been used to develop and upgrade the pest control decision-making technologies for implementation at the farmer level. This upgrading has required a major research effort to develop simulation and optimization models for the growth of the cotton crop, the dynamics of the major insect pests, and their impact on cotton production. Some of the results already have reached the stage where they have been adopted by farmers. These advances are reported in this chapter. The highly quantitative aspects of the developing technology are presented separately in Chapter 6.

THE NATURE OF
THE COTTON CROP PRODUCTION AND PROTECTION PROBLEM

The growth patterns of cotton are indeterminate, even of those varieties which are commonly thought to be determinate (see below). The patterns of growth are greatly influenced by weather, pest, and diseases, hence our discussion must consider these problems. Chapter 6 describes this growth and development in a highly mathematical form, but a more intuitive approach is developed here.

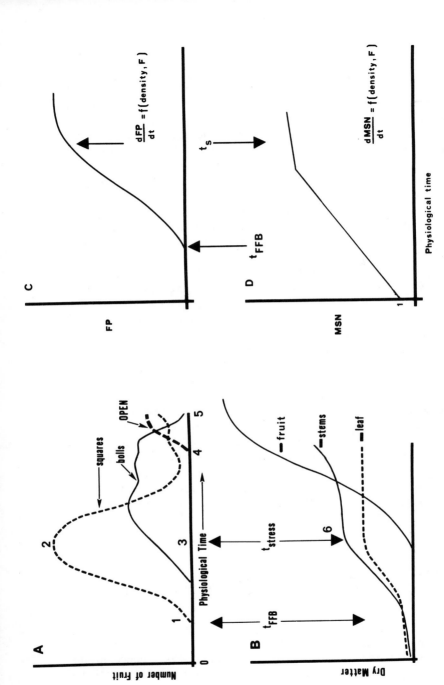

Figure 5.1. Diagramatic representation of crop growth and development: (A) phenology of fruit populations, (B) the patterns of dry matter allocation, (C) the pattern of fruit point production, and (D) the pattern of mainstem node production. The various symbols are described in the text.

Figure 5.1 depicts the stylized growth and development of a cotton crop. Several phenological events are commonly observed in the field (Fig. 5.1A): (1) the timing of the first fruiting branch (time $= t_{FFB}$), (2) peak squaring (l.g., t_s), (3) the onset of massive shedding of squares and small bolls ($> 80\%$; see Stockton et al., 1961), (4) the maturation of the crop (cotton crop), (5) if the season is long enough (e.g., Arizona), the initiation of a new or "top" crop, and (6) if dry matter data are also available, one would also notice that the accumulation of dry matter in leaf, root, and stem tissues coincides with the rapid increase in dry matter allocation to maturing fruits (Fig. 5.1B). Note that several of the events described above coincide with time (t_s), which is the time when the supply of carbohydrate is less than the demand by the growing plant parts (i.e., 2, 3, and 6 above, as well as a slowing of new fruit bud production (Fig. 5.1C) and mainstem node population (Fig. 5.1D).

Gutierrez et al. (1975), and later Wang et al. (1977) described and modeled these phenomena. Their reports should be consulted for greater biological and mathematical details (see also Chapter 6). They ascribe the evolution of the crop's growth and development to the interplay between the supply of photosynthate (S) and the plants' demands for growth (D). Figure 5.2 depicts a stylized representation of photosynthate supply and demand within the plant. Notice that the demand becomes larger than the supply at t_s, and this stress ($0 < S/D < 1$) causes the plant to shed the excess fruit (see McKinion et al., 1974), and alters the plant's priority scheme for cotton growth and development (Gutierrez, 1976).

Photosynthate stress ($S/D < 1$) may be induced by various factors such as cloudy weather combined with high temperatures, water stress, nitrogen

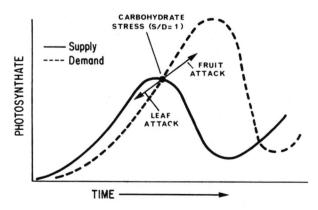

Figure 5.2. A stylized representation of carbohydrate sypply (S)–demand (D) within the cotton plant.

deficiency, and pest damage. The timing of this stress is variety-related (see Gutierrez, 1976), and as such is of very great importance to crop production and crop protection strategies. Most of the factors, except weather-related ones, can be managed. Cultural practices are easier to manipulate than are pests; hence we will stress pest management.

Pests may attack leaves, fruits, stems or roots, and the impact may cause S and/or D to vary. Figure 5.2 shows that leaf feeding affects the supply by reducing the photosynthetic area, causing wound healing losses and altering leaf age structure, while fruit loss affects demand for photosynthate. The destruction of leaves causes the stress point to be moved earlier in time resulting in stunted, unthrifty plants, while excessive shedding causes the plant to grow large and unproductive. It can be shown experimentally that small amounts of leaf feeding may not be deleterious (Gutierrez et al., 1975). Early leaf damage may be more severe than late damage. Also, it has been shown that moderate amounts of pest-induced fruit shedding may be beneficial (Gutierrez, Leigh, Wang & Cave, 1977), but see below.

In pest management, the problem is to judge whether it is economical to control a pest population, or to simply leave it alone. To do this effectively, one must be able to evaluate the potential impact of the pest on the crop and the influence of the control measures on the resurgence of the target pest, and noneconomic secondary pest populations, which may actually do more damage than the key pests if allowed to go uncontrolled (see Regev et al., 1976, for a discussion of the problem). These problems commonly create situations where the producer is placed on a season-long treadmill of pesticide treatments (van den Bosch, 1978).

The major cotton growing regions in the United States and the principal insect pests in each region can be characterized as follows:

1. Irrigated deserts of the Far West: the pink bollworm (*Pectinophora gossypiella* Saunders), lygus (*Lygus hesperus* Knight), bollworm (*Heliothis zea* (Boddie)), spider mites (Tetranychidae/spp.), and more recently, the tobacco budworm (*Heliothis virescens* Fab.)
2. The semi-arid regions of the Southwest: the boll weevil (*Anthonomus grandis* Boheman), fleahopper (*Pseudatomoscelis seriatus* Reuter), bollworm, and tobacco budworm
3. The humid regions of the mid-South and Southeast: the boll weevil, plant bugs (*Lygus* spp.), bollworm, and tobacco budworm

Currently, economic outbreaks of secondary pests are common in each of these regions. For example, in the eastern one-half of the United States cotton belt, routine insecticide applications are made to control the boll weevil, cotton fleahopper, and certain plant bugs (*Lygus* spp.), but these treatments also kill the natural enemies and unleash outbreaks of *Heliothis* spp. that

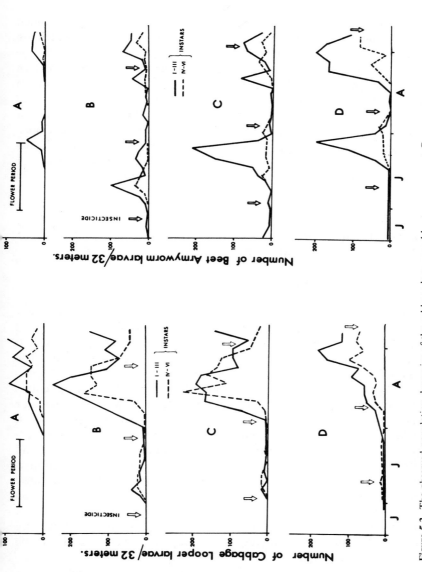

Figure 5.3. The observed population dynamics of the cabbage looper and beet armyworm at Corcoran, California during 1972 under natural control (A), and three different insecticide regimes (B, C, and D). The heavy downturned arrows indicate pesticide applications.

129

often cannot be economically controlled. Similarly, secondary pest outbreaks of *Heliothis* were common in California during the mid-1960s when farmers routinely applied pesticides for the control of lygus bug (Falcon et al., 1968; Gutierrez et al., 1975; see Fig. 5.3). In these areas, the problem is compounded because "indeterminate" flowering varieties of cotton are grown which are long fruiting and slow maturing, require heavy fertilization, and must be irrigated in areas of inadequate rainfall. The plant habitat and fruiting pattern are extremely suitable to late season damage by *Heliothis* spp. If maximum yields are to be produced from these cottons, the fruits must be set and matured during periods of the season when they are subjected to intense pest pressures. The solution of this problem requires that methods be developed so that key pests may be suppressed to subeconomic levels without inducing outbreaks by the secondary pest species, or leading to "flarebacks" or resurgence of the noxious species for which the treatments were originally applied.

LONG-TERM SOLUTIONS TO COTTON PEST MANAGEMENT PROBLEMS

Plant Breeding for Pest Resistance

For a number of years, scientists in California, Mississippi, Texas, and elsewhere have been investigating the possibilities of developing insect and disease resistant cotton varieties. The emphasis of the breeding programs at each of the cooperating NSF/IPM funded universities is different. The Mississippi Agricultural Experiment Station and associated U.S. Department of Agriculture–Agricultural Research Service (USDA–ARS) centers are screening cotton germ plasm for resistance to a range of pests with major emphasis on resistance to boll weevil, plant bugs, bollworms, spider mites, and whiteflies. The Texas Agricultural Experiment Station and ARS cooperators are breeding for bollworm, boll weevil, flea hopper, diseases, and nematodes resistance. The California Agricultural Experiment Station is cooperating with ARS, which is breeding cotton for resistance to lygus bugs, spider mites, and nematodes. All of these programs include emphasis on essential disease resistance, with some agencies emphasizing "multiple adversity resistance" breeding (MAR). These long-term programs are designed to develop cotton cultivars better adapted to the pest and climatic adversities of their region.

The NSF-EPA/IPM program enabled a closer coordination and intensification of this research effort to more fully exploit this line of defense against pests. The benefits from partial or complete resistance to a single key pest or from partial MAR are substantial. Gains from partial resistance may be enhanced by utilizing other pest control methods, and serve as a useful adjunct to cultural and natural biological control.

Beck and Maxwell (1976) summarized the known genetic sources of resistance in cotton to a number of pests up to 1974 (Table 5.1). It is seen from this table that certain characters which confer resistance to one pest species may confer high susceptibility to another. For example, cottons bearing the resistant factors glabrous, frego bract, and nectariless which are more resistance to boll weevil and *Heliothis* have been shown to have some sensitivity to fleahoppers. In Texas, all but one cultivar of the glandless or low gossypol seed cultivars tested were found to be highly susceptible to spider mites (J. K. Walker, personal communication). The glandless, frego bract, and nectariless characters have been incorporated into short-season backgrounds, and tests are currently being conducted on their agronomic qualities. Nevertheless, it is possible to develop varieties carrying multi-adversity resistance with respect to certain pests and climatic conditions (e.g., low temperatures, drought).

Plant Breeding for Pest Avoidance

Walker and Niles (1971) hypothesized that if an early, rapid fruiting, high yielding cotton could be developed (i.e., to "set" bolls during the first 30 to 40 days of fruiting), damage from boll weevil and bollworm could be minimized. This critical boll set period coincides with the F_1 and first part of the F_2 generations of the boll weevil. The strategy was based upon evidence that boll weevils do little damage to bolls that are 12 days old, or older. Currently available short-season cottons normally will set 50 percent or more of their fruits (12 days old or older) by the beginning of the F_2 generation (Walker et al., 1977). By contrast, the commonly grown indeterminate Deltapine 16 variety will set only 30 percent of its bolls during this time, and thus suffer severely from boll weevil attack.

In addition, short-season cottons have the added advantage of being harvested almost 1 month earlier than indeterminate varieties, well before environmental conditions induce diapause in boll weevil and pink bollworm. If this early harvest is coupled with area-wide stalk destruction before mid-September, which is possible, the overwintering populations of diapausing boll weevil (and pink bollworm) may be drastically reduced because the necessary food and breeding sites for their development would be destroyed. The size of the populations of these insects surviving the winter and available to infest cotton in the subsequent growing season are, as a result, much smaller (Adkisson, 1972). In addition, the absence of unnecessary pesticide applications reduces the possibility of secondary outbreaks of *Heliothis* spp.

Table 5.1.   Genetic Sources of Insect Resistance Found in Cotton and Currently Being Utilized in Breeding Programs—Texas, Mississippi, USDA, and Cooperating States of Louisiana and Missouri[a] [From Beck & Maxwell, 19

Morphological or Chemical Characters Identified	Major Cotton Insects					
	Boll weevil	Heliothis complex	Plant bugs Lygus lineolaris and hesperus	Cotton fleahopper	Spider mites	White
Frego	R(60–90% suppression)	N(insecticide coverage increased)	S	S	N	N
Nectariless	N	R(20–50% egg supression)	R	R	N	N
Smoothleaf (glabrous)	N	R(60% egg suppression)	S	S	R–	N
High gossypol	N	R	N	R–	N	N
Pilose (pubescence)	R–	S	R?	R	N	S
Okra leaf	N(better insecticide coverage increased kill in squares)	N	N	N	N	R–
Red color	R(choice situation)	N	N	N	N	N
"X" factor (*G. hirsutum* wild races)	N	R	N	N	N	N
Oviposition suppression factor (*G. hirsutum* wild races)	R(40% + suppression egg)	N	N	N	N	N

Table 5.1. (*Continued*)

Morphological or Chemical Characters Identified	Major Cotton Insects					
	Boll weevil	*Heliothis* complex	Plant bugs *Lygus lineolaris* and *hesperus*	Cotton fleahopper	Spider mites	Whiteflies
Plant bug suppression factor (Stoneville, wild races and other sources)	N	N	R	R	N	N
G. barbadense Pima S-2)	N	R–	N	N	R	N
Earliness of maturity	R(escape)	R(escape)	N	N	N	R?

R: resistant; N: no effect; S, increased susceptibility.

Subsequent to the original hypothesis, Dr. L. S. Bird of Texas A&M University developed a cotton variety, TAMCOT SP-37, which met the fruiting requirements proposed by Walker and Niles. In fact, this cotton produces 75 percent of its 12-day-old and older bolls within the first 30 days of flowering, producing 900 pounds of lint per acre (R. D. Parker & J. K. Walker, personal communication). These yields are equivalent to those of the full-season cottons grown under the same conditions without the associated heavy costs for pesticides, fertilizers, and water. A practical test of this system is discussed in a following section.

Biological Control

Natural control of pests in annual crops (e.g., cotton) has been largely overlooked, yet work by Falcon et al. (1968) showed that the application of pesticides against *Lygus hesperus* in California cotton caused very severe outbreaks of both primary and secondary insect pests and mites. Their observations were further verified by the later work of Ehler et al. (1973), Eveleens et al. (1973), and Gutierrez et al. (1975; see Fig. 5.3). Under the auspices of this IPM Project, very intensive investigations have just been completed in California on within-field (Byerly et al., 1978) and within-plant distribution of the various pests and natural enemies common to cotton (Wilson et al., in press). In addition, the group has intensively studied the biology and behavior of all of the major pests and natural enemies found in cotton (R. E. Jones, personal communication). As a result, they have elucidated in very fine detail the reasons for the occurrence of the pest outbreaks following pesticide applications, and in addition have quantified the impact of each predator species on the pest populations. This assessment is based upon detailed models of predator behavior (e.g., searching rates, position in the plant, host preferences, consumption rates, etc.) (R. E. Jones, personal communication). In general, cotton harbors many pests and natural enemies which migrate from other crops (e.g. alfalfa). While cotton may produce pests which migrate to other crops, their densities as pests are not usually great enough to contribute significantly to the buildup of natural enemies (i.e., it is a "sink" for natural enemies). The most important predators in California cotton are the green lacewing (*Chrysopa carnea* Stephens) and *Orius tristicolor* (White) (R. E. Jones, personal communication).

These kinds of findings are not restricted to California. Field experiments by G. Andrews (Mississippi State University, unpublished data) in the Mississippi Delta during 1973 and 1974 showed no detectable decreases in yield when cotton was left untreated for *Heliothis* spp. infestation. However, in 1975 under poor growing conditions and continuous tarnished plant bug attack, as well as a moderate bollworm infestation, untreated cotton yielded at least 100 pounds per acre less than an "as needed" insecticide-treated field.

Table 5.2.   Potential Parasites of Boll Weevil Recovered in Southern Mexico in 1976 (J. R. Cate, personal communication)

Parasite	Host Plant Association	Location
Pteromalidae		
Pteromalus grandis[a] (Burks)	*Hampea nutricia* Fryx.	Martinez de la Torre, Ver.; Cardenas, Tab.
	H. trilobata Standl.	Valladolid, Yuc.
	Cienfuegosia rosei Fryx.	Tehuantepec, Oax.
	Cissus sp.	Misantla, Ver.
	Phymosia umbellata (Cav.) Kearn.	Jilutepec, Ver.
Pteromalus hunteri (Crawford)[a]	*Sida cordifolia* L.	Mazátan, Chiapas
P. hunteri[c,d]	*Hibiscus tilaceus* L.	Tapachula, Chiapas
Pteromalid genus A[a]	*H. nutricia*	Huimanguillo, Tab.
	Mimosa pigra L.	Huimanguillo, Tab.
	Croton sp.	Mazátan, Chiapas

Table 5.2. *(Continued)*

Parasite	Host Plant Association	Location
Pteromalid genus B[a]	*H. nutricia*	Huimanguillo, Tab.
		Cardenas, Tab.
	M. pigra	Huimanguillo, Tab.
	Cissus sp.	Misantla, Ver.
Zatropis sp. nr.	*Gossypium hirsutum* (L.)	Tapachula, Chiapas
perdubius	(cultivated)	
(Girault)[c,d]		
Eurytomidae		
Aximopsis sp.[a]	*H. trilobata*	Valladolid, Yuc.
	Cissus, sp.	Misantla, Ver.
Eurytoma sp.[a]	*H. nutricia*	Cardenas, Tab.
	Croton sp.	Mazátan, Chiapas
Eulophidae		
Horismenus sp.[a,e]	*H. nutricia*	Jalapa, Tab.
	M. pigra	Huimanguillo, Tab.
	P. umbellata	
Eupelmidae		
Eupelmus sp. 1[a]	*H. nutricia*	Martinez de la Torre,
		Ver.; Huimanguillo, Tab.
	Cissus sp.	Misantla, Ver.
Braconidae		
Bracon sp. 1[a]	*H. nutricia*	Martinez de la Torre, Ver.
Bracon sp.	*H. tiliaceus*	Tapachula, Chiapas
Nealiolus sp.[c,d]	*G. hirsutum* (cultivated)	Tapachula, Chiapas
Apanteles sp.	*H. nutricia*	Huimanguillo, Tab.
	M. pigra	Huimanguillo, Tab.
	Croton sp.	Mazátan, Chiapas
Microchelonus sp.[a,e]	*H. nutricia*	Huimanguillo, Tab.
	H. trilobata	Francisco Escárcega, Camp.
		Vallidolid, Yuc.
	M. pigra	Huimangillo, Tab.
Chelonus sp.[a,e]	*H. nutricia*	Cardenas, Tab.

[a]Determined by Dr. Eric E. Grissell, Florida State Department of Agriculture, Gainesville, Florida.
[b]Determined by Dr. Paul M. Marsh, USDA/ARS, Systematic Entomology Laboratory, Beltsville, Maryland.
[c]Determined by Dr. Gordon Sordh, USDA/ARS, Systematic Entomology Laboratory, Beltsville, Maryland.
[d]Specimens obtained by Ing. René Bodegas, Centro de Investigaciones Ecologicas de Sureste, Tapachula (Chiapas), Mexico.
[e]Species suspected to be parasitic on insects other than boll weevil.

135

Efforts at biological control of the tarnished plant bug are currently under-way in Mississippi. Three species of Braconidae (*Peristenus stygicus* Loan, *P. rubicollis* Thompson and *P. digoneutis* Loan) have been released in the field. Three strains of *P. stygicus* have been introduced from Turkey, France, and Austria. *P. rubicollis* is a univoltine species which appears to be quite cold-tolerant, while *P. stygicus* and *P. digoneutis* are multi-voltine and heat-tolerant. Releases of these species have been made in areas containing the natural host plants of the plant bug, where it is hoped they will multiply and reduce the pest population before they invade cotton. All three species spin their coccoons in the soil, where they are attacked by fungi. The high rainfall common to the southeastern United States enhances the growth of the fungi and as such may limit the effectiveness of the parasites (M. Schuster, un-published data). Field evaluation of these parasites has not been completed.

Recent studies on the native and exotic natural enemies of boll weevil suggest that biological control is a viable method of control that is missing from the present integrated control programs for insect pests of cotton in Texas, Oklahoma, and the southeastern United States. Quantitative evalua-tions of biotic and abiotic mortality by Bottrell (1976; and unpublished data) show conclusively that in certain areas of Texas where insecticides are rarely used, parasites (predominately *Bracon mellitor* Say) contribute significantly to boll weevil mortality. *B. mellitor* was shown to exhibit a weak density-dependent response to boll weevil populations.

Parasites of boll weevil do not appear to provide economic control under the present production methods. For this reason, none of the present IPM programs for cotton take into account a natural enemy of the boll weevil. The early integrated control programs for boll weevil utilized this biotic component before insecticides came into general use. It is likely that they will work today, particularly under the low energy input production systems being developed. (See Smith et al., 1976, for an excellent review of early integrated control programs for boll weevil.)

Efforts to establish exotic natural enemies of boll weevil in the United States have been unsuccessful, and no new exotic natural enemies have been found. However, two recent efforts to locate new natural enemies appear to hold promise. Thirteen previously unrecorded parasite species have been reared from flower buds and fruits of various host plants infested by boll weevil (Table 5.2). Ten of these species belong to genera associated with boll weevil or other curculionids, and possibly may be previously undiscovered parasites of this pest. Further study is required to make positive parasite-host deter-minations.

The possibility exists that all of these parasites are from insect species other than the boll weevil. This is very likely the case for those parasites collected at Valladolid, Yucatan, since the boll weevil was not collected at this site. Instead, the new parasites were reared from a new anthonomine species, which is considered to be closely related to boll weevil (J. R. Cate & H. R. Burke, unpublished data), and feeds on *Hampea trilobata* Standley. Both weevils occur sympatrically at Francisco Escárcega, Campeche and in the area of Cardenas, Tabasco in Mexico.

The results of these recent studies suggest that biological control of boll weevil may be a viable tactic. They also suggest that further studies in biological control are urgently needed to develop this integrated control component.

SAMPLING PROGRAMS FOR PEST MANAGEMENT

The success of any IPM program depends upon the development of inexpensive, fast, and accurate sample methods for the pests and natural enemies found in the crop. Each cotton growing area has its unique pest problems, and sampling rules which may be suitable for one area may not be suitable for another.

In California, Byerly et al. (1978) compared the efficiency of the commonly used sampling methods in cotton (sweep net and "D-vac") for assessing population densities of pests and natural enemies with a newly devised absolute sampling method (whole plant bag samples, i.e., WPBS). The WPBS method involves placing organdy sleeves around the base of sample plants approximately 10 days before they are to be sampled. This presampling period allows the populations on those plants to recover from any disturbance caused by placement of the bag. Early in the morning on the sampling date, the bags are sealed at the bottom and quickly pulled up over the plant, capturing all of the organisms on it. The bag is sealed at the top, the stem cut, and the sample taken into the laboratory for counting.

The purpose of this study was to develop a correction factor (based on linear regression) for converting estimates of insect counts (pests and natural enemies) obtained using the sweep net and D-vac methods (both are relative methods) to estimates of absolute densities. The results were rather surprising, as analysis of the data showed that a consistent correction factor did not exist for most of the species. Consistent results were found only for *Lygus hesperus*; that is, during all years and all fields (e.g., Table 5.3). The results showed conclusively that the relative sampling methods were grossly inadequate for estimating the densities of most organisms in cotton. The lack of correspondence between the WPBS and relative methods occurs because the plant's growth and development not only influence the position of the insects in the canopy, but also the efficiency with which they can be sampled

Table 5.3. Regression of Whole Plant Bag Samples (WPBS) Against Sweep Net and D-vac Samples—Data collected during 1974 near Shafter, California

| Species | Stage | (A) Normal Least-Squares | | | | | | |
		a	SE_a	b	SE_b	$SE\,y.x$	r^2	$a = 0:+$ $a \neq 0:-$
		Relative Sampling Method—D-vac						
Geocoris spp.	A[a]	1.7802	0.4053	−3.1900	1.4300	0.3203	0.42	−
	I	0.3909	0.3416	4.2700	1.1600	0.6274	0.66	+
	T	0.4153	1.2140	3.7150	2.2600	1.2265	0.28	+
Orius spp.	A	0.1205	0.1777	2.3250	0.9200	0.2272	0.48	+
	I	0.3787	0.2041	4.2800	4.1650	0.3501	0.15	+
	T	0.3559	0.3649	3.2200	1.5600	0.4877	0.38	+
Chrysopa spp.	A	0.0203	0.0128	0.1650	0.5650	0.0338	0.01	+
	I	0.1115	0.0306	2.0000	3.0300	0.0408	0.06	−
	T	0.1657	0.0396	−0.4650	1.0750	0.0683	0.03	−
Nabis spp.	A	0.0342	0.0997	1.8350	1.8100	0.0872	0.13	+
	I	0.0050	0.0532	3.1200	0.9950	0.0795	0.59	+
	T	−0.0150	0.0827	2.7450	0.8600	0.1201	0.59	+
Spiders (All)	T	0.3946	0.0658	1.7176	1.6955	0.1359	0.11	−
Lygus spp.	A	0.0051	0.0305	1.1023	0.3409	0.0551	0.57	+
	I	0.0441	0.0464	3.7727	0.3864	0.0889	0.92	+
	T	0.0087	0.0724	2.8523	0.3686	0.1210	0.89	+
		Relative Sampling Method—Sweep-Net						
Geocoris spp.	A	1.5798	0.3012	−8.9144	4.2200	0.3393	0.22	−
	I	0.3320	0.4022	26.4950	8.3250	0.6869	0.59	+
	T	1.6364	1.3456	6.2052	11.9050	1.4165	0.04	+
Orius spp.	A	0.2307	0.1164	7.6300	2.4150	0.2028	0.59	+
	I	0.3176	0.1535	42.3850	26.2150	0.3416	0.27	+
	T	0.7155	0.2986	7.7700	5.7800	0.5519	0.21	−
Chrysopa spp.	A	0.0364	0.0184	−3.2900	3.4500	0.0320	0.12	+
	I	0.0967	0.0128	17.4950	5.0900	0.0257	0.63	−
	T	0.1604	0.0585	−0.6900	8.6750	0.0677	0.00	−
Nabis spp.	A	0.0574	0.0784	6.8000	6.7300	0.0872	0.13	+
	I	0.0546	0.0406	26.8350	6.5550	0.0786	0.71	+
	T	0.0052	0.1465	19.8350	8.6450	0.1760	0.43	+
Spiders	T	0.4968	0.1259	−9.8750	21.0900	0.1518	0.03	−
Lygus spp.	A	0.0061	0.0300	3.4860	0.9931	0.0522	0.64	+
	I	0.1198	0.0837	10.3577	2.1281	0.1546	0.77	+
	T	0.1339	0.1336	7.1139	2.0268	0.2117	0.64	+

[a]A: adults; I: immatures; T: total individuals.

by the two relative methods. This implies that population data reported in the literature, or in staff research files, using methods for the various insects in cotton, are highly suspect.

The alternative WPBS method for cotton, and a decision rule for determining the number of samples required for different levels of accuracy (after Iwao & Kuno, 1968) are reported by Byerly et al. (1978). The number of samples depends upon the degree of accuracy required (Fig. 5.4), the density of the organisms, and the degree of their aggregation. The decision formula, as given by Iwao and Kuno (1968), is

$$ n = \frac{t^2}{D^2} \frac{(\alpha + 1)}{\bar{x} + \beta - 1} $$

where

n = the number of required samples

t = Student's t, giving the frequency with which a sample of size n will give an estimate of precision D or better (e.g., if $t = 1$, frequency = 0.67).

D = the precision required as a fraction of the mean.

α, β = coefficients of regression of $\overset{*}{x} = x - S^2/\bar{x} - 1$ (see Iwao & Kuno, 1968). The value $\overset{*}{x}$ is called mean crowding by Lloyd (1967) and equals $\bar{x} + [S^2/\bar{x} - 1]$.

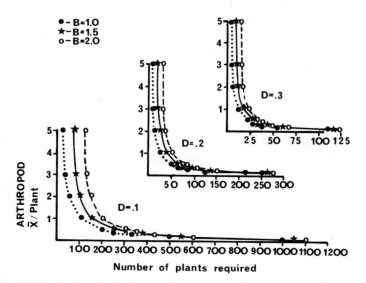

Figure 5.4. The decision rules for determining the number of Whole Plant Bag Samples (WPBS; see Byerly et al., 1978) required at different arthropod densities to achieve different levels of precision (D = 0.1, 0.2, and 0.3).

PRACTICAL TESTS OF IMPROVED PEST MANAGEMENT SYSTEMS

A Community IPM Study in Arkansas

One of the best cotton producing areas in Arkansas is in the lower Mississippi flood plain area about 25 miles east of Little Rock. The crops grown in the area are principally cotton, soybean, and rice. The bollworm, rather than the boll weevil, is the major pest in this area.

A community-wide IPM project on cotton was initiated on a limited scale in 1975 and expanded in 1976 to include an area of 50 square miles, of which 12,000 acres (640 acres/square mile) were planted in cotton. The cotton acreage of the 72 cooperators ranged from 70 to 1500 acres. An average of 10 applications of insecticides were directed against the bollworm during the 3 years prior to 1972 (i.e., prior to use of IPM). During 1973, 1974, and 1975, extensive data were collected three times weekly in an 18 acre experimental area near the center of the 12,000 acres under a "commercial" pest control program. These data were used to develop sampling and surveillance techniques, establish damage thresholds for bollworm, and to estimate bollworm development and mortality rates for constructing a model for short-range forecasting purposes. The experimental area unavoidably received some pesticide applications, which to some extent invalidates our intent to use it as a "check." Yet all of the IPM study fields were compared to it; hence we call it the Index field.

An analysis of the Index field data indicated that (1) long-range bollworm migration did not play a major role in the outbreak of this pest in the region, (2) estimates of within-field populations during June were more than sufficient to account for the population densities encountered later in the season, (3) the commonly used "economic injury thresholds" could safely be adjusted upward, and (4) the whole region (i.e., the community) could be considered as a single unit or field for the purpose of making recommendations on pest control procedures. In addition, considerable progress had been made during the period in developing a bollworm forecasting network.

The data taken in the Index area during the initial phases of the project suggested the bollworm problem had been aggravated by the pesticide use patterns common to the area. In general, the data indicated that the July bollworm populations did not cause economic loss, and where insecticides were not applied there was a buildup of natural enemies which greatly reduced the bollworm hazard during August.

Because an element of risk was involved in not treating the July population, the strategy of not treating unless essential was first explained to the growers (i.e., the IPM community), and then the growers were asked to decide whether an area-wide decision-making program should be undertaken. Over 90 percent of the growers voted, with some hesitancy, not to apply pesticides during July of 1976 unless advised by the IPM research group. Less than 50 of the 12,000 acres of cotton were treated during July that year. On August 6 the simulation model used for prediction, which had been updated with field sample data, projected that the bollworm would not be controlled by its natural enemies, and that a treatment would be necessary. A meeting of all of the cooperating growers was held and the recommendation given to treat with Chlordimeform®, plus a suitable larvicide to reduce the given population. Within 3 days, 90 percent of the 12,000 acres of cotton was treated. Index areas subsequently showed that failure to treat at the recommended time resulted in economic damage. A second final area-wide treatment (Chlordimeform® alone) was recommended on the 27th of August. In this case, the entire cotton acreage was treated within a 2-day period. Some 800 acres of late-maturing irrigated cotton required a further treatment during the first week of September.

The bollworm population encountered in both the untreated and IPM acreages during the season of 1976 are shown in Fig. 5.5. The striking differences between the populations of small and large larvae during July demonstrates the efficiency of the natural enemy complex in suppressing the young larvae during the early season. Prior to the start of the coordinated effort, bollworm populations (total larvae) in the area continued to increase during the season (e.g., 1974), despite the 10 applications of insecticides made by growers throughout the area (Fig. 5.5D). During 1975, low bollworm populations occurred through mid-August, with rapid increases only late in August. By contrast, the acreage under the 1976 IPM-managed program, as well as the Index plot, developed relatively minor populations (Fig. 5.5D). It appears that the area-wide coordination of insecticide applications produces better results than the more haphazard individual grower approach. The data do not suggest that insecticides are ineffective in controlling bollworm or other insect populations, but they do suggest that a mismanaged insecticide program is almost as ineffective as no program at all.

The 1976 results are not entirely conclusive because it was not a heavy bollworm year, but this may be because there were considerably more predators present than in the previous 2 years. Unfortunately, a large untreated check was not available (see below); thus we cannot be absolutely sure of cause and effect. The validity of these methods must be checked during other years. (Note: Later verified in 1977 and 1978.)

If the system is valid, the data indicate that further gains could be had by incorporating the short-season cottons into the production system. If short-season cottons had been utilized in the program, it is thought that only one application of insecticide (versus 10 under conventional programs) would have been necessary.

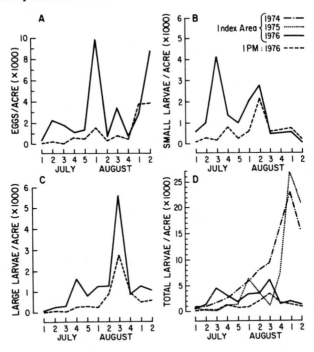

Figure 5.5 The average observed population trends in Arkansas cotton field under IPM, and the Index field. The Index field represents the population trends which would occur in the absence of IPM.

The Development of Cotton Pest Management
in the Central Valley of California

In the Central Valley of California cotton is now produced largely without the use of pesticides of any kind, though historically this has not been the case. The great majority of the pesticides used on cotton in this area are directed against lygus bugs and spider mites. (The pink bollworm is the key pest in the desert regions of California, e.g., in the Imperial Valley.)

In the early 1950s, 6 lygus/50 sweeps was the recommended economic threshold for this insect. From the mid-1950s until the early 1970s the economic threshold for *Lygus hesperus* infestations in Central Valley cotton had been arbitrarily established at 10 lygus bugs per 50 sweeps (Sevacherian & Stern, 1972) with each nymph being counted as two. Insecticide applications made for the control of this pest greatly aggravated secondary pest problems of cabbage looper, beet armyworm (Fig. 5.3), bollworm, and tetranychid mites

(Falcon et al., 1968). The importance of the individual pests varied from area to area within the valley. Large scale experiments by Falcon and his co-workers cast considerable doubt on the validity of the commonly used economic threshold and more importantly on the key pest status of *Lygus hesperus*. They demonstrated (1) the importance of assessing *L. hesperus* populations relative to the fruiting capacity of the plant and (2) that high lygus bug infestation rates in excess of 10 per 50 sweeps had little impact on yields. These studies set the framework for the later IPM work in California.

Under the IPM/NSF-EPA Project, a population model for cotton (Gutierrez et al., 1975) and a model for *L. hesperus* were developed. The critical parameter still missing was an estimate of *L. hesperus*-induced shedding of young squares. This information was difficult to obtain in the field because the damage is not visible and normal plant shedding ($> 80\%$ of all fruit points) tended to mask lygus effects. The plant model indicated that before lygus damage could be estimated, normal plant shedding must be greatly reduced or eliminated. The plant model indicated that large fruits must be removed before such a test, and that lygus bugs should be caged in small organdy bags only on the plant terminals so as to avoid any gross plant shading effects (see Gutierrez, Leigh, Wang and Cave, 1977, for complete details). Figure 5.6 shows the results of the experiments for lygus adults and nymphs, and indicates that a single lygus bug female causes 0.028 squares to shed per degree day above a threshold of 53.5°F, while a single male causes shedding at 40 percent of the female rate. Nymph-caused shedding rates are lower and are age-dependent, showing that the previous practice of equating one nymph with two adults was erroneous.

When these data were incorporated into the model, and observed numbers and phenology of lygus bugs were used (1974) to estimate field shedding, the model indicated that no differences in yield should be expected. When multiples of observed numbers were used, the results suggested that some increases in yield could result. Only feeding during the early squaring period should cause significant damage. These results were surprising, as they indicate that at certain population levels this key pest might actually be beneficial (increase yield). This confirms the observations of Falcon et al. (1971) who found no significant losses in yield, even though *L. hesperus* numbers exceeded the economic threshold of 10 per 50 sweeps on eight different sampling dates, reaching 34.3 on one date.

Concurrent with this later work, was a USDA/APHIS-sponsored IPM Cotton Pest Management Program in Kern, Kings, and Tulare Counties (H. Black, V. E. Burton, & G. A. Rowe, personal communication), which managed 6000, 18,000, and 22,500 acres during 1972, 1973, and 1974, respectively. The results of these analyses corroborated the earlier and concurrent work by Falcon et al. (1971) and Gutierrez, Leigh, Wang and Cave,

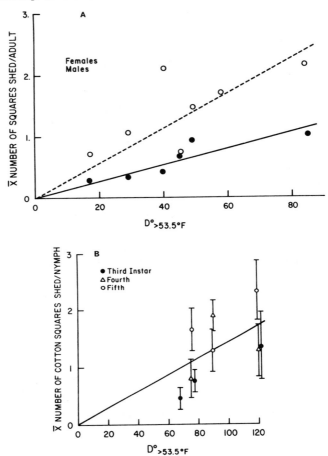

Figure 5.6. The relationship between *Lygus hesperus* feeding times in degree days (x axis) and the number of squares shed for (A) adult males and females, and (B) immatures.

1977. In addition, they developed a dynamic economic threshold for *Lygus hesperus* for the period before peak squaring. That threshold may be described as the economic number of lygus per 50 sweeps = 0.03 [squares/(0.001 acre)]. These data further show that the early squaring period is the critical time period, and it also confirms that under most circumstances, lygus infestations can be ignored after the period of peak squaring. Comparisons of the pesticide use (excluding sulfur dust) in project fields and adjacent fields in the three counties for 1973 and 1974 are shown in Table 5.4. The data show that the reductions in insecticide use were significant, varying from 6 to 57

Table 5.4. California Cotton Pest Management Project—A Summary of the Pounds of Insecticide Used (Active Ingredients) per Acre

	County	Combined Program Acreage	Average Use on Project Acreage	Average Use on Adjacent Acreage	Percent Change
1973	Kern		0.84	1.17	−28
	Kings	18,000	1.04	1.65	−37
	Tulare		0.85	1.17	−27
1974	Kern		0.90	0.96	−6
	Kings	22,500	0.78	1.82	−57
	Tulare		0.79	0.85	−7

percent less. A detailed model of lygus bug effects on cotton has been formulated, and optimization studies for its control are nearing completion (Gutierrez & Wang, in preparation).

As a result of all of the above studies (and those of others), the pest status of *Lygus hesperus* has been put into proper perspective, and as a result of reduced use of insecticides to control this pest, noctuiid pests of cotton (i.e., bollworm, cabbage looper, and beet armyworm) in the San Joaquin Valley are now rarely encountered in economic numbers.

Similar, but less dramatic, results have been revealed by investigations on other cotton pests. Currently, models for cabbage looper and beet armyworm phenology and dynamics are in progress (D. Hogg & A. P. Gutierrez, unpublished data), though the impact of their larval feeding on cotton had been estimated much earlier (Gutierrez et al., 1975). A model for pink bollworm-cotton-weather interaction has been developed (Gutierrez, Butler, Wang & Westphal, 1977), and is being used by Shoemaker and Gutierrez (in preparation) to develop optimal control strategies for this major pest of cotton in southern California. The results for pink bollworm already enable us to predict the phenology of this pest, to know rather precisely what its impact on the crop is expected to be, and to understand the nuances of its relationship with the plant (D. Westphal et al., in press). The model further tells us that it may be limited in the Central Valley by low winter temperatures. Though not yet adopted, these results have very immediate and practical implications.

A model for the bollworm, based upon detailed insect behavior is also in progress. Similar types of phenological relationships may be expected to emerge, and an assessment made of the within-season impact of this pest on cotton yields.

Extensive progress has been made in the field analysis and modeling of natural enemy impact on many of these Central Valley cotton insects. We now have a very good absolute sampling method for them (Byerly et al., 1978). We know the general distribution patterns of the various pests and their natural enemies within the field and within the plant canopy through time. Very simplified sampling methods based upon sampling main stem node leaves have been developed for cabbage looper, beet armyworm, bollworm eggs, and all stages of spider mites (Gutierrez et al., in preparation; Carey, in preparation; Wilson et al., in press). Dependable estimates of developmental times and temperature thresholds, food preferences, switching rates between different prey, and prey consumption rates can now be obtained and their behavioral patterns ascertained for all of the major pests and their predators in cotton in California. The causes of secondary pest outbreaks following pesticide applications are now well understood, and are increasingly being avoided as pest control consultants incorporate these concepts into their pest management programs.

Various methods for disseminating this information to field crop protection specialists are being devised. Among these are sets of simplified pest control recommendations based on estimates of potential yield loss expected from a prior assessment of crop phenology and pest abundance and detailed assessments based upon actual simulation runs. Computer systems for on-line delivery of the necessary data are well developed, even though the computer hardware is presently not available to the program for field use.

Use of Economic Thresholds versus Preventive Treatments in Mississippi

Smith et al. (1974) reported results of comparisons in the cost of insecticides for control of insects on conventional cottons in the Mississippi Delta. An average cost of $16.16/acre was incurred when pest populations were monitored and insecticides applied only when economic thresholds were exceeded. By comparison, a cost of $21.25/acre was incurred by growers using an intermediate program based on previous practices, while the weekly application of insecticides during the growing season cost $30.55/acre. The latter is a very common practice in Mississippi.

Simulated experiments have been performed in which bollworm and tobacco budworm damage was imposed on the cotton crop model. These experiments indicate that in arriving at appropriate economic thresholds for *Heliothis*, one must consider the developmental stage of the crop as well as the future environment and management practices the crop will experience. These models were also used to study the sensitivity of cotton to *Heliothis* damage at different times of the year, as well as the effects of nectariless and smooth leaf cotton on bollworm and tobacco budworm populations and yield (Harris et al., 1976). The models for these two species and one for the boll weevil (Jones et al., 1975) have been linked together and are currently being used for evaluating insect pest management options in Mississippi.

In 1975, the bollworm and tobacco budworm models were linked with a cotton model to study, by simulation, the effects of different insecticides and spraying patterns on cotton yield (Brown et al., 1976). The cotton model used in the 1975 simulated experiments was a modification of SIMCOT II (McKinion et al., 1974).

Short-Season Cotton-Pest Management Production System in Texas

The crux of the boll weevil pest management problem in most of Texas is to reduce the number of overwintering adults to such low levels that their progeny cause only minimal losses in yield, then delay insecticide applications sufficiently long such that natural enemies can be protected and secondary outbreaks of *Heliothis* avoided. Coupled with this was the recognition that the cotton bolls had to be matured before significant populations of boll weevil developed. This latter requirement was met by the new short-season variety TAMCOT SP-37.

In several pilot programs using the short-season cotton, one to three low dosage applications of insecticides, which are effective against boll weevils but not harmful to their natural enemies (e.g., azinphosmethyl), were used well before the cotton began to flower. The treatments were made only when needed and based upon field monitoring of weevil populations. More toxic wide-spectrum pesticides, such as methyl parathion, were avoided.

Under this program, infestations of bollworm or tobacco budworm were allowed to increase beyond the formerly used "economic thresholds" of 8 to 10 percent damaged squares, because natural enemies quickly suppressed populations to sub-economic levels within a few days with no pest influence on final yields (Baldwin et al. 1974). Thus, the old concept of economic thresholds for *Heliothis* still in prevalent use in the United States on conventional indeterminate cottons has little meaning for the new short-season pest management/crop production system for cotton as developed in Texas. In this system, the boll weevil is controlled before it causes crop damage, while *Heliothis* spp. are generally ignored because natural biological control provides better suppression than any of the insecticides currently registered for use on cotton. Many of the insecticides are largely ineffective against resistant strains of these pests.

The economic benefits of the short-season pest management/crop production system based upon an experiment in Frio County, Texas are summarized in Table 5.5 (Sprott et al., 1976). The costs for producing an acre of typical, indeterminate cotton under irrigation in an area heavily infested with boll weevil was about $278, but the gross return on a production of 1 bale per acre was $340, yielding a net profit of $62 per acre. One cooperating producer used more fertilizer, water, and pesticides, with an average cost of $326 per acre, but this increased yields and returned a gross income of $435 per acre, with a net profit of $109 per acre. In these tests, the short-season cotton was grown under the new production system, using 80 percent less fertilizer, 50 percent less water, and 75 percent less insecticides. Production costs for short-season cotton grown at 40-inch row spacings were approximately the same as those incurred by the typical producer. The yield, however, was well above that of the typical producer and slightly increased over

Table 5.5.   A Per Acre Comparison of Cotton Production with Alternative Cotton Pest Management Practices—Frio County, Texas

Item	Unit	Typical Input Producer[a]	High Input Producer	Short Season (40″)	Short Season (26″)
			Production Technique		
Inputs					
Fertilizer	lb.	80–40–0	116–62–0	24–24–24	24–24–24
Irrigation	ac. in.	20	18	12	12
Pesticides	lb.	9.6	16.9	6.6	6.6
Total	1000 kcal	3624	3645	2445	2445
Cost	$/ac.	278	326	281	279
Cost	¢/lb.	47.60	42.56	33.84	26.90
Production					
Yield	lb./ac.	500	625	649	765
Gross[b]	$/ac.	340	435	452	532
Net[b]	$/ac.	62	109	170	252

[a]Average inputs of nitrogen, water, and pesticides were used. These costs were based on enterprise budget published by the Texas Agricultural Extension Service. Greater than average rates of nitrogen, water, and pesticides were used by the high-input producers.
[b]Based on a cotton price of $0.60/lb. for lint and $120/ton for seed.

that of the high-input producer. The net profit was more than double that of the typical producer and 55 percent greater than that of the high-input producer. When the short-season cotton was planted in narrow rows (26-inch row spacing), yields were greatly increased, costs were held at the same level, and net profit per acre was increased to $242 per acre. This is a profit increase of $143 per acre over that of the high-input cotton producer who used the conventional pest control system, and $190 greater than that of the typical producer of the area.

Thus, the short-season system has highly attractive features of decreasing costs and greatly increasing both yields and profits. The short-season cotton production system also requires less fossil fuel for production, because less cultivation, pesticides, fertilizer, and irrigation water are used. Because of these advantages, cotton producers in boll weevil infested areas of Texas are shifting to this new production system which lowers the cost of production by 30 percent or more, but gives the same or greater yields and increased profits.

Pest Management Systems Using Conventional Cotton in the Brazos River and Trinity River Regions of Texas

The increasing insecticide resistance of the bollworm and tobacco budworm has had a severe impact on yields and costs in the Brazos and Trinity River valleys of Texas, and has been the major impetus for the practical development of an IPM program in those areas. The strategy requires that (1) boll weevils be controlled with a fall "diapause" control program, (2) fleahoppers be controlled with low dosages of insecticide applied early in the season, (3) fleahopper treatments be terminated as early as possible to enhance natural enemy populations for control of late season bollworm and tobacco budworm populations, and (4) the insecticide applications are begun only after the pest populations have exceeded economically unacceptable levels.

In the Brazos area, Sterling and Haney (1973) compared the cotton production costs and benefits of the traditional insecticide treatment program with those of the above IPM strategy. Insecticide use under the new system was reduced from 12.9 to 6.4 lbs./acre (50% reduction), while lint yield increased about 60 lbs. Estimated net returns would be expected to increase about $22/acre if the new strategy were adopted generally in the region.

In the Trinity area, a 50-percent reduction in use of insecticides (from 10.8 to 5.6 lbs./acre) resulted in higher yields and a net profit increase of $37/acre (Sterling & Haney, 1973).

CONCLUSIONS

The development of successful insect pest management programs for any crop is an exercise not only in entomology and ecology, but also in agronomy, economics, environmental improvement, and many other disciplines. The farm manager must place the costs and benefits of pest control in the context of total long-term production costs and benefits. Farmers are becoming aware that their agronomic practices during a given year may have long-term consequences to their businesses in future years, and to the environment, as well. They know that constraints may be placed upon the use of insecticides if they are used indiscriminately. Successful farmers, of course, must manage the total system so that profits adequately compensate their investment and management efforts. Because insect pests and the crop varieties often undergo a continual state of evolution, pest-crop management systems must also evolve. Old economic thresholds for pests are continually being discarded and new ones established. In fact, the old utilization of a static economic threshold (Stern et al., 1959) is being replaced by a dynamic one. The knowledge acquired in the evolution of the various IPM studies reported above clearly illustrates this point. The basic principles of integrated control hold for these new systems: the entire crop ecosystem must be closely considered, economic thresholds must be used, and inputs of chemicals that disrupt biological controls should be used only as economically and ecologically justified.

Smith et al. (1974) reported that integrated insect control programs of some sort are now in use by a majority of cotton farmers in the United States. The level of sophistication of these programs varies greatly, but as the complex systems models developed by scientists working in the present cotton IPM program (Chapter 6) are simplified and reduced to provide better guidelines for monitoring and decision-making, IPM will gain wider farmer acceptance.

LITERATURE CITED

Adkisson, P. L. 1972. Use of cultural practices in insect pest management, in Implementing Practical Pest Management Strategies. *Proc. Nat. Ext. Insect Pest Management Workshop. Purdue Univ., LaFayette, Ind. (Mar. 14–16, 1972),* pp. 37–50.

Baldwin, J. L., J. K. Walker, J. R. Gannaway, and G. A. Niles. 1974. Bollworm attack on experimental semidwarf cottons. *Tex. Agric. Expt. Stn. Bull.* 1144.

Beck, S. D., and F. G. Maxwell. 1976. Use of plant resistance, in C. B. Huffaker and P. S. Messenger (eds.), *Theory and Practice of Biological Control.* Academic, New York, pp. 615–636.

Bottrell, D. G. 1976. Biological control agents of the boll weevil, in *Proc. Conf. Boll Weevil Suppression, Management and Elimination Technology, Memphis, Tenn. (Feb. 13–15, 1974)*, pp. 22–25.

Bottrell, D. G., and P. L. Adkisson. 1977. Cotton insect pest management. *Ann. Rev. Entomol. 22*:451–481.

Brown, L. G., J. W. Jones, and F. A. Harris. 1976. A simulation study of insect pest management alternatives by integration of a *Heliothis* spp. model and a cotton crop model. *Am. Soc. of Agric. Engineers Paper* 76–5026.

Byerly, K., A. P. Gutierrez, R. E. Jones, and R. F. Luck. 1978. Comparison of sampling methods for some arthropod populations in cotton. *Hilgardia 46*:257–282.

Carey, J., A. P. Gutierrez, and R. E. Jones. In preparation. The within plant distribution and population dynamics of tetranychid mites in cotton.

Doutt, R. L., and R. F. Smith. 1971. The pesticide syndrome diagnosis and suggested prophylaxis, in C. B. Huffaker (ed.), *Biological Control*. Plenum, New York, pp. 3–15.

Ehler, L. E., K. G. Eveleens, and R. van den Bosch. 1973. An evaluation of some natural enemies of cabbage looper in cotton in California. *Environ. Entomol. 2*:1009–1015.

Eveleens, K. G., R. van den Bosch, and L. E. Ehler. 1973. Secondary outbreak induction of beet armyworm by experimental insecticide applications in cotton in California. *Environ. Entomol. 2*:497–503.

Falcon, L. A., R. van den Bosch, J. Gallagher, and A. Davidson. 1971. Investigation of the pest status of *Lygus hesperus* in cotton in Central California. *J. Econ. Entomol. 64*:56–61.

Falcon, L. A., R. van den Bosch, C. A. Ferris, L. K. Stromberg, L. K. Etzel, R. E. Stinner, and T. F. Leigh. 1968. A comparison of season-long cotton-pest-control programs in California during 1966. *J. Econ. Entomol. 61*:633–642.

Gutierrez, A. P. 1976. Management of cotton pests. *Joint EPPO/IOBC conference on systems modelling in modern crop protection. Paris, France (Oct. 12–14, 1976)*.

Gutierrez, A. P., G. D. Butler, Jr., Y. Wang, and D. Westphal. 1977. The interaction of pink bollworm (Lepidoptera: Gelichiidae), cotton, and weather: a detailed model. *Can. Entomol. 109*:1457–1468.

Gutierrez, A. P., L. A. Falcon, W. Loew, P. A. Leipzig, and R. van den Bosch. 1975. An analysis of cotton production in California: A model for Acala cotton and the effects of defoliators on its yields. *Environ. Entomol. 4*;125–136.

Gutierrez, A. P., T. F. Leigh, Y. Wang, and R. Cave. 1977. An analysis of cotton production in California: *Lygus hesperus* injury—an evaluation. *Can. Entomol. 109*: 1375–1386.

Gutierrez, A. P., and Y. Wang. 1977. Applied population ecology: Models for crop production and pest management in G. A. Norton and C. S. Holling (eds.), Proceedings of a conference on pest management, 25–29 Oct. 1976. CP 77–6. Int. Inst. Appl. Syst. Analy. (IIASA), Laxenburg, Austria, pp. 255–280.

Gutierrez, A. P., and Y. Wang. In preparation. Optimal pest control for *Lygus hesperus*.

Gutierrez, A. P., L. T. Wilson, and D. B. Hogg. In preparation. The within plant distribution of some pests of cotton; with special emphasis on *Trichoplusia ni* (Hubner).

Harris, F. A., L. G. Brown, J. W. Jones, G. L. Andrews, and M. W. Parker. 1976. Use of cotton plant-insect interaction in insecticide, plant resistance and economic threshold studies. *Proc. 1976 Beltwide Cotton Prod. Res. Conf. (Natl. Cotton Counc., Memphis, Tenn.)*, pp. 144–149.

Huffaker, C. B., and B. A. Croft. 1976. Integrated pest management in the U.S.: Progress and promise. *Environ. Health Perspect. 14*:167–183.

Iwao, S., and E. Kuno. 1968. Use of the regression of mean crowding on mean density for estimating sample size and the transformation of data for analysis of variance. *Res. Popul. Ecol. 10*:210–214.

Jones, J. W., H. D. Bowen, R. E. Stinner, J. R. Bradley, Jr., and J. S. Bacheler. 1975. Simulation of boll weevil populations as influenced by weather, crop status and management practices. *Am. Soc. Agric. Engineers Paper 75-4515*.

Lloyd, M. 1967. Mean crowding. *J. Anim. Ecol. 36*:1–30.

McKinion, J. M., D. N. Baker, J. D. Hesketh, and J. W. Jones. 1974. SIMCOT II: A simulation of cotton growth and yield, in *Computer Simulation of a Cotton Production System—A User's Manual*. ARS-S-52, pp. 27–82.

McKinion, J. M., J. W. Jones, and J. D. Hesketh. 1974. Analysis of SIMCOT: photosynthesis and growth. *Proc. 1974 Beltwide Cotton Prod. Res. Conf. (Natl. Cotton Counc., Memphis, Tenn.)*, pp. 117–124.

Regev, U., A. P. Gutierrez, and G. Feder. 1976. Pest as a common property resource: A case study in the control of the alfalfa weevil. *Am. J. Agric. Econ. 58*:187–197.

Sevacherian, V., and V. M. Stern. 1972. Spatial distribution patterns of *Lygus* bugs in California cottonfields. *Environ. Entomol. 1*: 695–703.

Shoemaker, C., and A. P. Gutierrez. In preparation. An optimization model for pink bollworm control.

Smith, R. F. 1976. Phases in the development of integrated pest management. *Third Pest Contr. Conf., Cairo, Egypt. (Oct. 2–5, 1976)*. Mimeo.

Smith, R. F., J. L. Apple, and D. G. Bottrell. 1976. The origins of integrated pest management concepts for agricultural crops, in J. Lawrence Apple and Ray F. Smith (eds.), *Integrated Pest Management*. Plenum, New York, pp. 1–16.

Smith, R. F., C. B. Huffaker, P. L. Adkisson, and L. D. Newsom. 1974. Progress achieved in the implementation of integrated control projects in the USA and tropical countries. *EPPO Bull. 4*:221–239.

Sprott, J. M., R. D. Lacewell, G. A. Niles, J. K. Walker, and J. R. Gannaway. 1976. Agronomic, economic and environment implications of short-season, narrow-row cotton production. *Tex. Agric. Expt. Stn.* MP-1250c.

Sterling, W. L., and R. L. Haney. 1973. Cotton yields climb, costs drop through pest management systems. *Tex. Agric. Progr. 19*:4–7.

Stern, V. M., R. F. Smith, R. van den Bosch, and K. S. Hagen. 1959. The integration of chemical and biological control of the spotted alfalfa aphid. The integrated control concept. *Hilgardia 29*:81–101.

Stockton, J. R., L. D. Doneen, and V. T. Walhood. 1961. Boll shedding and growth of the cotton plant in relation to irrigation frequency. *Agronomy J. 53*:272–275.

Van den Bosch, R. 1978. The Pesticide Conspiracy. Doubleday, New York.

Walker, J. K., J. R. Gannaway, and G. A. Niles. 1977. Age distribution of cotton bolls and damage from the boll weevil. *J. Econ. Entomol. 70*:5–8.

Walker, J. K., and G. A. Niles. 1971. Population dynamics of the boll weevil on modified cotton types. Implications for pest management. *Tex. Agric. Exp. Stn. Bull.* 1109.

Wang, Y., A. P. Gutierrez, G. Oster, and R. Daxl. 1977. A population model for plant growth and development: coupling cotton-herbivor interactions. *Can. Entomol. 109*:1359–1374.

Westphal, D., A. P. Gutierrez, and G. D. Butler, Jr. In preparation. Pink bollworm and cotton—a study of co-adaptation.

Wilson, L. T., A. P. Gutierrez, and T. L. Leigh. In press. The within plant distribution of bollworm immatures on cotton. *Hilgardia.*

6

THE SYSTEMS APPROACH TO RESEARCH AND DECISION MAKING FOR COTTON PEST CONTROL

A. P. Gutierrez

Division of Biological Control
University of California, Berkeley, California

D. W. DeMichele

Department of Industrial Engineering
Texas A&M University, College Station, Texas

Ying Wang

Division of Biological Control
University of California, Berkeley, California

G. L. Curry

Department of Industrial Engineering
Texas A&M University, College Station, Texas

Ronald Skeith

Department of Industrial Engineering
University of Arkansas, Fayetteville, Arkansas

L. G. Brown

Department of Agricultural Engineering
Mississippi State University, Mississippi State, Mississippi

INTRODUCTION

Pest management is a catchy term used mostly by entomologists who naturally focus on the invertebrate pest, but it may be a misleading term. The experience of the IBP initiated Integrated Pest Management Project (NSF/EPA supported) in the United States has shown that what entomologists often mean when they discuss pest management (PM) is crop production and protection, and that pest management is but one aspect, though a key aspect, like plant nutrition or some other discipline.

A realistic analysis (economic, ecological, etc.) of the cotton (*Gossyppium hirsutum* L.) production system has proved to be extremely difficult, primarily because of the highly indeterminate nature of its growth. Cotton growth and development, perhaps more than that of most other crops, is highly variable, and is greatly influenced by many factors (e.g., weather, insect damage, agronomic practices, etc.). This indeterminate attribute of cotton restricts the utility of the commonly used statistical procedures for assessing the impact of the various factors affecting cotton growth and development and yield. This fact has forced the agriculturist to use various methods from engineering and systems science (i.e., systems analysis). Systems analysis has considerable application to cotton production, provided the essence of the crop production problems can be distilled out in a

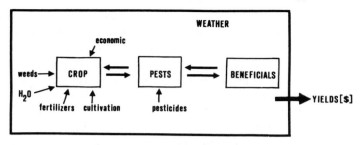

Figure 6.1. A simplified representation of an agroecosystem.

mathematically tractable form. As we shall see, cultivated cotton is an extremely tractable experimental system.

The methods of systems analysis are quite diverse (Watt, 1966), but whichever method is chosen, it must be adequate for dealing with problems in population ecology: populations of plants, pests and natural enemies as modified by weather, and by man's various agronomic inputs. Figure 6.1 depicts a trivial representation of this complex problem from the farmer's point of view. Because crops are grown for profit, the output is necessarily a monetary one.

If the crop production system is to be analyzed, it must first be put into some simplified conceptual framework. Given that this framework exists, the results of the analysis (producing understanding) can then be used to develop strategy optimization for pest control, plant breeding, or some other aspect of crop production. This chapter shows how the traditional concepts of population ecology can be applied to the study of real life crop production and pest management problems. The chapter further summarizes the progress to date and projects into the future how these findings may be implemented into the farmer's crop production and protection decision process. Most of the practical results of this work were presented in Chapter 5. The models are, after all, designed to examine field problems, not esoteric relationships. The mathematics employed, while difficult for many biologists, is trivial by mathematical standards. Gutierrez and Wang (1977) review the conceptual framework for much of this chapter.

MODELING—A GENERALIZED APPROACH

The plant ecologists Harper and White (1974) stated that populations of plants (e.g., crop plants) are composed of individuals of different ages, and each plant is further composed of populations of plant parts (e.g., root, stem, leaf, and fruit tissues) of different ages. Our terminology is different from theirs, but the concept remains valid. In a crop, the ages of individual plants are more uniform, but the second supposition holds.

The population processes for plants appear to be analogous to those of animal populations. The production of new photosynthetic material by the plant and its allocation to leaves, stems, and roots (growth) or seeds (births) are, in all respects, analogous to the growth and birth processes in animal populations. The plant grows by trapping solar energy via photosynthesis, while all higher trophic levels use this trapped energy in lower ones until all of the energy is dissipated (i.e., via entropy). The death processes are similarly analogous. Whole plants or plant parts may die. In the latter case, the parts die when leaves fall off, the stems and roots become woody, or parts may be killed by parasites or predators, which at this level of abstraction we may call herbivores for nomenclatural convenience. Plants also exhibit density-dependent responses. For example, crowding produces smaller, less hardy, and less fecund individuals just as with animals.

The Mathematical Framework

The classic population models (e.g., Thompson, 1924; Nicholson-Bailey, 1935; Lotka, 1925; Volterra, 1931; etc.), ignored age structure, and considered all individuals equal. These models are not suitable for most field problems. From laboratory studies, Slobodkin (1954), Frank (1960), and Auslander et al. (1974), and from a field study, Hughes and Gilbert (1968) added other population variables (e.g., age structure) in their models. Using matrix algebra, Leslie (1945) provided a convenient mathematical form for examining population models that include age structure (i.e., time varying life tables; Hughes, 1963). The major conceptual weakness of the Leslie Matrix model is that it also considers that all individuals in a cohort (i.e., those born at the same time) develop at the same rate.

The Leslie model can be represented as follows:

$$
\begin{bmatrix} N_{0,\,t+\Delta t} \\ N_{1,\,t+\Delta t} \\ N_{2,\,t+\Delta t} \\ \\ \\ \\ \\ \\ \\ N_{n,\,t+\Delta t} \end{bmatrix}
=
\begin{bmatrix}
mx_0(\cdot) & mx_1(\cdot) & mx_2(\cdot) & \cdot & \cdot & \cdot & \cdot & \cdot & mx_n(\cdot) \\
1x_0(\cdot) & & & & & & & & \cdot \\
0 & 1x_1(\cdot) & & & & & & & \cdot \\
& & 1x_2(\cdot) & & & & & & \cdot \\
& & & \cdot & & & & & \cdot \\
& & & & \cdot & & & & \cdot \\
& & & & & \cdot & & & \\
& & & & & & \cdot & & \cdot \\
0 & & & & & & & 1x_{n-1}(\cdot) & 0
\end{bmatrix}
\begin{bmatrix} N_{0,\,t} \\ N_{1,\,t} \\ N_{2,\,t} \\ \\ \\ \\ \\ \\ \\ N_{n,\,t} \end{bmatrix}
$$

where the age-dependent birth $[mx_i(\cdot)]$ and survivorship $[1x_i(\cdot)]$ rates are complex functions incorporating much of the complexity described above. For example: $1x_{i,t}(\cdot) = f$(age, nutrition, crowding, predation, . . .), and is the product of the survivorship probabilities of the k factors [i.e.,

$$\left(\prod_{j=1}^{k} 1x_{i,j} = 1x_{1,j} \cdot 1x_{2,j} \cdot 1x_{3,j} \ldots 1x_{n-1,j} \right) \cdot mx_{i,t}(\cdot)$$

and in this case the factors (e.g., age) scale the potential birth rate from some maximum.

A continuous form of this discrete model is the von Foerster model (von Foerster, 1959), which was originally developed to describe the growth of cell populations:

$$\frac{\partial N}{\partial t} + \frac{\partial N}{\partial a} = -\mu(\cdot) \cdot N(t, a)$$

In this model, $N(t, a)$ is the number density function (note: it could be mass or energy units), t is time, a is age, and $\mu(\cdot)$ is a complex death function incorporating $1x$ and mx factors. The model requires two conditions: $N(0, a)$ is the initial age distribution, and $N(t, 0)$ is the density of newborn. This model can be solved either numerically or analytically (see Wang et al., 1977). It can be generalized to several variables (see Streifer, 1974); work by Frank, 1960, on *Daphnia pulex*; and Slobodkin, 1954, on *Daphnia obtusa*). Slobodkin's model is of particular interest because, in addition to age and time, he considered that size should be taken into account in defining classes of physiologically identical organisms. Sinko and Streifer (1967) proposed a continuous form of a general model. Let $N(t, a, m)$ be the population density function depending on t, a, and m ($=$mass or size). If assume that $dt = da$, then

$$\frac{\partial N}{\partial t} + \frac{\partial N}{\partial a} + \frac{\partial}{\partial m_j} [g(\cdot)N(t, a, m)] = -\mu(\cdot)N(t, a, m)$$

where $g(\cdot)$ and $\mu(\cdot)$ are the mass growth and death rates, respectively, and are specific to the problem in hand. The equation is not limited to these variables. In general, suppose that there are k physical characteristics, $m_1, m_2, \ldots, m_k$, then the balance equation is

$$\frac{\partial N}{\partial t} + \frac{\partial N}{\partial a} + \sum_{j=1}^{k} \frac{\partial}{\partial m_j} [g_j(\cdot)N] = -\mu(\cdot)N$$

where $g_j = dm_j/dt =$ the growth rate of characteristic m_j.

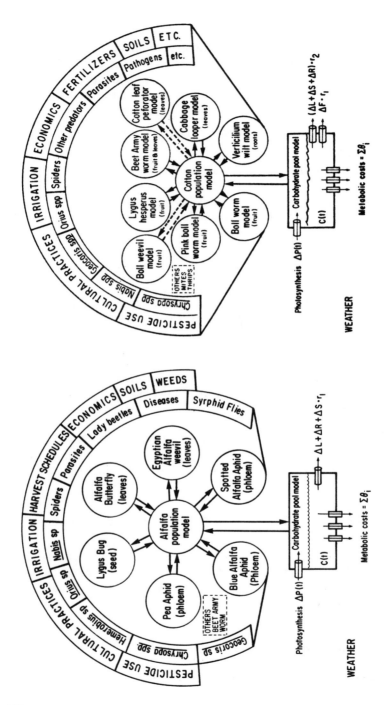

Figure 6.2. Some of the potential interactions in an alfalfa ecosystem. Notice that all interactions ultimately impinge on the carbohydrate pool model.

Figure 6.3. Some of the potential interactions in a cotton ecosystem. Notice that all interactions ultimately impinge on the carbohydrate pool model.

As implied above, this general population model has considerable application to modeling populations of organisms in higher trophic levels. Population modeling of all trophic levels has three distinct but essential parts: (1) developing the analytical framework, (2) collection of field and laboratory data, and (3) model testing against data sets not used in the construction of the model. The aim of simulation model building should first be understanding and not merely prediction; though in our work a very good blend of the theory and practical application has been achieved. Figures 6.2 and 6.3 depict some of the components of the alfalfa and cotton agroecosystems.

THE COTTON ECOSYSTEM MODEL

Plant Population Model

A recent book by Evans (1975) discussed the general state of development of the earlier plant modeling and crop physiology literature, while Thornley (1976) provides a detailed description of submodels of the physiology of plant growth and development. In recent years, several workers have developed models describing the growth and development of single plants (De Wit et al., 1970; Hesketh et al., 1971; Duncan et al., 1971; Hesketh et al., 1972; McKinion et al., 1974; Stapleton et al., 1973; Fick et al., 1974). For the most part, these models are complex algorithms. Some of them have been good simulators of individual plant growth, but have been less than satisfactory for use in agroecosystem studies of crop plant populations. The emphasis has been more on the intricacies of physiology than on functional population dynamics in a more practical sense. A more critical shortcoming is that in most cases the models have lacked a definable mathematical form and this greatly reduces their usefulness. Other mass flow models of a more analytical nature (McKinion et al., 1975) distill much of the essence of the problem, but are unrealistically structured (e.g., they ignore age structure). Those models are basically of the following form:

$$\frac{dW}{dt} = \frac{dP}{dt} - G_r \cdot \frac{dQ}{dt} - R_0 W - R_s \cdot \frac{dC}{dt}$$

where

W = total plant weight [plant weight (Q) + carbohydrate reserves (C)]

P = photosynthate

G_r = growth respiration

R_0 = maintenance respiration

R_s = respiration associated with stored starch

t = time in day degrees

Miles et al. (1973) developed a similar model for alfalfa growth and development, but it was less analytical in nature.

Briefly, the cotton model [Gutierrez et al., 1975 (a discrete form); Wang et al., 1977] is composed of balance equations for plant, leaf, stem, root, and fruit populations and the nutrient pool (= photosynthate). The equations and initial conditions for these models can be written as

$$\frac{\partial p}{\partial t} + \frac{\partial p}{\partial a} = - \mu_p(\cdot) \cdot p(t, a) + I_p \qquad \begin{array}{l} p(t, 0) = \beta_p(t) \\ p(0, a) = \gamma_p(a) \end{array} \qquad (1)$$

$$\frac{\partial L}{\partial t} + \frac{\partial L}{\partial a} = - \mu_L(\cdot) \cdot L(t, a) \qquad \begin{array}{l} L(t, 0) = g_L(t) \\ L(0, a) = \alpha_L(a) \end{array} \qquad (2)$$

$$\frac{\partial S}{\partial t} + \frac{\partial S}{\partial a} = - \mu_S(\cdot) \cdot S(t, a) \qquad \begin{array}{l} S(t, 0) = g_S(t) \\ S(0, a) = \alpha_S(a) \end{array} \qquad (3)$$

$$\frac{\partial R}{\partial t} + \frac{\partial R}{\partial a} = - \mu_R(\cdot) \cdot R(t, a) \qquad \begin{array}{l} R(t, 0) = g_R(t) \\ R(0, a) = \alpha_R(a) \end{array} \qquad (4)$$

$$\frac{\partial F}{\partial t} + \frac{\partial F}{\partial a} = - \mu_F(\cdot) \cdot F(t, a) \qquad \begin{array}{l} F(t, 0) = \beta_F(t) \\ F(0, a) = \gamma_F(a) \end{array} \qquad (5)$$

$$\frac{\partial M}{\partial t} + \frac{\partial M}{\partial a} = - \mu_M(\cdot) \cdot M(t, a) \qquad \begin{array}{l} M(t, 0) = g_M(t) \\ M(0, a) = \alpha_M(a) \end{array} \qquad (6)$$

where the independent variables t and a are time and age, respectively, measured in physiologic time units (see biophysical models section). Note that in the simplest deterministic form $dt = da$. In fact, $dt \neq da$ in nature because individuals in the population tend to age at different rates (see biophysical models section). $L(t, a)$, $S(t, a)$, $R(t, a)$, and $M(t, a)$ are mass density functions for leaf, stem, root, and fruit tissues, respectively, while $p(t, a)$ and $F(t, a)$ are number density functions for whole plants and fruits. The term I [Equation (1)] is a net immigration rate (in a general sense, it could be new seeds or other plant propagation methods). This term is not applicable to domestic cotton crops which are planted, but is important for wild cotton where new seeds may be blown into or carried into the area encompassing the study population. It is included in Equation (1) to make the model general for any population of plants. The various $\mu(\cdot)$ are complex death rate functions. The model is basically a model of the cotton canopy encompassing a population of plants, but if $p(0, a) = 1$ in Equation (1), the model collapses to a single plant model. All of these models [(1) to (6)] are coupled via the carbohydrate pool model.

The carbohydrate pool submodel (Fig. 6.4) differs radically in form from Equations (1) to (6). The equations for estimating photosynthesis used in the

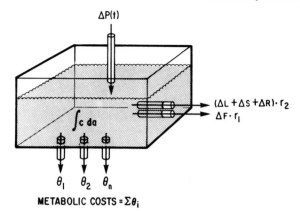

Figure 6.4. The metabolic pool model for allocating photosynthate (P), carbohydrate reserves (C) to meet metabolic needs (θ_i) of the plant or growth of plant parts [e.g., leaves (L), stem (S), root (R), or fruit (F)]. Note that the levels of the outflows indicates a priority scheme, and r_1 and r_2 indicates the fraction of the maximum growth achieved during some $\Delta t (0 \le r_i \le 1)$.

model are derived from McKinion et al. (1974). Let $\phi(t)$ be the carbohydrate present in the pool at time t. Then ϕ satisfies the following equation:

$$\frac{d\phi}{dt} = \frac{dP}{dt} - \left(\sum_{i=1}^{3} \theta_i(t) + \dot{R} + \dot{S} + \dot{M} + \dot{L} \right)$$

If $d\phi/dt$ is approximated by $[-\phi(t) + \phi(t + \Delta t)]/\Delta t$, the difference representation of Equation (3) is then

$$\phi(t + \Delta t) = \phi(t) + \frac{dP}{dt} \cdot \Delta t - \left[\sum_{i=1}^{3} \theta_i(t) + \dot{R} + \dot{S} + \dot{M} + \dot{L} \right] \cdot \Delta t$$

where, dP/dt is the rate of photosynthate production, $\phi(t)$ is the reserve carbohydrate available, the θ_i values are the metabolic costs associated with plant growth and development (c.f. McKinion et al., 1974), and $\dot{R}$, $\dot{S}$, $\dot{M}$ and $\dot{L}$ are the realized plant growth rates. The cotton growth is determined principally by photosynthate availability [= supply (Q)] some or all of which is used to meet daily age dependent demands (= D) by the various plant parts (L, S, R, M) and the θ_i's.

The computation of Q and D, as well as the priority scheme for allocating dry matter can be described algebraically as follows, while the process can be visualized as shown in Figure 6.4 (the levels of the taps for dry matter allocation can be viewed as a priority scheme):

The total demand (D) at time t equals $\sum_{i=1}^{3} \Delta\theta_i + \Delta L + \Delta R + \Delta S + \Delta M$,

where the $\Delta\theta_i$ are respiratory losses and ΔR, ΔS, ΔL, and ΔM are the maximum growth rates by the plant parts (roots, stems, leaves and fruits respectively). If $D_t \le Q_t$ (i.e., $r = Q/D > 1$), then $\Delta C_t = Q_t - D_t$; otherwise, the following priority scheme applies (Gutierrez et al., 1975):

1. $Q_1 = Q_t - \sum\limits_{i=1}^{2} \Delta\theta_1$, where $\theta_3 \cdot \Delta t = (\theta_M + \theta_L + \theta_S + \theta_R) \Delta t$

2. $r_1 = Q_1/(M + \theta_M) \cdot \Delta t$, where $0 \le r_1 \le 1$, $Q_1 > 0$ and $\overset{*}{\theta}_3 = \theta_3 - \theta_M$

3. $Q_2 = Q_1 - (\theta_M + \dot{M}) \cdot \Delta t \cdot r_1$ (i.e., for fruits = M)

4. If $Q_2 > 0$, $r_2 = \dfrac{Q_2}{(\dot{S} + \dot{R} + \dot{L} + \dot{\theta}_3) \cdot \Delta t}$ for $Q_2 > 0$ and $0 \le r_2 \le 1$

5. $\Delta C_t = Q_2 - [\overset{*}{\theta}_3 + \dot{S} + \dot{R} + \dot{L}] \cdot \Delta t \cdot r_2$ (i.e., for all other plant parts)

The second $0 \le Q/D = r_1 < 1$ ratio is used to scale the rates of mainstem node and fruit point production, as well as to estimate the rate and ages of fruits (squares and small bolls) to be shed.

The time when $r < 1$ occurs may vary depending upon several factors (e.g., plant density, weather, etc.), but its impact can be readily observed in the field as reduced rate of dry matter accumulation in leaves, stems, and roots, reduced mainstem node and fruit point production, and the occurrence of peak squaring and rapid growth of large bolls (see Fig. 6.5). The model has been a good simulator of cotton growth and development in the San Joaquin Valley of California, where only Acala cotton is grown. The parameter modifications required to convert the Acala cotton model to a form to simulate the Delta pine variety of cotton grown in Arizona are given by Gutierrez, Butler, Wang and Westphal (1977).

1. Developmental phenology
2. Estimates of photosynthesis under various conditions of leaf age and light intensity
3. Estimates of dry matter partitioning amongst plant parts
4. Carbohydrate sink-source relationships (i.e., when and where photosynthate goes)
5. Density-dependent growth relationships
6. Mortality of plants and plant parts due to various causes
7. The effects of weather on plant growth and development, especially temperature, light, and moisture
8. The effects of agronomic practices on plant growth and development

It is certain that submodels for more accurately estimating photosynthetic efficiency (e.g., ones utilizing light penetration) and water use by plants will be forthcoming soon, and this model is formulated with these developments in mind. It could accommodate their inclusion. These recent advances are discussed in a later section.

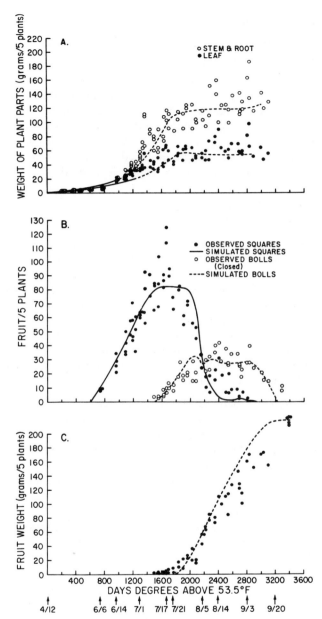

Figure 6.5. The phenology and development of a cotton crop in Corcoran, California during 1975: (A) dry matter accumulation in stems + roots and leaves, (B) the observed fruit patterns, and (C) the accumulation of dry matter in fruit.

165

MODELS FOR HERBIVORES AND
HIGHER TROPHIC LEVEL ORGANISMS

In the past, zoologists have studied herbivore populations without adequately considering the dynamics of their food. As we shall see, this feedback is very important, not only to understanding the ecological aspects of the problem, but also the economic implications. The herbivore and higher trophic level organisms can also be modeled in the following form, and in fact, become submodels of the higher trophic level models:

$$\frac{\partial N_i}{\partial t} + \frac{\partial N_i}{\partial a} = -\mu(\cdot) \cdot N_i(t, a) + I_{N_i}(\cdot)$$

The model structure is the same as that for modeling the plant except that $I_{N_i}(\cdot)$ is a complex emigration or immigration rate function of the pest species into or out of the field.

Extensive field and laboratory studies have been done to examine the species-specific biologies of each of the pest species (see Figs. 6.2 and 6.3). Relevant data from the literature have, in general, been sparse because the data were only partially reported, were inconsistent or in error, or taken in a manner which was not useful for developing various ecological plant–pest–predator–weather relationships. The following types of information must be gathered for the insect species:

1. Temperature-dependent developmental rates
2. Estimates of developmental times for each life stage, as well as their age dispersion of cohorts through time
3. Estimates of $1x$ and mx (see Andrewartha & Birch, 1954) under optimal and suboptimal conditions
4. Data on their seasonal phenology, population dynamics, and age structure
5. Overwintering characteristics
6. Site and age preference for plant parts (plant feeders), or for prey (predators)
7. Feeding rates
8. Sex ratios
9. Detailed observation on behavior
10. Detailed field life table data for use in developing various mortality submodels (see next section)

Population models for several of the important pests of cotton have been developed, and the computer algorithms are by and large discrete approximations of the models proposed here. A partial list of these models follows:

1. Beet armyworm—cabbage looper/cotton (Gutierrez et al., 1975); Hogg & Gutierrez, in progress.
2. Bollworm/cotton (*Stinner et al., 1974; Wilson & Gutierrez, in progress; Brown et al., in progress; Skeith, in progress)
3. Boll weevil/cotton (Wang et al., 1977; *Jones et al., 1974; Curry & DeMichele, personal communication)
4. *Lygus hesperus*/cotton (Gutierrez, Leigh, Wang and Cave, 1977)
5. Pink bollworm/cotton (Gutierrez, Butler, Wang and Westphal, 1977)

A model for spider mite dynamics on cotton is currently in progress (*J. Carey, in progress).

Coupling Models of Organisms of Different Trophic Levels

Coupling of plant and herbivore processes or activities is not conceptually too difficult, given that the appropriate data are at hand (see above). First, in coupling the two models we must recognize that the time step for the plant may not equal the time step for the herbivore(s). The developmental rates of the herbivore life stages may also vary or they may be influenced differently by the age or quality of the plant part(s) attacked (e.g., pink bollworm— Gutierrez, Butler, Wang and Westphal, 1977). Next we must integrate age preferences for plant parts, prey searching success, hunger, attack rates, and so on, in a predation model relevant for each pest.

A fundamental understanding of the attack processes of a pest on a crop is very germane to the development of sound economic thresholds or to formulate economic models to assess the damage (Na = numbers or mass attacked) of the pest on a wide scale. For plants, Na may be fruit numbers, leaf, stem, fruit or root mass photosynthate, and/or wound-healing losses, or disease recovery costs (w). Figure 6.4 depicts the nature of the problem, as the plant's ability to compensate for damage is to a large degree dependent on available reserves (see Gutierrez et al., 1976). If $\int c/da$ is the reserves, $D_{max} = \Delta L + \Delta S + \Delta R + \Delta M$ = the maximum demand for plant growth, ΔP is the new photosynthate, and $\sum \theta_i$ represents the various metabolic costs, then reserves are utilized when $D_{max} + \sum \theta_i > \Delta P$. If maximum growth is to occur, the demand must be less than the supply of photosynthate:

$$D_{max} + \sum \theta_i \le P + \lambda \int cda$$

*Those preceded by an asterisk have not been successfully coupled with a suitable cotton plant model.

where λ is some rate of reserve availability. It is possible that under stress conditions (e.g., insect damage $= Na$) the ability of the plant to compensate is severely impaired. Reduced yields can occur because of herbivore injury (Na), reduced demand (D^*_{max}) or reduced photosynthesis (P^*):

$$Na + D^*_{max} + \sum\theta_i > P^* + \lambda \int cda$$

A similar model exists for the herbivores and predators (Gutierrez & Wang, 1977) (Fig. 6.6). The important component is hunger as it affects the predator's search behavior, though hunger may not affect some parasitoids which have their full complement of eggs when they emerge from the pupa and do not host feed. The animal has some age-dependent maximum gut capacity (Ω) which may be full or may be partially empty. If the gut is partially empty, the insect is hungry (i.e., it can eat), but the amount it eats (Na) is the mass of prey it captures. The model for hunger is

$$0 \leqslant h_{t+1} = h_t - Na_t + mx_t + \Delta\Omega(a) + g_t + \sum\psi_{i,t} \leqslant \Omega$$

where

$h =$ the mass of prey the animal can eat at time t

$Na =$ the mass of prey captured

$mx =$ the mass of progeny produced

$g =$ the mass growth of the animal

$\sum\psi_i =$ metabolic costs

The priority of nutrients for g and mx change with age and h (e.g., adults stop growing but do produce progeny), while survivorship ($1x$) is also a function of h (e.g., Gutierrez et al., in prep.).

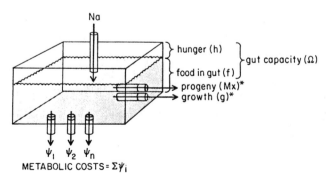

Figure 6.6. The metabolic pool model for herbivores and predators. Na is the mass of prey captured. *The levels of the outflows indicate priority levels which depend on age and hunger.

"PREDATION" MODELS, OR ESTIMATING ATTACK SUCCESS

Herbivore/Predation

An herbivore is a plant predator, hence the relevant notions of predation models apply equally well to herbivores as to the higher trophic level organisms. If an herbivore attacks a population of leaves, it alters their age structure (by differentially consuming different age classes and amounts of tissue), and causes wound healing losses. Because individual leaves may be only partially eaten, and still be alive and functioning at reduced levels, mass rather than numbers are the units of attack (Na). Many herbivores attack fruits and cause them to abscise or be unavailable for other herbivores. In such cases, Na will refer to numbers, though its relationship to mass may also be useful.

Cotton and alfalfa have many of the same pests and also some of the same predators. The principal general entomophagous predators in both cotton and alfalfa in the San Joaquin Valley are *Geocoris pallens* Stål, *G. punctipes* Say, *Chrysopa carnea* Stephens, *Nabis americoferus* Carayon and *Orius tristicolor* (White). In cotton, predatory insects are much more important than the entomophagous parasitoids whose behavior does not fit this pattern. The latter's omission in this discussion is therefore not serious. The migration of insect pests from alfalfa to other crops, including cotton, is well documented [e.g., lygus bug (Sevacharian & Stern, 1972)], but the movement of their natural enemies from alfalfa, and into cotton, is less well understood (Ehler et al., 1973; Eveleens et al., 1973).

Predators of plants or animals may be polyphagous or oligophagus or have some other feeding habit. In some cases, the predators may feed on both plants and animals (e.g., *Lygus*, an herbivore; and *Orius*, a predator). The problem becomes more complex as we consider other behavioral components such as prey preference, age preferences, switching (Murdoch, 1969), age-dependent searching rates, predator interference, and so on. Intensive field and laboratory studies are being conducted on the feeding behavior of all predator species listed above in San Joaquin Valley, Mississippi, and Texas cotton. Less is known concerning the activities and roles of insect parasitoids and disease pathogens, but preliminary evidence indicates that these natural enemies are effective only at moderately high host densities for some of the pests, at least in cotton crop systems in California.

These behavioral problems create difficulties in developing models to accurately estimate Na. Lotka (1925) and Volterra (1926, 1931) described an equation for estimating the number of prey [e.g., animals, (here extrapolated to plant parts)] attacked (Na) when the two species (prey and predator) are moving randomly in space, while Thompson (1924) described what is basically a random search model. The Nicholson and Bailey (1935) model, like the others, is overly simplistic as it ignores hunger, age class, and so on. The models proposed by Holling (1959, 1966) were adequate to describe the activities of a single predator feeding on a population of prey, but not for a population of predators feeding on a population of prey, which is what we see in nature. Griffith and Holling (1969) attempted to expand this model to a complex of predators, but the result is inadequate.

Other models by Hassell and Varley (1969), Royama (1970), Hassell and May (1973), Rogers and Hassell (1974), and others, consider such factors as mutual interference among predators, influence of prey density, nonrandom distribution of prey and predator searching, and so on. While these models add complexity, they do not provide adequate estimates of Na for field utilization (Huffaker et al., 1977).

Frazer and Gilbert (1976) and Gilbert et al. (1976) derive a more tractable, or at least a more field applicable, model which overcomes most of the difficulties.

$$Na = N_0[1 - e^{-(bTP/N_0)}\ 1 - e^{-(aN_0/b)}]$$

where

Na = the number of prey attacked at time t

N_0 = the number prey present

b = the predator demand rate

a = the predator search rate

$T = \Delta t$

P = predator numbers

These authors incorporated Holling's hunger component, not on an individual predator basis, but rather, averaged for the predator population. This model was used with good success by Frazer and Gilbert to model coccinellid predation at low aphid densities, and by Gutierrez, Wang, and Daxl (1979) to model boll weevil (an herbivore) attack on cotton squares using Nicaraguan cotton data. This model provides more adequate estimates of Na at all host-predator densities, and estimates survivorship from predation ($1x_s$) in population models. The model needs to be expanded to incorporate relevant stochastic processes such as predator search rates, attack success, and hunger. This model has many interesting properties. For example, it can be seen from a Taylor expansion that as N_0 becomes small, the term

$$1 - e^{-aN_0/b} \to aN_0/b \quad \text{as} \quad aN_0/b \to 0,$$

or

$$1 - e^{-aN_0/b} = 1 - 1 + aN_0/b - \frac{(aN_0/b)^2}{2!} + \frac{(aN_0/b)^3}{3!} - \cdots$$

$$- (-1)^{n-1} \frac{(aN_0/b)^{n-1}}{(n-1)!} + (-1)^{n+1} \frac{(aN_0/b)^n}{n!}$$

which when substituted into the model yields

$$Na = N_0(1 - e^{-aTP})$$

As $P \to 0$, the Nicholson-Bailey model becomes a Lotka-Volterra model because

$$1 - e^{-aTP} = 1 - 1 + aTP - \frac{(aTP)^2}{2!} + \frac{(aTP)^3}{3!} - \cdots$$

$$- (-1)^{n-1} \frac{(aTP)^{n-1}}{(n-1)!} + (-1)^{n-1} \frac{(aTP)^n}{n!},$$

and the model reduces to

$$Na = N_0 aTP$$

The definitions of the parameters a and b are important to the model, and are described below:

a is derived from the Nicholson-Bailey model and equals

$$- \ln\left(\frac{N_0 - Na}{N_0}\right) \Big/ P\, T$$

If $P = 1$, and $T = t = 1$ (i.e., standard conditions for estimating a empirically), then

$$a = -\ln\left(\frac{N_0 - Na}{N_0}\right) = -\ln\left(1 - \frac{Na}{N_0}\right) = -\ln(1x)$$

In the Nicholson-Bailey model it is a negative log survivorship, while in the Lotka-Volterra model it equals the mortality rate (i.e., effective part of A searched). A is the area available for search.

To estimate a empirically (for each species and life stage), a realistic field or laboratory experimental arena must be devised, and values of A and T assigned such that the behavior which underlies a is not greatly influenced [e.g., large changes in hunger (h)].

In its simplest form, single predators would be allowed to search in the experimental arena for different densities of hosts (mass and numbers). Because hunger as well as N_0 greatly influence a, h_0 at $T = 0$ must be specified, hence $a = f(h_0, N_0)$. The effects of N_0 on a have been described above, and the effects of h_0 can be incorporated as follows:

$$a_{t+1} = a_t \cdot \frac{h_t}{h_0}$$

where h_0 is the hunger near satiation.

The parameter b can be estimated from simpler laboratory experiments and depends on the level of h (see hunger equation), hence if $h_t \geq b_{max} \cdot \Delta t$, then $b = b_{max}$. If $h_t < b_{max} \cdot \Delta t$, $b = h_t/\Delta t$.

In nature, T and A as well as P and N_0 rarely remain constant. The model as formulated now accommodates P and N_0, but not T and A. A is usually some habitat (the surface of cotton leaves, etc.), so the dynamics of its growth must be accommodated. For example, in alfalfa, A might be some unit of length, specifically the number of stems per ft^2 × their respective lengths. In the Frazer-Gilbert model, the observed value (A_0) would be compared to the standard

$$A_t = \frac{A_0}{A}$$

and the new value A_t used.

A similar procedure could be used for estimating T_t, where $T_0 = \Delta t_0$ may be variable (i.e., day degrees), and compared with the standard Δt used in the experiment:

$$T_t = \frac{T_0}{T}$$

A general model following work of Frazer and Gilbert (1976) for the attack of a complex of predators on several prey species (also several ages) has been formulated by R. E. Jones (unpublished).

Biophysical Models

Accuracy and further advancement of population modeling requires that relevant stochastic processes (e.g., Barr et al., 1973) and appropriate behavior be included (Gilbert et al., 1976), and that the underlying physiologic processes affecting age-specific survivorship $[1x(\cdot)]$ and natality $[mx(\cdot)]$ be described in rigorous mathematical terms. Considerable progress has been made in this area.

The basic premise in much of the work is that intrinsic developmental rates, and many other factors of plant and insect population dynamics are controlled or significantly affected by the thermal environment. For example, much of the earlier work on developmental rates of insects and plants used the notion of day degrees. This concept assumes that their developmental rates are exactly proportional to temperature throughout much of the range favorable to their environment. A good review of the effects of temperature on organism development is given by Precht et al. (1973). There are numerous empirical and physiochemical model formulations of the development process, but only two are relevant: (1) the day degree or temperature summation rule put forward by de Candolle (1855), and (2) the nonlinear temperature inhibition model derived by Johnson and Lewin (1946). The day degree summation rule is widely used in applied zoology, entomology, and agronomy due to its ease of application and the fact that within certain temperature limits it approximates observed values. This rule has been used successfully in the crop productivity model described previously, only because the observed temperatures were within the favorable range. Figure 6.7 shows the typical S-shaped curve of development, including the linear region of the development curve where the day degree concept is valid. Problems with the day degree model include: (1) the developmental zero, "Ta," is not the true threshold for development; and (2) gross errors in predicted development rates occur at the temperature extremes (Stinner et al., 1974). The results of the poikilotherm study provide a functional form and conceptual basis whereby problems (1) and (2) have been eliminated (Sharpe and DeMichele 1977).

Johnson and Lewin (1946) developed a nonlinear mathematical function to describe the inhibition of microbial growth at high temperatures. A similar derivation for low temperature inhibition of enzymatic processes has been formulated by Hultin (1955). Each of these equations described the inhibiting effect of *either* high *or* low temperatures on organism development. However, what was needed was a mechanistically based formulation on poikilotherm development that yields a linear response in mid-temperatures and nonlinear temperature inhibition at both high and low temperatures. The extent of the temperature inhibition can be shown by an Arrhenius

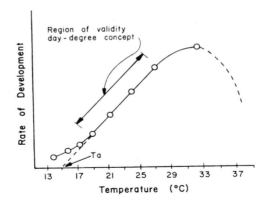

Figure 6.7. Typical relationship between temperature and rate of insect development (Wigglesworth, 1965). Ta represents the theoretical "developmental zero."

plot (Ingraham, 1973). Typical Arrhenius plots of organism development are shown in Figure 6.8. It can be seen that both high and low temperature inhibition of process rates are significant factors.

The poikilotherm analysis resulted in a derivation of a general model for organism development, based on biological process rates incorporating both high and low temperature inactivation of control enzymes. Six thermodynamic constants appear to characterize many organisms' enzymatic response to the full range of temperature (see Sharpe & DeMichele, 1977). This universal characteristic allows the use of the same concept and mathematical structure for most poikilotherms, and is easily incorporated in the previously described population models.

Closely related to the work on temperature dependent development is the work that has been completed on organism emergence from one state of development to another. It has long been recognized that individuals of a population do not respond in precisely the same way to a given stimulus or environment. This somewhat random spread in performance results in a distribution of development rates rather than a single deterministic value. This type of variability can create havoc when attempting to simulate a population with a deterministic model. This variability probably results from metabolic and genetic differences among the individuals of the population. It is a common assumption that this variability is symmetrically distributed. Symmetrical distributions have been proposed by Glass and Grebenik (1954), Menke (1973), Jones (1974), and Stinner, Gutierrez and Butler (1974). while a considerable number of other researchers have represented variability

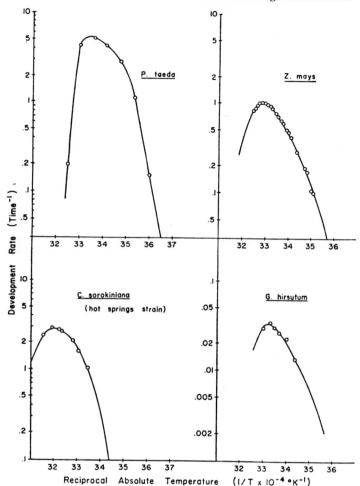

Figure 6.8. Arrhenius plots of development for other organisms. Comparison of the Sharpe and DeMichele (1977) model with constants from Table 6.2 and experimental data (a) *Pinus taeda*, (b) *Zea mays*, (c) *Chlorella sorokiniana*, and (d) *Gossypium hirsutum*.

in development time by a mean and standard deviation. This latter representation, however, contains insufficient information for the purposes of estimating the probability distribution of developmental times.

Observations on variability in development times between individuals within a population of homeotherms or poikilotherms reared at constant temperatures show a definite asymmetry (see Fig. 6.9). Anderson (1971) for

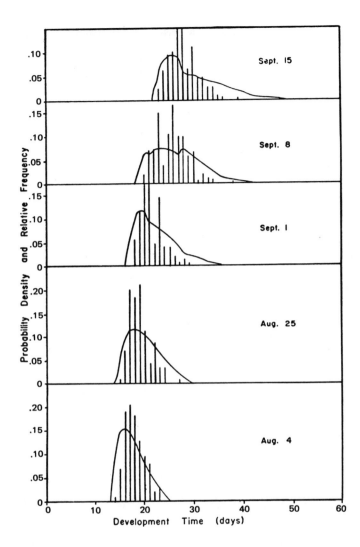

Figure 6.9. Theoretical probability density functions of emergence from Sharpe et al. (1977) for boll weevil for variable temperature regimes typical of Lubbock, Texas. Also shown on same scale are relative frequency histograms of observed development times for same given variable temperature regimes.

plants, and Hoffard and Coster (1976) and Stinner et al. (1975) for insects, report a skew in the distribution of development times. Stinner et al. (1975), working with populations of *Heliothis zea*, included the skew in the development times in their simulation model of *Heliothis* populations by the utilization of a cumulative beta distribution.

It is shown in the emergence model of Sharpe et al. (1977) that the distribution of poikilotherm development times for constant and variable temperatures can be described by a model based upon one simple assumption. This assumption is that the concentration of enzymes which are rate controlling for development are symmetrically distributed about some genetically determined mean concentration. It then follows mathematically that the

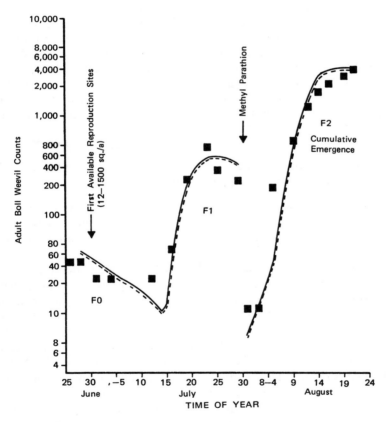

Figure 6.10. Walker's 1968 College Station boll weevil field population data (black square) and Curry et al. (1978) stochastic simulation model results: the two lines show the model sensitivity to initial population estimates.

skew in the distribution in development times, observed by Stinner et al. (1975), and others, results naturally from the transformation from development rates to emergence times. The model enables predictions of the distribution of emergence times for organisms reared under any set of variable temperature field conditions. For example, a preliminary boll weevil model was reported in the 1973 IPM annual report which utilized the fractional development approach of Fye et al. (1969) and Butler and Watson (1974). While the model clearly demonstrated the need to include the physiologic state of the plant in the insect population dynamics model, it could accurately simulate only the F_1 generation in the field. A sensitivity study of this model coupled with insight gained from a review of the *Heliothis* model developed by Stinner, Rabb, and Bradley (1974) suggested that stochastic variability in the rates of development, reproduction, and lengths of life of individuals was required. Sensitivity analysis also showed the need for a stochastic reproduction profile model for the boll weevil. The boll weevil reproduction data of Isely (1932) and Cole (1970) were used to develop the reproduction profile model. The stochastic population model was compared with several years of field data with good success (e.g., Fig. 6.10). The results indicate that the model is sensitive to the accuracy of the initial population count. The model further emphasized the strong interplay between plant development and weevil biology (see also Wang et al., 1977).

WHAT USE ARE MODELS IN COTTON PRODUCTION AND PLANT PROTECTION?

Probably the most important use of a model is that it forces the researcher to specify the components. Because of this, the process enables the worker to quickly define what is known about the problem and as such is most useful in guiding the research.

Most of the above models are designed for use in developing crop production-crop protection strategies and management systems; hence, they must be examined for accuracy in the field under a variety of conditions. In a practical sense, the only check of a model's "validity" is its explanatory value for "check data" not used in the construction of the model itself (i.e., independent testing), but the test must include a wide range of variables. For example, the cotton models (Gutierrez et al., 1975; Wang et al., 1977; see Fig. 6.5) both require that the model mimic the phenology of crop growth and development, dry matter accumulation in the various plant parts, and the age structure of the fruit population. When discrepancies arise, the researcher must explain them, so the model may be immensely useful, not only for telling us what we know, but by telling us what we do not know so that we can design the appropriate experiments.

Some groups stress crop yield or pest population trend forecasting over short periods of time (i.e., simulation models) in order to determine if a chemical treatment must be made or can be delayed until the next sample is taken (Chapter 5). Because the driving parameter for these models is the weather, accurate predictions are, under the best circumstances, reliable only for a few days into the future (NAS, 1976).

Other groups (California) stress the development of general strategy models based upon a good understanding of the system (i.e., optimization models). These will be discussed here in greater detail.

Optimization Models

Optimization models for various aspects of crop production and protection are beginning to appear in the literature, and in many cases they are extensions of simulation models (e.g., Regev et al., 1976; see also Chapter 2). For example, if we wish to establish the optimum pest control strategy among several possibilities, we must first decide what our objective is (i.e., our objective function). It might be a biological objective (e.g., maximum light penetration) or it might be maximum profit (π).

Farmers usually want to maximize profit (π), hence their objective function is

$$\max \pi = B(x) - C(x)$$

where $B(x)$ are all the benefits of using some decision variable (e.g., $x_i = $ pesticides, or biological control, etc.), while $C(x)$ are all of the costs of using the procedures. (N.B. It could be a zero amount.) Also note that $B(\cdot)$ and $C(\cdot)$ could be complex functions. The optimal decision when pesticides are applied occurs when

$$\frac{dB(x)}{dx} = \frac{dC(x)}{dx}$$

The validity of the optimization results depends on the validity of the model used, as well as the validity of the questions asked of the model. Optimization models to establish sound economic thresholds and control strategies are currently in progress for pink boll worm (*Pectinophora gossypiella* (Saunders)) (Shoemaker & Gutierrez, personal communication) and *Lygus hesperus* Knight (Gutierrez & Wang, 1977). Both of these species are key pests in various areas of the United States cotton belt.

The optimization problem may be more biological than economic in nature (as in the immediate goals in plant breeding). Crop genotypes are selected by man to meet specific needs, and the standard plant breeding experiments usually involve planting many cultivars and assessing various attributes such as yield and quality (or pest resistance) at the end of the season. This is an arduous, time-consuming procedure. It is possible that a good weather driven plant model could help predetermine the characteristics crop plants should have for any particular area. The cotton model tells us that the dry matter partitioning coefficient, the timing of fruit development, and so on, determine to a large extent the growth and development of the crop (i.e., determine yields). The weather inputs (e.g., for Fresno, California) are part of the cotton system model, and they characterize the region. The optimization model seeks to maximize yields (Y), hence the objective function is

$$\max_{x_{i,\,i=1,2,\ldots,n}} Y = \max \int_0^\infty \int_0^\infty mF(t^*, a, m, x_1, x_2, \ldots, x_n, \cdot)\, da\, dm$$

where $F(t^*, a, m, x_1, x_2, \ldots, x_n, \cdot)$ satisfied the system of equations which describe the cotton model. Note that the x_1 are decision variables (see below), m is the mass of cotton fruit, F is the number of fruits, t^* is the end of the season, and a is age. The decision variables (x_i) that have been used in the model are

1. Planting density
2. Timing of the first fruiting branch
3. Rate of fruit point production
4. Age of transition for the growth rates of single fruits
5. Growth rate of young fruits
6. Growth rate of older fruits

Other decision variables might be the leaf-to-stem ratio during both the vegetative and fruit maturation phases. Powell's method (Powell, 1964) was used in this work so that much of the complexity could be retained. Other optimization procedures are currently being investigated for use on this problem. The value of this type of analysis to plant breeding are obvious, as the x_i values represent all characters which can be easily observed and probably be selected. Preliminary results indicate that achieving a global optimum is difficult, but that several local optima of approximately the same Y are commonly found. That is to say that the plants may be engineered in several different ways to produce the same crop.

If, say, we now wish to use the program to minimize insect damage, we could incorporate the insect model into the optimal (or observed) plant program, and use the same objective function. The decision variables in this case might be quantity and timing of pesticides, planting date for the crop, plant density, or some attribute of plant growth and development.

IMPLEMENTATION

The next step in the study of pest management in cotton is the implementation of these models into the decision process of the producer. It would seem that we have proceeded far enough with the research to begin thinking of its implementation. Efforts to begin implementation are currently underway in Arkansas, California, Mississippi, and Texas.

Arkansas

This research group has developed a phenology-population dynamics model for *Heliothis zea* (R. Skeith, IPM Project Report, 1976), and have been actively organizing and educating growers in a large community about the benefits of using the model in forecasting population flights. The effort successfully coupled field scouting efforts with computer evaluations of current damage potential of the pest.

The results of the Arkansas effort have been very dramatic in that pesticide treatments were reduced from 10–12 to approximately 2 in 1976. This reduction was achieved uniformly across the experimental area (see Chapter 5 for greater detail). The research group is currently engaged in coupling a plant model with their bollworm model.

California

Considerable work is in progress in California on developing the computer technology required to put these models into actual field practice. Several simulation models which couple the plant and pest(s) have been developed and others are under development, (e.g., cotton, pink bollworm, lygus, beet armyworm) and they have been written so that they may be accessed from remote terminals. The costs of obtaining the computer hardware and implementing the programs have been somewhat limiting. In addition, the emphasis has been on developing a very extensive data base on cotton growth and development, the dynamics of the pests and their natural enemies, and developing sampling strategies and computer software for data management. The results of these investigations are being passed on to the growers and the number of pesticide applications have been reduced (e.g., see Chapter 5, the Kern County pest management program).

Mississippi

Smith et al. (1974) reported results of comparisons in the cost of insecticides for control of insects on conventional cottons in the Mississippi Delta. An

average cost of $16.16/ acre was incurred when pest populations were moni-
tored, and insecticides applied only when the economic thresholds were
exceeded. By comparison, a cost of $21.25/acre was incurred by growers
using an intermediate program based on previous practices, while the weekly
application of insecticides during the growing season cost $30.55/acre. The
latter is a very common practice in Mississippi (see further Chapter 5).

Texas

A major effort is underway in Texas to develop a mini-computer network to
implement many of the research results developed by the NSF/EPA cotton
research group in IPM. This pilot system is known as BUGNET.

The computerized system will be designed for use by district or county
agents or with their supervision. If the system grows, we should be able to
provide individual growers with the assistance and analysis to enable them
to use the system on their own.

While the present uses are largely biased toward entomological and
specifically cotton entomology applications, other crops, insects, and pests
will eventually be added. Already several other disciplines have shown
active interest in the computer based delivery system. Agricultural Economics
and Meteorology have applications which could be implemented. BUGNET
is intended to make the latest information affecting management decisions
available to farmers. This information will primarily be in the form of plant
growth and insect forecasting model simulations as well as limited management recommendations based on current biological and economic conditions.

The on-line capabilities in pest management are dependent on four main
components: (1) research and modeling of the pest and crop, (2) biological
monitoring, (3) environmental monitoring, and (4) an information delivery
system to educate and communicate. These four components are being
developed to be jointly used in accomplishing the goals of this program
(Chapters 1 and 5).

LITERATURE CITED

Anderson, W. K. 1971. Emergence and early growth response of cotton to controlled
    temperature regimes. *Cotton Grow. Rev. 48*:104–115.

Andrewartha, H. G., and L. C. Birch. 1954. *The Distribution and Abundance of Animals.*
    University of Chicago Press.

Auslander, D. M., G. F. Oster, and C. B. Huffaker. 1974. Dynamics of interacting
    populations. *J. Franklin Inst. 297*:345–376.

Barr, R. O., P. C. Cota, S. H. Gage, D. L. Haynes, A. N. Kharkar, H. E. Koenig, K. Y. Lee, W. E. Ruesink, and R. L. Tummala. 1973. Ecologically and economically compatible pest control. *Mem. Ecol. Soc. Aust. 1*: 241–264.

Butler, G. D., Jr., and F. L. Watson. 1974. A technique for determining the rate of development of *Lygus hesperus* in fluctuating temperatures. *Florida Entomol. 57*: 225–230.

Candolle, A. de. 1855. *Géographique Botanique Raisonée*. Masson, Paris.

Cole, C. L. 1970. Influence of certain seasonal changes on the life history and disapause of the boll weevil, *Anthonomous grandis* Boheman. Ph.D. thesis. Department of Entomology, Texas A&M University, College Station.

Curry, G. L., R. M. Feldman, and K. C. Smith. 1978. A stochastic model of a temperature dependent population, *J. Theor. Biol. 13*: 197–204.

DeWit, C. T., R. Brouwer, and F. W. T. Penning De Vries. 1970. The photosynthetic systems, in *Prediction and Measurements of Photosynthetic Productivity*. PUDOC, Wageningen, The Netherlands, pp. 47–78.

Duncan, W. G., D. N. Baker, and J. D. Hesketh. 1971. Simulation of growth and yield in cotton. III. A computer analysis of the nutritional theory. *Proc. Beltwide Cotton Prod. Res. Conf. (Natl. Cotton Counc.,* Memphis, Tenn.) p. 78.

Ehler, L. E., K. G. Eveleens, and R. van den Bosch. 1973. An evaluation of some natural enemies of cabbage looper in cotton in California. *Environ. Entomol. 2*: 1009–1015.

Evans, L. T. (ed.). 1975. *Crop Physiology: Some Case Histories*. Cambridge University Press.

Eveleens, K. G., R. van den Bosch, and L. E. Ehler. 1973. Secondary outbreak induction of beet armyworm by experimental insecticide applications in cotton in California. *Environ. Entomol. 2*: 497–503.

Fick, G. W., R. L. Loomis, and W. A. Williams. 1974. Sugar beet, in L. T. Evans (ed.), *Crop Physiology: Some Case Histories*. Cambridge University Press, pp. 259–295.

Frank, P. W. 1960. Prediction of population growth form in *Daphnia pulex* cultures. *Am. Nat. 94*: 357–372.

Frazer, B. D., and N. E. Gilbert. 1976. Coccinellids and aphids: A qualitative study of the impact of adult lady birds (Coleoptera: Coccinellidae) preying on field populations of pea aphids (Homoptera: Aphididae). *J. Entomol. Soc. B.C. 73*: 33–56.

Fye, R. E., R. Pantana, and W. C. McAda. 1969. Developmental periods for boll weevils reared at several constant and fluctuating temperatures. *J. Econ. Entomol. 62*: 1402–1405.

Gilbert, N. E., and A. P. Gutierrez. 1973. A plant–aphid–parasite relationship. *J. Anim. Ecol. 42*: 323–400.

Gilbert, N. E., A. P. Gutierrez, B. D. Fraser, and R. E. Jones. 1976. *Ecological Relationships*. Freeman, London.

Glass, D. V., and E. Grebenik. 1954. The trend and pattern of fertility in Great Britain. *Pap. R. Comm. Popul., London, H.M.S.O. 6*(1): 255.

Griffith, K. J., and C. S. Holling. 1969. A competitive submodel for parasites and predators. *Can. Entomol. 101*: 785–818.

Gutierrez, A. P., G. D. Butler, Jr., Y. Wang, and D. Westphal. 1977. A model for pink bollworm in Arizona and California. *Can. Entomol. 109* : 1457–1468.

Gutierrez, A. P., J. B. Christensen, C. M. Merritt, W. B. Loew, C. G. Summers, and W. R. Cothran. 1976. Alfalfa and the Egyptian alfalfa weevil. *Can. Entomol. 108* : 635–648.

Gutierrez, A. P., L. A. Falcon, W. B. Loew, P. A. Leipzig, and R. van den Bosch. 1975. An analysis of cotton production in California: A model for Acala cotton and the effects of defoliators on its yields. *Environ. Entomol. 4* : 125–136.

Gutierrez, A. P., T. F. Leigh, Y. Wang, and R. Cave. 1977. An analysis of cotton production in California: *Lygus hesperus* injury—an evaluation. *Can. Entomol. 109* : 1375–1386.

Gutierrez, A. P., and Y. Wang. 1977. Applied population ecology: Models for crop production and pest management, in G. A. Norton and C. S. Holling (eds.), *Proceedings of a Conference on Pest Management*, 25–29 Oct. 1976. CP 77–6. Int. Inst. Appl. Syst. Anal. (IIASA), Laxenburg, Austria, pp. 255–280.

Gutierrez, A. P., Y. H. Wang, and R. Daxl. 1979. The interaction of cotton and boll weevil—a study of co-adaptation. *Can. Entomol. 111* : 357–366.

Harper, J. L., and J. White. 1974. The demography of plants. *Annu. Rev. Ecol. Syst. 5* : 419–463.

Hassell, M. P., and R. M. May. 1973. Stability in insect host parasite models. *J. Anim. Ecol. 42* : 693–726.

Hassell, M. P., and G. C. Varley. 1969. New inductive population model for insect parasites and its bearing on biological control. *Nature* (Lond.) *223* : 1133–1137.

Hesketh, J. D., D. N. Baker, and W. G. Duncan. 1971. Simulation of growth and yield in cotton: respiration and the carbon balance. *Crop Sci. 11* : 394–398.

Hesketh, J. D., D. N. Baker, and W. G. Duncan. 1972. II. Simulation of growth and yield in cotton—environmental control of morphogenesis. *Crop. Sci. 12* : 436–439.

Hoffard, W. H., and J. E. Coster. 1976. Endoparasitic nematodes of *Ips* bark beetles in eastern Texas. *Environ. Entomol. 5* : 128–132.

Holling, C. S. 1959. The components of predation as revealed by a study of small-mammal predation of the European pine sawfly. *Can. Entomol. 91* : 292–320.

Holling, C. S. 1966. The functional response of invertebrate predators to prey density. *Mem. Entomol. Soc. Can. 48* : 3–86.

Huffaker, C. B., R. F. Luck, and P. S. Messenger. 1977. The ecological basis for biological control. *Proc. XV Int. Congr. Entomol.* (*Washington D.C., 1976*), pp. 560–598.

Hughes, R. D. 1963. Population dynamics of the cabbage aphid, *Brevicoryne brassicae* (L.). *J. Anim. Ecol. 32* : 393–426.

Hughes, R. D., and N. Gilbert. 1968. A model of an aphid population—a general statement. *J. Anim. Ecol. 37* : 553–563.

Ingraham, J. L. 1973. Genetic regulation of temperature responses, in H. Precht, J. Christophersen, H. Hensel, and W. Larcher (eds.), *Temperature and Life*. Springer-Verlag, Berlin, Heidelberg, pp. 60–77.

Iseley, D. 1932. Abundance of the boll weevil in relation to summer weather and to food. *Ark. Agric. Expt. Stn. Bull.* 271.

Johnson, F. H., and I. Lewin. 1946. The growth rate of *E. coli* in relation to temperature, quinine and coenzyme. *J. Cell Comp. Physiol. 28* : 47–97.

Jones, J. W. 1974. A simulation model of boll weevil population dynamics as influenced by the cotton crop status. Ph.D. thesis, North Carolina State University, Raleigh.

Jones, J. W., H. D. Bowen, J. R. Bradley, and R. E. Stinner. 1974. Modeling interactions of boll weevils on cotton crops by analysis of behavioral patterns. *Proc. Beltwide Cotton Prod. Res. Conf. (Natl. Cotton Counc., Memphis, Tenn.)*, pp. 102–103.

Leslie, P. H. 1945. On the use of matrices in certain population mathematics. *Biometrika 35* : 213–245.

Lotka, A. J. 1925. *Elements of Physiological Biology.* Williams & Wilkins, Baltimore.

McKinion, J. M., J. W. Jones, and J. D. Hesketh. 1974. Analysis of SIMOT: Phytosynthesis and growth. *Proc. Beltwide Cotton Prod. Res. Conf. (Natl. Cotton Counc. Memphis, Tenn.)*, pp. 117–124.

McKinion, J. M., J. W. Jones, and J. D. Hesketh. 1975. A system of growth equations for the continuous simulation of plant growth. *Am. Soc. Agric. Eng. Paper 18*(5): 975–979, 984.

Menke, W. W. 1973. A computer simulation model: The velvet bean caterpillar in the soybean agroecosystem. *Fla. Entomol. 56* : 92–102.

Miles, G. E., R. J. Bula, D. A. Holt, M. M. Schreiber, and R. M. Peart. 1973. Simulation of alfalfa growth. *Am. Soc. Agric. Eng. Paper 73*–4547.

Murdoch, W. W. 1969. Switchings in general predators: Experiments on predator specificity and stability of prey populations. *Ecol. Monogr. 39* : 335–354.

National Academy of Sciences. 1976. Climate and food. National Academy of Sciences, Washington, D.C.

Nicholson, A. J. and V. A. Bailey. 1935. The balance of animal populations. *Proc. Zool. Soc. Lond. Pt. 1*, pp. 551–598.

Powell, M. J. D. 1964. An efficient method for finding the minimum of a function of several variables without calculating derivatives. *The Computer J. 7* : 155–162.

Precht, H., J. Christopherson, H. Hensel and W. Larcher (eds.). 1973. *Temperature and Life.* Springer-Verlag, Berlin, Heidelberg.

Regev, U., A. P. Gutierrez, and G. Feder. 1976. Pest as a common property resource: A case study in the control of the alfalfa weevil. *Am. J. Agric. Econ. 58* : 187–197.

Rogers, D. J., and M. P. Hassell. 1974. General models for insect parasite and predator searching behavior: Interference. *J. Anim. Ecol. 43* : 239–253.

Royama, T. 1970. Factors governing the hunting behavior and selection of food by the great tit, *Parsus major* L. *J. Anim. Ecol. 39* : 619–668.

Sevacherian, V., and V. M. Stern. 1972. Spatial distribution patterns of lygus bug in California cotton fields. *Environ. Entomol. 1* : 695–703.

Sharpe, P. J. H., G. L. Curry, D. W. DeMichele, and C. L. Cole. 1977. Distribution model of organism development times. *J. Theor. Biol. 66* : 21–38.

Sharpe, P. J. H., and D. W. DeMichele. 1977. Reaction kinetics of poikilotherm development. *J. Theor. Biol. 64*:649–670.

Sinko, J. W., and W. Streifer. 1967. A new model for age-size structure of a population. *Ecology 48*:910–918.

Slobodkin, L. B. 1954. Population dynamics in *Daphnia obtusa* Kurz. *Ecol. Monog 24*:69–89.

Smith, R. F., C. B. Huffaker, P. L. Adkisson, and L. D. Newsom. 1974. Progress achieved in the implementation of integrated control projects in the USA and tropical countries. *EPPO Bull. 4*:221–239.

Stapleton, H. N., D. R. Buxton, F. L. Watson, D. L. Nolting, and D. N. Baker. 197 Cotton: A computer simulation of cotton growth. *Ariz. Agric. Expt. Stn. Tec Bull.* 206.

Stinner, R. E., G. C. Butler, Jr., J. S. Bacheler, and C. Tuttle. 1975. Simulation of temperature-dependent development in population dynamics models. *Can. Entomol. 107*:1167–1174.

Stinner, R. E., A. P. Gutierrez, and G. D. Butler, Jr. 1974. An algorithm for temperature dependent growth rate simulation. *Can. Entomol. 106*:519–524.

Stinner, R. E., R. L. Rabb, and J. R. Bradley. 1974. Population dynamics of *Helioth zea* (Boddie) and *H. virescens* (F.) in North Carolina: A simulation model. *Environ Entomol. 3*:163–168.

Streifer, W. 1974. Realistic models in population ecology. *Adv. Theor. Ecol. 8*:199–26

Thompson, W. R. 1924. La theorie mathématique de l'action des parasites entomophage et le facteur du hasard. *Ann. Fac. Sci. Marseille, Ser. 2, 2*:69–89.

Thornley, J. H. M. 1976. *Mathematical Models in Plant Physiology.* Academic, Londor

Volterra, V. 1926. Variazioni e fluttuazioni del numero d'individui in species anima conviventi. *Mem. accad. Lincei 2*:31–113.

Volterra, V. 1931. *Lecons sur la Théorie Mathématique de la Lutte pour la Vie.* Gauthier Villars, Paris.

Von Foerster, H. 1959. Some remarks on changing populations in Frederick Stohlmar Jr. (ed.). *The Kinetics of Cellular Proliferation.* Grune & Stratton, New York.

Wang, Y., A. P. Gutierrez, G. Oster, and R. Daxl. 1977. A general model for plant growth and development: Coupling cotton–herbivore interactions. *Can. Entomol 109*:1359–1374.

Watt, K. E. F. 1966. The nature of systems analysis, in K. E. F. Watt (ed.), *System Analysis in Ecology.* Academic, New York, pp. 1–14.

Wigglesworth, V. B. 1965. *The Principles of Insect Physiology.* Methuen, London.

7

GENERAL ACCOMPLISHMENTS TOWARD BETTER INSECT CONTROL IN ALFALFA

E. J. Armbrust

Illinois Natural History Survey
  and
Illinois Agricultural Experiment Station, Urbana, Illinois

B. C. Pass

Department of Entomology, University of Kentucky, Lexington, Kentucky

D. W. Davis

Department of Biology, Utah State University, Logan, Utah

R. G. Helgesen

Department of Entomology, Cornell University, Ithaca, New York

G. R. Manglitz

Agricultural Research Service, USDA
  and
University of Nebraska, Lincoln, Nebraska

R. L. Pienkowski

Department of Entomology
Virginia Polytechnic Institute and State University, Blacksburg, Virginia

C. G. Summers

Department of Entomological Sciences
University of California, Berkeley, California

INTRODUCTION

Alfalfa is the world's most valuable cultivated forage crop and it is recognized as providing the best food value for all classes of livestock. In the United States there are nearly 30 million acres of alfalfa which provides more than 50 percent of the total hay crop. It is exceeded in total acreage by corn, wheat, and soybeans, but its protein production potential per acre far exceeds that of each of these three crops. Because alfalfa is a nationally grown crop that is adapted to a wide variety of geographic and climatic conditions, the Integrated Pest Management Project for Alfalfa was begun in seven areas representing different agricultural and biotic regions. Each of these regions has its own unique agroecosystem and problems, but each contains certain alfalfa insect pests common to the other. The entire scope of problems is represented by those in California, Utah, Nebraska, Illinois, Kentucky, Virginia, and New York.

A principal investigator was designated for each region to coordinate the team research effort at that respective institution. At the beginning of this project avenues of communication and representation were established with various other alfalfa research groups on both a national and regional scale. In several instances, research was coordinated with outside projects. Several states that were not funded by this project supplied field data for some of the studies reported herein. This coordination and exchange of data assisted in making this a successful and productive project.

The alfalfa ecosystem is unique among field crops in that it represents a relatively long-lasting, well-established perennial system. This makes it of prime importance ecologically. It is further unique in combining long- and short-term cycles of drastic environmental change with a mutualistic soil–plant relationship, and levels of resistance to certain pests which permit great management flexibility.

Since the early 1940s the major method of insect pest control in crop ecosystems has been the use of chemical insecticides against the damaging stages of a pest at the time of population outbreaks. This unilateral approach has often produced disasterous side-effects on nontarget organisms and the environment. Because of the undesirable side-effects of this method of insect control, many workers have strongly advocated a more fundamental approach—one which considers the basic relationships among host plant, pest populations, related abiotic and biotic factors, and the combined management practices man uses to produce the crop.

The main objective of the Integrated Pest Management Project for Alfalfa has been to gain a better understanding of the behavior of the alfalfa ecosystem and develop multidisciplinary insect pest management programs for alfalfa which will improve production efficiency and environmental quality. The following subobjectives have assisted in fulfilling this goal: (1) to ascertain the ecological interactions of the components of the alfalfa ecosystem, (2) to develop simulation models of the alfalfa ecosystem, (3) to ascertain the basic biological parameter values necessary to quantify the above models, and (4) verification and field testing of the models and the indicated management systems.

The Integrated Pest Management Project for Alfalfa brought together a broad diversity of talent, not available at any single institution, to contribute to a common goal—that of developing a new concept in alfalfa insect pest management which truly integrates biological, chemical, and cultural methods of control, while contributing a fundamental explanation of the underlying biological phenomena. These various experts include specialists in the areas of environmental modeling, systems analysis, biometrics, computer programming, insect ecology, insect pest management, insect behavior, insect pathology, biological control, population dynamics, pesticide toxicology, animal nutrition, plant physiology, plant breeding, plant ecology, and information retrieval. Many of these persons have contributed in past years various findings which relate to this project. Eventually these components may have been utilized in a pest management system but this would have evolved slowly. The funding provided by the U.S. National Science Foundation and Environmental Protection Agency has served as a catalyst to bring these experts together in a multidisciplinary team to develop a new holistic (comprehensive and unified) system of alfalfa pest management for immediate utilization.

The alfalfa weevil, *Hypera postica* (Gyllenhal), and the closely related species, the Egyptian alfalfa weevil, *Hypera brunneipennis* (Boheman), have been the most important pests of alfalfa in the United States. The alfalfa weevil was first discovered near Salt Lake City, Utah, in 1904 and for near 50 years it remained confined to 12 western states. However, in 1952 it was discovered in Maryland and from there it spread rapidly throughout the remainder of the United States. The Egyptian alfalfa weevil was first discovered in the United States in 1939, and it has remained confined mainly to California and Arizona.

Even though one might consider the alfalfa weevil complex as a single national alfalfa pest problem, various geographic regions have their own unique problems because the weevils' biologies and control requirements are extremely dependent on climatic factors. Because the weevil complex, and more specifically *H. postica* and its widely distributed parasite, *Bathyplectes curculionis* (Thomson), is of concern to growers on a national scale, it has been the key insect pest under investigation in this project. Other insects, both pest and natural enemy species, were also considered in the total arthropod complex on alfalfa, and whatever knowledge as could be gained from specific studies on them has been incorporated into the program. Because of the diversity of alfalfa production and management and the biological differences in the weevil complex in relation to crop development in a specific region, many aspects of the research were conducted in each of several regions.

SIGNIFICANT ACCOMPLISHMENTS

Analysis of Existing Information

Because there is a vast amount of available information on the biology, ecology, and control of alfalfa insect pests, an immediate effort was made to search, categorize, analyze, and utilize the existing literature and unpublished data. By the end of the present project there will be published indexed bibliographies on the spotted alfalfa aphid, *Therioaphis maculata* (Buckton),

(Davis et al., 1974), pea aphid, *Acyrthosiphon pisum* (Harris), (Harper et al., 1977), *Sitona* species (Morrison et al., 1974), potato leafhopper, *Empoasca fabae* (Harris), the weevil complex, and weevil parasites. This represents over 9000 literature references of which two-thirds have been indexed and entered into a computerized literature retrieval system.* This accomplishment has been of valuable assistance to all the biologists as well as those involved in modeling efforts. In many instances it was found that parasites had been established, cultural practices and timing of insecticide applications had been investigated, and considerable effort had been put forth in developing resistant varieties. These data have been integrated into a system that uses each factor in a complementary fashion.

There are over 2000 literature citations dealing with the weevil complex and its parasites, but it was apparent that certain aspects of alfalfa weevil and parasite biology needed special research attention. The modeling effort (see Chapter 8) helped to distill out these unknowns and point to specific research needs. Especially needed was a better understanding of the mortality factors influencing fluctuations in the insect pest and natural enemy populations in order to develop a better insect management system for alfalfa, a system that leans increasingly on improved insights gained from the systems approach (Chapters 2 and 8, specifically).

Standardization and Improvement of Research Techniques

In order for researchers to be able to compare results from field studies of the alfalfa weevil for each life stage, standardized research and sampling techniques that were based on a unit area were agreed upon. It was further agreed that the sweepnet should not be used for population studies until sweepnet catch could be related to absolute densities.

Where necessary, research results were obtained to refine current methods of sampling for each life stage and it was soon discovered that sampling intervals would best be based on physiologic time as measured by degree days.

Adult Alfalfa Weevil Sampling

Estimates of absolute density (weevils per square foot) for adult alfalfa weevils have been notoriously difficult to obtain. Even° under the best conditions about 1 man-hour is needed in the field plus 3 days of extraction time and use of 20 Berlese funnels in the laboratory to estimate adult density for one field. A more typical case would require about three to four times this effort.

*Illinois Natural History Survey and INTSOY, 163 Natural Resources Building, Urbana, Illinois 61801.

A study was undertaken in Illinois* to obtain information about the relationship between sweepnet catch and absolute density. It was hypothesized that the ratio (M) of absolute density to catch per sweep would depend largely on alfalfa height, temperature, wind speed, relative humidity, and solar radiation intensity. Data were gathered between April 1973 and November 1975 from several fields in Illinois and Indiana, resulting in 29 data sets.

Multiple regression analysis was used to obtain an equation for predicting the ratio M from environmental conditions. The best equation, based on analysis to date is

$$M = 0.2 + 10^{7.3595 - 0.001343(T + 3L)^2 - 0.3744H + 0.0058TH}$$

Where T is temperature in degrees Fahrenheit, H is crop height in centimeters, and L is solar radiation in Langley units (cal/cm^2/min). Use of this equation is facilitated by using tables† with which one can estimate M in about 15 seconds.

This equation has not been field tested using independent data, and until that time it is premature to place much confidence in its predictions. It is expected that absolute density can be estimated within 20 percent of the true mean 90% of the time based on 200 to 500 sweeps (depending on catch per sweep), and that about 15 to 30 minutes for one man will be required in each field.

Comparison of Two Sweepnet Techniques for Egyptian Alfalfa Weevil

Cothran et al. (1975) found that, on the average, the 180° sweep captured approximately 1.8 times as many Egyptian alfalfa weevil larvae as the pendulum type sweep (Table 7.1). Documentation of a significant difference between the two techniques emphasizes the importance of describing all major aspects of the sampling method(s) used in a given study. Such qualifications often have not appeared in published works which include sweepnet data for the weevil complex.

When average counts are low, the standard deviations are low and when average counts are high, the standard deviations are high. Also, there were no significant differences between the regression coefficients of the two methods. This has the implication that if the number of weevils in the field or plot is high, the pendulum sweep is as reliable as the 180° sweep.

*Unpublished data: W. G. Ruesink, E. J. Armbrust, and D. P. Bartell, Illinois Natural History Survey and Illinois Agricultural Experiment Station, Urbana.

†Unpublished data: W. G. Ruesink, E. J. Armbrust, and D. P. Bartell, Illinois Natural History Survey and Illinois Agricultural Experiment Station, Urbana.

Table 7.1.   Comparison of 180° and Pendulum (P) Sweepnet Sampling Techniques for Recovery of Egyptian Alfalfa Weevil Larvae—Yolo Co., Calif., 1972 (Cothran et al., 1975)

		Mean No. Larvae/Sweep for Indicated Sweepnet Techniques[a]					
	March 11				March 17		
SC^b	180° swp.	SC^b	P swp.	SC^b	180° swp.	SC^b	P swp.
6	80.6 a	6	49.1 a	3	254.0 a	6	152.1 a
5	74.5 a	4	40.9 ab	9	217.1 ab	3	118.3 ab
4	72.0 ab	9	39.7 ab	6	216.3 ab	8	108.2 abc
3	69.7 ab	5	38.5 ab	5	209.3 abc	2	105.0 bc
10	67.3 abc	3	37.8 abc	10	189.5 abc	5	99.1 bcd
9	67.2 abc	7	34.7 bc	2	166.6 bc	9	95.6 bcd
2	67.0 abc	10	34.3 bc	4	164.8 bc	4	91.9 bcd
1	53.5 bcd	8	31.9 bc	8	163.4 bc	10	83.9 bcd
7	49.1 cd	2	31.3 bc	7	153.1 bc	7	73.1 cd
8	45.7 d	1	28.5 c	1	144.8 c	1	69.7 d

[a]Means followed by the same letter are not significantly different at the 5% level of probability. Duncan's multiple range test.
[b]Individual sampler's code number.

Comparison of Sampling Methods for Adult Potato Leafhoppers

Adult potato leafhoppers were sampled from alfalfa under varying field conditions using a newly designed emergence trap, absolute density determinations, and the sweepnet.

Efficiency of the emergence trap in collecting adults was not significantly affected by wind, temperature, or alfalfa height. Decreasing solar intensity, however, resulted in a significantly decreased efficiency as predicted by the equation, $E = 0.87S^{0.36}$ [E = efficiency, S = Langleys (cal/cm^2/min)]. At solar intensities greater than 0.8 Langley units, the emergence gave a good estimate ($\geqslant 80\%$) of absolute density. Wind velocity, and to a much lesser extent, temperature were the two important variables affecting sweepnet catches. When pendulum sweeps were used, the relationship of sweepnet catches to wind velocity was predicted by the equation, $M = 0.075(1.205)^W$ [M = sweepnet conversion factor, W = wind speed (m/sec)] (Table 7.2). The ratio for pendulum versus 180° sweeps for adult E. fabae was 0.48 : 1. Absolute density ranged from 17.6 to 66.0/m^2 and the overall seasonal sex

Table 7.2. Sweepnet Conversion Factor (M) as a Function of Wind Speed (W) in Meters/second and Temperature (T) in Degrees Celsius as Predicted by the Equation $M = 2.70 (1.498)^W (T)^{-1.135}$—Multiplying catch per sweep by M produces an estimate of leafhoppers/m^2 (Cherry et al. 1977)

| Wind Speed | | Temperature | | | | |
m/sec (mph)	(°C) (°F)	14 (57)	19 (66)	24 (75)	29 (84)	34 (93)
0 (0)		0.14	0.10	0.07	0.06	0.05
1 (2.2)		0.20	0.14	0.11	0.09	0.07
2 (4.5)		0.30	0.21	0.16	0.13	0.11
3 (6.7)		0.45	0.32	0.25	0.20	0.17
4 (9.0)		0.68	0.48	0.37	0.30	0.25
5 (11.2)		1. 02	0.72	0.55	0.45	0.37
6 (13.4)		1.53	1.08	0.83	0.67	0.56

ratio of males to females was 1.5:1. Female leafhoppers were caught more frequently in sweeps, and less frequently in emergence traps than expected, probably due to lessened flight activity of females relative to males.

Sampling Egg Populations of Clover Root Curculio

Sampling techniques for estimating egg populations for the clover root curculio, *Sitona hispidula* (F.), were developed in alfalfa and red clover fields in Kentucky (Ng et al., 1977). An appropriate sample unit was a 3 in. × 3 in. (7.6 cm × 7.6 cm) area of plants and the soil beneath it to a depth of 4 to 5 in. (10.2–12.7 cm). The sample was processed by a flotation technique. The spatial disposition of the eggs conformed to a clumped pattern in both fields; 88 to 93 percent of the variation in log variances was accounted for by the variation in log means (Figs. 7.1 and 7.2). There were no significant differences in dispersion pattern between the two crops. Variance analysis revealed that block and plot differences were rarely significant but the interquadrat variance predominated. The number of samples required to estimate the population with specified levels of significance and margin of error was inversely proportional to population density.

Sampling Plans for the Weevil Complex

Latheef and Pass (1974a) conducted research to determine the spatial distribution pattern of eggs, larvae, and pupae of the alfalfa weevil, and they further developed a sampling plan for age-specific weevil populations (Latheef

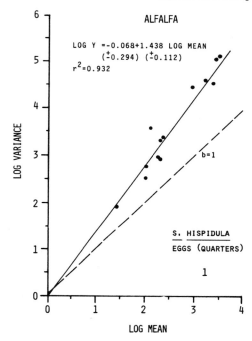

Figure 7.1. Relationship of log variance to log mean for 14 samples of *S. hispidula* eggs in alfalfa. Regression fits and Poisson expectations are shown as solid and dotted lines, respectively. Sample unit was a 3 × 3 in. (7.6 × 7.6 cm) area of plants and 4 to 5 in. (10.2–12.7 cm) of soil beneath it.

& Pass, 1974b). Sampling variations were investigated using analysis of variance procedures for nested random models. Interquadrat variance was the major source of population variation in two contiguous plots in which the spatial distribution conformed to an overdispersed pattern. Interblock and interplot differences were rarely significant. There was no single major source of population variance at a third plot where the field counts approached the Poisson series. The variance components associated with blocks, plots, and samples were significantly greater than zero in most cases. A least-square foot analysis of the form $\log Y = \log a + b\sqrt[3]{X}$ revealed a high correlation between sample size (log Y) and mean density (log X) for eggs and larvae when the data points from the three plots were pooled. An exponential function of the form $Y = ae^{bx-1/3}$ was derived from the regression equation to plot the predicted relationship between sample size and mean density in the original scale. The number of samples required was inversely proportional

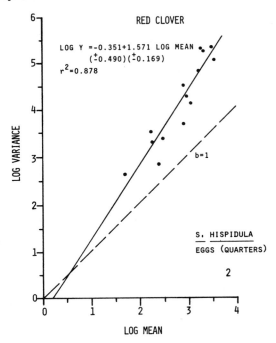

Figure 7.2. Relationship of log variance in log mean for 14 samples of *S. hispidula* eggs in red clover. Regression fits and Poisson expectations are shown as solid and dotted lines respectively. Sample unit was a 3 × 3 in. (7.6 × 7.6 cm) area of plants and 4 to 5 in. (10.2–12.7 cm) of soil beneath it.

to population levels for the low-density plots. However, when population increased and the distribution approached the Poisson series at high density the graph leveled off asymptotically; this relationship was graphed for two levels of sampling precision. Populations of prepupae, pupae, and adults of the insect may be estimated with a 10 percent standard error of the mean from ten 1-ft^2 quadrat areas of alfalfa plants.

Christensen et al. (1977) also investigated the within-field spatial pattern of Egyptian alfalfa weevil larvae. The changing dispersion patterns with time were found to be influenced by both density induced competition for food and varying spatial distribution of host plants. Mean crowding ($\overset{*}{x}/\bar{x}$) values characteristic for both years quickly stabilized. This made possible a method for calculating leaf consumption rates by larval populations with an aggregated distribution, k being derived from $\overset{*}{x}/\bar{x}$.

Collection of Field Samples for Life Table Construction

Using standard sampling techniques for each life stage of the weevil, population data for the weevil complex and for *B. curculionis* have been collected over a period of 3 to 4 years from one or more locations in Kentucky, California, Illinois, Indiana, Virginia, New York, Ohio, Utah, Michigan, and Nebraska. These data have been extremely valuable for determining age-specific survival rates and key mortality factors, quantifying the effects of population dynamics and impact of parasites on population regulation and field validation of models (Chapter 8).

The life tables were based on 1-ft^2 samples of alfalfa collected at random within each field. Eight age intervals were used to trace the course of each generation. Graphic key factor analysis was used to study the factors responsible for changes in population density. By this method, the killing power, or *k*-value, of each mortality factor was estimated by taking the difference between the logarithm of population density before and after its action. Using these values, the data from Kentucky were plotted for 11 life tables. The results indicated that k_g, mortality of summer adults (to emigration), followed the same fluctuating course as did changes in generation mortality (K). Thus total generation mortality was dominated by the magnitude of summer adult mortality (cause not specific) and this represented the key factor of importance in explaining generation survival of *H. postica* on alfalfa. At the same time, mortality of larvae due specifically to parasitism tended to compensate for the changes in adult mortality by reducing generation to generation* variation in K, and thus contribute a density dependent regulatory component.

Quantitative Biology Studies

In the construction of insect models for the alfalfa weevil and *B. curculionis*, it soon became apparent that there were certain research areas that needed special attention for both alfalfa weevil and *B. curculionis*: diapause rate, diapause duration and termination, adult survival, oviposition rate, and dispersal between habitats. A detailed quantitative understanding of these features as affected by different, or changed environmental conditions, was needed.

A literature search at the beginning of this project revealed few detailed studies concerning the biology of the parasites, and the impact of insecticides on them. Even more limited were studies dealing with predators in alfalfa

*Unpublished data; M. A. Latheef and B. C. Pass, University of Kentucky, Lexington.

and their impact on the alfalfa weevil or its parasites. Hundreds of species of organisms are commonly found in alfalfa, with the role of many being as yet poorly understood. In these studies, in order to safeguard against disrupting whatever natural stability as may exist in this ecosystem, the system was disturbed as little as possible. Thus investigations of the natural control complex in relatively undisturbed situations became an essential part of this project. There has been wide variation, however, in the emphasis of effort on specific members of the natural enemy complex.

Competition Between Species of Parasites

Because in some areas there are two or more species of important weevil parasites, it was important to investigate competition between them. Laboratory studies were conducted in Kentucky with *B. curculionis* and *Bathyplectes anurus* (Thomson) and it was found that the latter is more efficient at finding the host larvae. In general, weevil larval mortality was higher when the two species were present than when only one species was present.*

Diapause of B. curculionis

The population dynamics of *B. curculionis* is greatly affected by diapause or the lack of it in certain generations, and this has an effect on the necessary synchronization of peak parasite populations with peak host larval populations. A better understanding of the ecological and biological factors regulating diapause has provided more accurate values for our insect model (Chapter 8).

B. curculionis exhibits two types of development under field conditions. Most individuals have one generation per year, spending about 10 months within cocoons as diapausing prepupae. Some individual parasites, primarily during the earlier part of the season, do not enter diapause and have a second or rarely a third generation during the same season. Parrish and Davis (1978) found that under laboratory conditions it was possible to rear successive generations of nondiapausing parasites with no tendency to exhibit either inherited or alternating developmental types. The major effect of environmental manipulations of temperature, light condition, and relative humidity was observed from exposures during adult parasite development between mating and oviposition. Some effects were noted during the time after parasite emergence but before mating, and some after the weevil larvae had been parasitized. When light and temperature regimes were synchronized, using 9-hour photophase at 25°C and 15-hour scotophase at 7°C, over 95 percent of *B. curculionis* were of the nondiapausing type (Table 7.3). Any conditions

*Unpublished data; T. R. McKinney and B. C. Pass, University of Kentucky, Lexington.

Table 7.3. Effect of Uninterrupted Light or Dark and a 9-hour Daily Light Phase on the Prevention of Diapause in *B. cucurlionis* Progeny (Parrish & Davis, 1978)

Photoperiod	Nondiapausing (percent)[a]
Continuous dark	0.0 c
Continuous light	10.8 b
9 hr light + 15 hr dark	96.5 a

[a]Analysis by LSD test; values followed by the same letter are not significantly different at the 5 percent level.

which varied greatly from these, commonly resulted in less than 15 percent, or even 0 percent, nondiapausing parasites. Fewer nondiapausing individuals were also noted with relative humidity below 20 percent or above 80 percent, when older adult parasites were used, or when instar IV weevil larvae were parasitized.

Reproduction and Attack Rate Potential of B. curculionis

Even though *B. curculionis* is one of the most widespread and successful parasite species of the alfalfa weevil, few researchers in the past had dealt with the quantitative aspects of the host-parasite interactions and the results of these interactions in terms of the population dynamics of the alfalfa weevil. Research was conducted in Illinois to determine the effects of host density, temperature, and parasite age on the parasite's reproduction potential. The numerical values obtained from these studies are essential for developing descriptive models of the host/parasite interaction (Chapter 8). Besides providing values for the above models, the data are also being used to calculate the searching efficiency of the parasite and its relationship with parasite density, Latheef et al. (1977). The effect of *H. postica* density on attack rate by *B. curculionis* was tested using an open-choice experimental design. Host density had no significant effect on the proportion of hosts parasitized (Figs. 7.3, 7.4, and 7.5).

Yeargan and Latheef (1976) (Figs. 7.6, 7.7, and 7.8) found that host mortality not attributable to normal parasite development was, nevertheless, positively correlated with parasite density, suggesting some form of host injury by the parasite. Parasite densities of 1, 2, 4, 8, and 16 parasites per

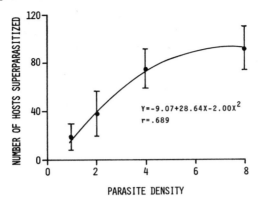

Figure 7.3. Relationship between number of superparasitized hosts and number of parasites. The vertical lines represent the standard error associated with each mean.

experiment (12.25 sq ft searching area) had no significant effect on the average success per parasite, indicating that little or no mutual interference (Hassell & Varley, 1969) occurred at these densities.

Mortality Factors Affecting *B. curculionis*

Applications of pesticides in the alfalfa ecosystem interrupt the normal relationship between alfalfa and its associated arthropod fauna. Cocoons of

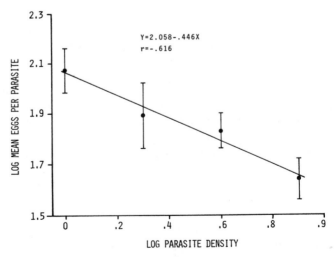

Figure 7.4. Relationship between log number of eggs laid per parasite and log number of parasites. The vertical lines represent the standard error associated with each mean.

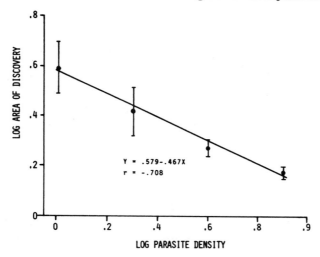

Figure 7.5. Relationship between log area of discovery and log number of parasites. The vertical lines represent the standard error associated with each mean.

B. curculionis can be found in the litter for a period of 1 to 2 weeks for non-diapausing individuals and 10 to 11 months for diapausing ones. Bartell et al. (1976) demonstrated that the construction of the cocoon, and the metabolic state of the individual within it, influences the penetration of insecticides (Tables 7.4 and 7.5).

Because the cocoons of diapausing *B. curculionis* remain in the litter from one spring to the next they are vulnerable to insecticides over a very long period of time, and to a variety of other factors such as weather, predators, parasites, pathogens, and cultural practices, Cherry and Armbrust (1975, 1977) and Cherry et al. (1976) found that invertebrate predation caused the heaviest mortality in these diapausing larvae. Specific predators were identified by exposing *B. curculionis* larvae to various surface dwelling invertebrates found in alfalfa fields. In addition, field plantings of parasite larvae in modified screen mesh cages were used to determine the size of the predators involved and also to determine if litter density affected rates of predation or the species involved.

Results from feeding studies showed that spiders, Cicindelidae, Formicidae, and small species of Staphylinidae did not prey on *B. curculionis* larvae in cocoons. The two groups of predators which did consume them were field crickets, *Gryllus pennsylvanicus* Burm., and various species of Carabidae (Table 7.6). The predators of the parasite larvae planted in the field were mainly insects of moderate size and their use of *B. curculionis* was not significantly affected by litter density (Table 7.7). The greatest number of total predators (*G. pennsylvanicus* plus carabids) caught/day/pitfall trap (Table 7.7), and the greatest predation on field planted *B. curculionis* larvae, occurred concurrently during September and October. These data suggest that predation during fall (September and October) may be significant in reducing field populations of diapausing parasite larvae. Based on feeding studies and pitfall trap catches, *G. pennsylvanicus* and the carabids, *Abacidus permundus* (Say),

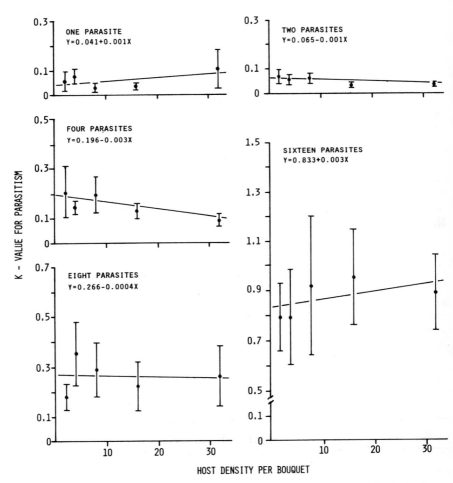

Figure 7.6. Relationship between *Hypera postica* density and the rate of attack by *Bathyplectes curculionis* (H_o:b = 0, P > 0.10 for all graphs).

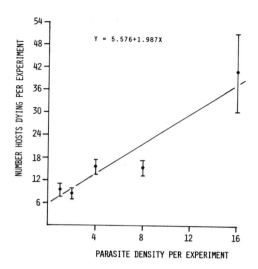

Figure 7.7 Relationship of *Bathyplectes curulionis* density to "indirect" mortality of *Hypera postica* larvae during the 72-hour period following exposure to the parasite(s) ($H_o:b = 0$, $P < 0.01$).

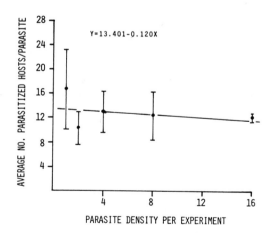

Figure 7.8. Relationship of *Bathyplectes curculionis* density to the average number of *Hypera postica* larvae parasitized by each parasite ($H_o:b = 0$, $P > 0.10$).

Table 7.4.   Summary of Results from Topical Applications of Insecticides on *Bathyplectes curculionis* Cocoons (Bartell et al., 1976)

Treatment[a]	Diapausing		Nondiapausing	
	No. Treated	No. Dead	No. Treated	No. Dead
Carbofuran	30	0	30	30
Phosmet	30	0	26	26
Methyl parathion	30	1	30	30
Methoxychlor	30	0	30	30
Total	120	1	116	116
Control (acetone)	40	1	25	11
Control (no acetone)			10	2

[a]Disregarding different concentrations.

Evarthrus sodalis LeConte, *Harpalus pennsylvanicus* DeGeer, and *Scarites subterraneus* F., were the most significant specific predators on *B. curculionis* larvae. These and other invertebrate predators cause a far greater mortality to diapausing *B. curculionis* than the combined effects of weather, hyperparasites, diseases, and insecticide usage.

Economic Thresholds and Compensation of Alfalfa to Insect Damage

Early in the project some states, especially California ('Koehler & Rosenthal, 1975), began economic injury studies for the weevil complex and certain

Table 7.5.   Mean Percent of Labeled Carbofuran That Penetrated Cocoons Containing Diapausing (*D*) and Nondiapausing (*ND*) *Bathyplectes curculionis* (Bartell et al., 1976)

Time	Mean Percent of Radioactive Label					
	On Outside		In Cocoon		In Parasite Tissues	
	D	*ND*	*D*	*ND*	*D*	*ND*
2nd day	93.7	60.7	4.7	3.41	1.7	5.4
4th day	93.8	75.6	5.8	20.7	0.4	3.7
8th day	95.9	73.8	3.4	22.2	0.7	4.0
Average for 8 days[a]	94.5	70.0	4.6	25.6	0.9	4.3

[a](*ND*) Significantly different from (*D*) in all three comparisons.

Table 7.6. Feeding Tests to Determine the Predators of *B. curculionis* Larvae in Cocoons (Cherry & Armbrust, 1977)

Taxa	N	Body Length (mm) Range	Predation Response
Araneae		2–18	−
Agelinidae	6		−
Araneidae	7		−
Clubionidae	1		−
Lycosidae	8		−
Micryphantidae	1		−
Salticidae	5		−
Tetragnathidae	7		−
Theridiidae	2		−
Thomisidae	3		−
Carabidae		2–25	variable[a]
Abacidus permundus (Say)	32		+
Colosoma spp.	7		−
Evarthrus sodalis LeConte	5		+
Harpalus pennsylvanicus DeGreer	16		+
Scarites subterraneus	12		+
Unidentified	82		variable[a]
Cicindelidae	34	10–12	−
Formicidae	61	2–10	−
Gryllidae			
Gryllus pennsylvanicus Burm.	21	9–23	variable[b]
Staphylinidae	23	2–10	−

[a]Generally only carabids > 10 mm body length fed on the larvae.
[b]Generally only *G. pennsylvanicus* > 15 mm body length fed on the larvae.

other participating institutions cooperated with researchers at Purdue University to determine these levels under various biological and geographical conditions (Hintz et al., 1976). Also, in order to evaluate management practices and control measures, quantitative knowledge was required of the effects of insect defoliation on yield and quality. In New York, Liu and Fick (1975) compared yield and quality in systems cut twice and three times during the season, where insect pests were controlled and not controlled (Tables 7.8 and 7.9). The greatest effect for a single harvest was in the second cutting of the three-cut system where feeding occurred in the early stages of regrowth. Indirect plant responses to insect feeding may partially compensate for direct losses or, on the other hand, cause further reductions in yield and feed quality.

Table 7.7. Field Predation on *B. curculionis* Larvae in Predator Exclosure Cages and at Different Litter Densities—Washington County, IL, Sept.–Oct. 1975 (Cherry & Armbrust, 1977)

Cage Treatment	$\bar{x} \pm SE$ Larvae Preyed Upon/Month/Cage/5 Cocoons
Open top (control)	2.65 ± 0.24
Open top-heavy litter[a]	2.77 ± 0.52
Open top-little litter[b]	3.44 ± 0.62
$\frac{1}{4}$ inch screen top	1.70 ± 0.49
$\frac{1}{8}$ inch screen top	0.33 ± 0.33

[a] $\bar{x} \pm SE = 5.60 \pm 0.64$ gm dry weight/cage.
[b] $\bar{x} \pm SE = 0.72 \pm 0.23$ gm dry weight/cage.

Table 7.8. Yield of Alfalfa Herbage Components with a Three-Cut Management and Sprayed (*W*0) or Natural (*W*1) Alfalfa Weevil Populations—Metric Tons/Ha (Liu & Fick, 1975)

Herbage Component[a]	1st Cut			2nd Cut			3rd Cut			Season Total		
	*W*0	*W*1	P[b]	*W*0	*W*1	P[b]	*W*0	*W*1	P[b]	*W*0	*W*1	P[b]
High Natural Weevil Population (1971)												
DM	2.79	2.39		3.63	2.50	×	3.67	3.47		10.09	8.36	**
IVTDDM	2.31	1.95		2.72	1.85	*	2.91	2.78		7.95	6.58	**
Leaves	1.39	1.18		1.92	1.29	*	2.16	2.14		5.47	4.61	*
CP	0.58	0.51		0.70	1.48	*	0.75	0.74		2.94	1.73	**
Low Natural Weevil Population (1972)												
DM	4.30	4.25		3.34	2.98		2.82	2.80		10.46	10.03	
IVTDDM	3.38	3.34		2.44	2.20		2.14	2.16		7.97	7.70	
Leaves	2.05	2.07		1.67	1.50	*	1.55	1.58		5.27	5.51	
CP	0.93	0.94		0.66	0.63		0.65	0.62		2.24	2.19	

[a] DM = dry matter; IVTDDM = in vitro true digestible dry matter; CP = crude protein.
[b] Probability level for a significant difference between sprayed (*W*0) and natural (*W*1): × $\leq$ 10%; * $\leq$ 5%; ** < 1%; all others, not significant.

Table 7.9.   Yield of Alfalfa Herbage Components with a Two-Cut Management and Sprayed or Natural Alfalfa Weevil Populations—Metric Tons/Ha (Liu & Fick, 1975)

Herbage Component[a]	1st Cut			2nd Cut			Season Total		
	W0	W1	P[b]	W0	W1	P[b]	W0	W1	P[b]
	High Natural Weevil Population (1971)								
DM	2.85	3.27	×	3.57	4.45	×	6.42	7.72	
IVTDDM	1.77	2.11		2.81	3.48		4.58	5.59	
Leaves	1.03	1.18		2.06	2.52		3.09	3.70	
CP	0.38	0.45		0.76	0.95		1.14	1.40	
	Low Natural Weevil Population (1972)								
DM	5.43	5.25		3.70	3.39		9.13	8.64	
IVTDDM	3.50	3.33		2.67	2.58		6.16	5.91	
Leaves	1.75	1.69		1.71	1.74		3.46	3.43	
CP	0.83	0.79		0.77	0.76		1.60	1.55	

[a]DM = dry matter; IVTDDM = in vitro true digestible dry matter; CP = crude protein.
[b]Probability level for a significant difference between sprayed (W0) and natural (W1); × ≤ 10%; all others not significant.

Fick and Liu (1976) found that there are many indirect effects of insect defoliation, and these have to be considered in evaluating insect damage to alfalfa and these indirect effects have to be included in plant and insect models (Chapter 8).

Various harvest methods used in Utah* have resulted in up to 90 percent reduction in alfalfa weevil numbers over the duration of the season, with striking differences among methods. In addition, spring rains and overhead irrigation cause up to 50 percent mortality to first instar larvae. Summers (1976) found that many other species of arthropods, both pests and natural enemies of the pests, experienced a similar mortality. In order to increase predation and parasitism of some of the other insect pests in alfalfa, researchers in California have investigated the practice of leaving uncut strips of alfalfa because early cutting removed most of the larval hosts required for

*Unpublished data: D. W. Davis, Utah State University, Logan.

parasite oviposition, whereas a late cut kills most of the parasitized larvae. The uncut strips permit continuous parasite production and they can move into the cut areas when those areas develop again.

Behavior and Mobility of Alfalfa Insects

The behavior and mobility of insects are important considerations in the development of pest management systems within a given crop system and between closely related systems. Many alfalfa insects, including the alfalfa weevil, are extremely mobile. The response of newly hatched, unfed alfalfa weevil larvae to several stimuli was measured by Pienkowski.* First instar alfalfa weevil larvae demonstrate a strong positive phototaxis and do not show a definite geotaxis or response to the odor of alfalfa. The ability of the newly hatched weevil larvae to travel over various surfaces was studied to determine the significance of weed hosts at field margins as a source of early season weevil infestations in alfalfa fields.†

Pienkowski (1976) and Pienkowski and Golik (1969) have studied the orientation of adult alfalfa weevils to various stimuli in the laboratory. Adult weevils exhibit a strong negative photo- and hygrotaxis, and a moderate positive thigmotaxis while in diapause (Tables 7.10 and 7.11).

Many spring adults of both the alfalfa weevil and the Egyptian alfalfa weevil leave alfalfa fields and disperse by flight to diapause-aestivation sites. This dispersal is often associated with harvest and in an attempt to study the physical factors that affect fall migration of the Egyptian alfalfa weevil, Christensen et al. (1974) monitored temperature, wind, cloud cover, evaporation, and precipitation. The number of these weevil adults migrating during autumn from aestivation sites to alfalfa was correlated with large differences in daily max–min temperatures (Gutierrez et al., 1976). It was found that the rate of movement could be accurately predicted from the cumulative day degrees (D°) below their metabolic threshold. The rate of movement into the field is reasonably constant; hence the rate and maximum number of adults entering a field can be estimated. Immigration into a field is also influenced by the condition of the field. Fall planted alfalfa without a cover crop represents a field essentially free of litter cover and protected sites in the winter. Alfalfa which was seeded into an existing grass pasture suffered much greater injury and attracted or retained much higher populations of Egyptian alfalfa weevil adults than nearly clean-planted alfalfa. These data are extremely important if models are to be used over more than a 1-year period.

*Unpublished data: R. L. Pienkowski, Virginia Polytechnic Institute and State University, Blacksburg.
†Unpublished data: R. L. Pienkowski, University of Kentucky, Lexington (presently at Virginia Polytechnic Institute and State University, Blacksburg).

Table 7.10.   Orientation to Light of Adult Alfalfa Weevils in Diapause (Pienkowski, 1976)

Portion of Arena	Males[a]	Females[a]	Mixed Sexes
Under dark refuge			
over black surface	6.5 a	6.5 a	8.0 a
over white surface	6.5 a	6.5 a	8.0 a
	0.5 b	3.5 ab	1.5 b
Under clear refuge			
over black surface	0.0 b	0.0 b	0.0 b
over white surface	0.0 b	0.0 b	0.0 b
Not under refuge			
over black surface	1.5 b	0.0 b	0.25 b
over white surface	0.5 b	0.0 b	0.25 b
Percent aggregated	85.0	100.0	92.5

[a]Means in a column with the same letter are not significantly different (Duncan's multiple range test, $P < 0.01$).

Alfalfa Variety Studies

A high level of resistance in alfalfa to *Hypera* spp. has not been found, but varieties with low levels of resistance have been developed (Sorensen et al., 1972). These previously developed varieties are of the dormant type and were tested exclusively against *Hypera postica*. Summers and Lehman (1976) demonstrated that resistance can be found in nondormant type alfalfas and

Table 7.11.   Percent Orientation of Alfalfa Weevil Adults in Diapause to Wet or Dry Surfaces, and Dark or Light Hiding Places (Pienkowski, 1976)

	Chamber w/Dark Refuge					Chamber/Clear Refuge				
	Under refuge		In open			Under refuge		In open		
Hours	Dry	Wet	Dry	Wet	Other[a]	Dry	Wet	Dry	Wet	Other[a]
1–3	76.3	12.5	5.0	0.0	6.3	47.5	11.3	25.0	1.3	15.0
20–29	73.0	13.5	4.5	3.0	6.0	36.5	11.5	24.0	7.0	21.0
40–49	85.0	8.7	2.5	1.3	2.5	51.3	10.0	13.7	3.7	21.3
70–79	87.5	7.5	1.3	1.3	2.5	42.5	8.7	20.0	3.7	25.0
90–99	78.4	3.1	8.8	3.4	6.3	32.2	5.0	31.3	12.8	18.7
Avg.	80.0	9.1	1.8	4.4	4.7	42.0	9.3	22.8	5.7	20.2

[a]Refers to weevils on the top or side of the chamber and on top of the refuge.

that varieties developed for resistance to *H. postica* also resist *H. brunnei-pennis* (Table 7.12). These data verify the importance of the tolerance component in resistance to *Hypera* spp. but also indicate the possible presence of other components of resistance.

Much emphasis in the area of alfalfa resistance to insects was centered at the University of Nebraska with insect pests other than the weevil complex. Kehr and Manglitz (1976) examined the performance and resistance to pea aphid, spotted alfalfa aphid, and bacterial wilt of all available certified seed lots of Dawson alfalfa. Field and greenhouse studies showed no major significant differences between any of the seed lots tested. In studies with the potato leafhopper, Kehr et al. (1975) found that the rank of four varieties from resistant to highly susceptible was MSB, N.S. 16, Vernal, and Buffalo. In general, insecticide applications increased the yield and quality of susceptible varieties more than that of varieties with resistance.

Manglitz and Gorz (1974) obtained information on *Medicago* spp. as hosts of the yellow clover aphid complex. One of the aphids of this complex, the spotted alfalfa aphid, has been an important pest of alfalfa in many parts of the United States.

Mueke et al. (1978) tested six insecticides—carbofuran, methoxychlor, Mobam® [benzo (*b*) thien-4-yl methylcarbamate], leptophos, methidathion, and dialifor as control for pea aphid on four alfalfa varieties (Cody, Dawson,

Table 7.12.   Larval Populations, Damage Ratings, Tolerance Ratings, and Leaf Weight in Seven Alfalfa Cultivars—Fresno Co., California (Summers & Lehman, 1976)

Cultivar	Mean No. Larvae/ Sweep[a]	Damage Rating[a,b] Whole Plot	Damage Rating[a,b] Single Stem	Tolerance Rating[a,c]	Dry Wt. (mg) of Leaves/cm Stem[a]
Team	40.0 ab	3.6 a	3.5 a	4.84 a	13.81 a
Weevilcheck	38.9 ab	3.5 a	3.6 a	4.26 b	14.28 a
UC 73	35.9 a	3.3 a	3.3 a	4.27 b	14.07 a
UC Cargo	59.7 bc	5.3 b	5.7 a	2.84 d	10.58 b
UC Salton	60.3 bc	6.0 b	5.2 b	3.45 c	11.68 b
UC 67	69.8 c	5.5 b	4.7 b	3.48 c	11.71 b
Moapa	65.4 c	5.0 b	5.4 b	3.60 c	11.92 b

[a]Means followed by the same letter(s) are not significantly different at the 5% level of probability. Duncan's multiple range test.
[b]1 = no defoliation, 10 = complete defoliation.
[c]Higher values indicate greater tolerance to feeding damage.

Team, and Vernal) differing in resistance to the pea aphid. They found that aphid numbers often increased following leptophos treatment. These increases were noted most frequently, but not exclusively, on resistant varieties. The other insecticides either reduced or had no effect on aphid numbers. The resistant varieties Dawson and Team were generally very effective in reducing pea aphid numbers.

Kugler et al. (1977) studied the effects of combining ability for characters associated with resistance to the spotted alfalfa aphid, the pea aphid, and two mirids [the alfalfa plant bug, *Adelphocoris lineolatus* (Goeze), and the tarnished plant bug, *Lygus lineolaris* (Palisot de Beauvois)]. General combining ability effects were far more important than specific combining ability effects among single crosses for forage yield, stem length, and leaf malformations under mirid infestation, and for pea aphid and spotted alfalfa aphid resistance. Such information is useful in the design of breeding programs to develop resistance to these alfalfa pests.

Practical Achievements Toward Improved Weevil Control

Detailed descriptions of the population dynamics of the alfalfa weevil and its parasite are making it possible to develop management models. For example, defining the phenology of the weevil and its parasites has led to adjustments in management practices (insecticides and cutting) to maximize parasite survival and minimize weevil survival. Furthermore, the development of a simulation model for alfalfa growth has added another dimension to understanding the weevil/alfalfa interaction. Predictions of plant growth and phenology should allow for improved scheduling of management practices such as harvesting.

In addition to the basic biology of the alfalfa plant, weevil, and parasite, an understanding of such phenomena as plant compensation to insect damage, disruption of ecological balance in the alfalfa crop system, and the economics of the weevil/alfalfa interaction will lead to a more realistic structuring of the alfalfa ecosystem. The results should suggest new approaches to management of the key pest, the alfalfa weevil. For example, this past season, IPM alfalfa researchers, in cooperation with Extension staff and alfalfa growers, continued an alfalfa weevil management program in alfalfa fields in Illinois, Kentucky, New York, Iowa, Utah, Virginia, and other cooperating states. Each program made use of one scout, who was furnished with a decision-making chart. The method for monitoring alfalfa weevil populations and making recommendations as to the timing of insecticide applications was developed as an adjunct to the modeling efforts of the Integrated Pest Management Alfalfa Subproject. The complexities of the models, weather factors, biological interactions, and so on, were simplified in a series of charts that an individual with a minimum of specialized training can use efficiently. Three key factors were incorporated into the chart and used to alert the scout to potential of and/or recognition of the existing problems: (1) degree day accumulation since January 1, (2) number of large larvae present, and (3) crop height.

Although all the information from the pilot program has not been reviewed, the following conclusions can be made: (1) the charts are designed such that an individual with no entomological knowledge can be easily and quickly trained to use them, (2) none of the 50 fields scouted in Illinois required a second foliar or stubble treatment, thus the real world system appears to be realistically represented in the charts, (3) more work needs to be done on timing of applications to maximize parasite and predator survival and incorporate this into the charts, and (4) the growers were pleased to see someone taking an active interest in their alfalfa, and responded by making their applications as recommended shortly after they received their recommendations.

Other Insect Pests

Alfalfa Aphids

Relatively few aphid species colonize alfalfa in sufficient numbers to pose a serious economic threat to the crop. Until recently the spotted alfalfa aphid and the pea aphid were the only species causing significant economic loss in alfalfa. The confrontation of entomologists and the spotted alfalfa aphid led to the strategy and methodology of integrated control (Hagen et al., 1971). The utilization of imported hymenopterous parasites, spotted alfalfa aphid-resistant varieties, and the conservation of native predators through the selective use of insecticides (choice of material, dosage rate, and timing) has largely eliminated the spotted alfalfa aphid as a serious pest of forage alfalfa.

Success in the management of pea aphid populations has been less spectacular than with the spotted alfalfa aphid. Although the same predator species that are so reactive to the spotted alfalfa aphid are also active against the pea aphid, changes in aphid phenology during recent years have led to the presence of destructive population levels during the summer as well as in the spring and fall. Hagen et al. (1971) have discussed two possible explanations for such a change in behavior and its impact on population regulation by naturally occurring biological control.

The introduction of a closely related aphid pest, *Acyrthosiphon kondo* Shinji, has further complicated the aphid situation in alfalfa. Since its discovery in California in 1974 it has spread into Nevada, Arizona, New Mexico and Kansas in the United States, and Baja, Sonora, and Coahuila in Mexico. *A. kondoi* is far more destructive than the pea aphid and is capable of causing economic loss at a much lower population level than is the pea aphid.

Efforts have thus been underway in California for the past 2 years to develop simulation models of the pea aphid, *A. kondoi*, and the spotted alfalfa aphid. These models are being developed in such a way that they can be easily coupled with the model recently developed by Gutierrez et al. (1976) for the Egyptian alfalfa weevil and alfalfa.

CONCLUSIONS

This research project has provided a wealth of information which has contributed to a better understanding of the alfalfa agroecosystem. Specifically, studies have dealt with ascertaining the ecological interactions of the components of the alfalfa ecosystem, developing simulation models, determining basic biological parameter values, and verifying and field-testing models and management programs. This has resulted in developing multidisciplinary insect pest management programs for alfalfa which will improve production efficiency and environmental quality.

Much supportive research has been performed using funds from sources other than this IPM project, including state experiment stations, Hatch projects, USDA grants and contracts, and other NSF grants. One of the unique features of this project is that it provides funds to support research in very specialized basic areas, such as the behavior of insect parasites and predators and the effects of suboptimal conditions on developmental rates. Such research is essential to the accomplishment of our more applied objectives since it is essential for the completion of the component models. Yet this research typically would not be financed by more applied interests.

Table 7.13 shows a topic breakdown of the 160 alfalfa subproject publications or submitted manuscripts to date; it reveals the research scope and emphasis of the project.

There has been a good balance of basic biological and ecological research with the modeling efforts (Chapter 8). The project has been successful because the research team coordinated their major efforts. Detailed descriptions of the population dynamics of the alfalfa weevil complex and their parasites have made it possible to improve the realism of the insect models, as well as to understand the factors contributing to alfalfa pest management. For example, defining the phenology of the weevil and its parasites has led to adjustments in management practices such as harvest, insecticide applications, and so on, that have increased parasite survival and reduced weevil survival. Without such biological research, the models would not be realistic

Table 7.13.   Alfalfa Subproject Publications Categorized by Topic—1977

Topics	No. of Papers
Weevil Complex	
Parasites, predators, pathogens	32
Chemical control	7
Cultural control or plant resistance	9
Biology	25
Economic damage levels and sampling	16
Economics	11
Modeling	21
Other Insects	24
Miscellaneous	15
TOTAL	160

enough for pest management purposes, the assumptions of the models would have been too spurious or unproven, and no entomologist would be willing to accept the results.

LITERATURE CITED

Bartell, D. P., J. R. Sanborn, and K. A. Wood. 1976. Insecticide penetration of cocoons containing diapausing and nondiapausing *Bathyplectes curculionis*, an endoparasite of the alfalfa weevil. *Environ. Entomol. 5* :659–661.

Cherry, R. H., and E. J. Armbrust. 1975. Field survival of diapausing *Bathyplectes curculionis*, a parasite of the alfalfa weevil. *Environ. Entomol. 4* :931–934.

Cherry, R. H., E. J. Armbrust, and W. G. Ruesink. 1976. Lethal temperatures of diapausing *Bathyplectes curculionis* (Hymenoptera: Ichneumonidae), a parasite of the alfalfa weevil (Coleoptera: Curculionidae). *Great Lakes Entomol. 9* :189–193.

Cherry, R. H., and E. J. Armbrust. 1977. Predators of *Bathyplectes curculionis*, a parasite of the alfalfa weevil. *Entomophaga 22* :323–329.

Cherry, R. H., K. A. Wood, and W. G. Ruesink. 1977. Emergence trap and sweep net sampling for adults of the potato leafhopper from alfalfa. *J. Econ. Entomol. 70* : 279–282.

Christensen, J. B., W. R. Cothran, C. E. Franti, and C. G. Summers. 1974. Physical factors affecting fall migration of the Egyptian alfalfa weevil, *Hypera brunneipennis* (Coleoptera: Curculionidae): A regression analysis. *Environ. Entomol. 3* :374–376.

Christensen, J. B., A. P. Gutierrez, W. R. Cothran, and C. G. Summers. 1977. The within field spatial pattern of the larval Egyptian alfalfa weevil, *Hypera brunneipennis* (Coleoptera: Curculionidae): an application of parameter estimates in simulation. *Can. Entomol. 109* :1599–1604.

Cothran, W. R., C. G. Summers, and C. E. Franti. 1975. Sampling for the Egyptian alfalfa weevil: A comparison of 2 standard sweep-net techniques. *J. Econ. Entomol.* 68:563–564.

Davis, D. W., M. P. Nichols, and E. J. Armbrust. 1974. The literature of arthropods associated with alfalfa. 1. A bibliography of the spotted alfalfa aphid, *Therioaphis maculata* (Buckton) (Homoptera: Aphididae). *Ill. Nat. Hist. Survey Biol. Notes* 87.

Fick, G. W., and B. W. Y. Liu. 1976. Alfalfa weevil effects on root reserves, developmental rate, and canopy structure of alfalfa. *Agron. J.* 68:595–599.

Gutierrez, A. P., J. B. Christensen, W. B. Loew, C. M. Merrit, C. G. Summers, and W. R. Cothran. 1976. Alfalfa and Egyptian alfalfa weevil. *Can. Entomol.* 108:635–648.

Hagen, K. S., R. van den Bosch, and D. L. Dahlsten. 1971. The importance of naturally occurring biological control in the western United States, in C. B. Huffaker (ed.), *Biological Control.* Plenum, New York, pp. 253–293.

Harper, A. M., J. P. Miska, G. R. Manglitz, E. J. Armbrust, and B. I. Irwin. 1977. The literature of arthropods associated with alfalfa. 3. A bibliography of the pea aphid. *Acyrthosiphon pisum* (Harris) (Homoptera: Aphididae). *Univ. of Ill. Spec. Publ.* 50.

Hassell, M. P., and G. C. Varley. 1969. New inductive population model for insect parasites and its bearing on biological control. *Nature (Lond.)* 223:1133–1137.

Hintz, T. R., M. C. Wilson, and E. J. Armbrust. 1976. Impact of alfalfa weevil larvae feeding on the quality and yield of first cutting alfalfa. *J. Econ. Entomol.* 69:749–754.

Kehr, W. R., and G. R. Manglitz. 1976. Performance of certified seed lots of Dawson alfalfa. *Nebr. Agric. Exp. Stn. Res. Bull.* 277.

Kehr, W. R., R. L. Ogden, and J. D. Kindler. 1975. Management of four alfalfa varieties to control damage from potato leafhoppers. *Nebr. Agric. Exp. Stn. Res. Bull.* 275.

Koehler, C. S., and S. S. Rosenthal. 1975. Economic injury levels of the Egyptian alfalfa or the alfalfa weevil. *J. Econ. Entomol.* 68:71–75.

Kugler, J. L., W. R. Kehr, and R. L. Ogden. 1977. Combining ability effects for resistance to four insects in selected alfalfa (*Medicago sativa* L.) clones. *Crop. Sci. 17*:621–624.

Latheef, M. A., and B. C. Pass. 1974a. A sampling plan for age-specific population estimates of *Hypera postica* in Kentucky alfalfa fields. *Environ. Entomol.* 3:872–875.

Latheef, M. A., and B. C. Pass. 1974b. Spatial distribution patterns of *Hypera postica* in Kentucky alfalfa fields. *Environ. Entomol.* 3:866–871.

Latheef, M. A., K. V. Yeargan, B. C. Pass. 1977. Effect of density on host–parasite interactions between *Hypera postica* and *Bathyplectes anurus. Can. Entomol.* 109:1057–1062.

Liu, B. W. Y., and G. W. Fick. 1975. Yield and quality losses due to alfalfa weevil. *Agron. J.* 67:828–832.

Manglitz, G. R., and H. J. Gorz. 1974. Additional hosts of the "yellow clover aphid complex." *J. Econ. Entomol.* 67:453–454.

Morrison, W. P., B. C. Pass, M. P. Nichols, and E. J. Armbrust. 1974. The literature of arthropods associated with alfalfa. II. A bibliography of the *Sitona* species (Coleoptera: Curculionidae). *Ill. Nat. Hist. Surv. Biol. Notes* 88.

Mueke, J. M., G. R. Manglitz, and W. R. Kehr. 1978. Pea aphid: Interaction of insecticides and alfalfa varieties. *J. Econ. Entomol. 71*: 61–65.

Ng, Y. S., M. A. Latheef, and B. C. Pass. 1977. Estimation of egg populations of the clover root curculio, *Sitona hispidula* (Coleoptera: Curculionidae) in Kentucky alfalfa and red clover fields. *Can. Entomol.* (in press).

Parrish, D. S., and D. W. Davis. 1978. Inhibition of diapause in *Bathyplectes curculionis*, a parasite of the alfalfa weevil. *Ann. Entomol. Soc. Am. 71*: 103–107.

Pienkowski, R. L. 1976. Behavior of the adult alfalfa weevil in diapause. *Ann. Entomol. Soc. Am. 69*: 155–157.

Pienkowski, R. L., and Z. Golik. 1969. Kinetic orientation of the alfalfa weevil to its host plant. *Ann. Entomol. Soc. Am. 62*: 1241–1245.

Sorensen, E. L., M. C. Wilson, and G. R. Manglitz. 1972. Breeding for insect resistance, in C. H. Hanson (ed.), *Alfalfa Science and Technology*. Am. Soc. Agron., Madison, Wis., pp. 371–390.

Summers, C. G. 1976. Population fluctuations of selected arthropods in alfalfa: Influence of two harvesting practices. *Environ. Entomol. 5*: 103–110.

Summers, C. G., and W. F. Lehman. 1976. Evaluation of non-dormant alfalfa cultivars for resistance to the Egyptian alfalfa weevil. *J. Econ. Entomol. 69*: 29–34.

Yeargan, K. V., and M. A. Latheef. 1976. Host–parasitoid density relationships between *Hypera postica* (Coleoptera: Curculionidae) and *Bathyplectes curculionis* (Hymenoptera: Ichneumonidae). *J. Kansas Entomol. Soc. 49*: 551–556.

8

THE SYSTEMS APPROACH TO RESEARCH AND DECISION MAKING FOR ALFALFA PEST CONTROL

W. G. Ruesink

Illinois Natural History Survey
Urbana, Illinois

C. A. Shoemaker

Department of Environmental Entomology
Cornell University, Ithaca, New York

A. P. Gutierrez

Division of Biological Control
University of California, Berkeley, California

G. W. Fick

Department of Agronomy
Cornell University, Ithaca, New York

INTRODUCTION

The modeling and systems analysis portion of the International Biological Program's NSF-EPA sponsored integrated pest management (IPM) subproject for alfalfa has concentrated on the alfalfa weevil, *Hypera postica* (Gyllenhal), in the East and Mid-west states and on the Egyptian alfalfa weevil, *Hypera brunneipenis* (Boheman), in the West. Chapter 7 explained the reasons why other pests were excluded from consideration and they will not be repeated here. The goals of the modeling effort have been mainly, (1) to improve our scientific understanding of the alfalfa ecosystem and (2) to develop improved practical pest management systems by integrating the biological and economic considerations, using relevant improved systems science technology.

Work on the alfalfa weevil had been divided into three parts: (1) development of a descriptive model of the alfalfa weevil and its key parasite, *Bathyplectes curculionis* (Thomson), directed by W. G. Ruesink, University of Illinois, (2) development of a descriptive model of the alfalfa plant directed by G. W. Fick, Cornell University, and (3) development of models for optimizing (optimization models for) alfalfa weevil management directed by Christine Shoemaker, Cornell University. The Egyptian alfalfa weevil effort has had basically two divisions: (1) development of descriptive models of the alfalfa plant and of the Egyptian alfalfa weevil directed by A. P. Gutierrez, University of California, and (2) development of models for optimizing Egyptian alfalfa weevil management, directed by U. Regev* and A. P. Gutierrez, University of California.

THE PLANT MODEL OF FICK

The name ALSIM is an acronym for ALfalfa SIMulation. ALSIM 1 predicts dry matter yields of various parts of an alfalfa crop as a function of time. The

*Ben Gurion University, Beersheba, Israel.

ALSIM I   (Level I)

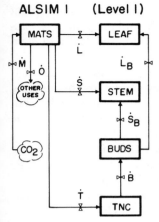

Figure 8.1 ALSIM 1 (Level 1) is based on this model of material flow in the alfalfa crop. Rectangles represent the parts of the system modeled; arrows, the pathways of material flow; valve symbols, the rates controlling material flow; cloud symbols, parts of the system not treated in the model. Variable names are defined in the description of the model.

block diagrams (Fig. 8.1) identify the parts of the system (i.e., state variables) for which predictions are made. They also show pathways of material flow between parts and identify the rates that control the flow of material. The material flowing in ALSIM 1 is fixed carbon expressed as plant dry matter (DM). For compatibility with the alfalfa insect models, ALSIM 1 was given a time step of 1 day. The effort to find a simple model structure capable of handling regrowth following a simulated harvest resulted in five state variables in the core of the model (Fig. 8.1).

Material (MATS) available for top growth and storage defines the supply of photosynthetic DM (carbon) that can be used for growth of the parts of the alfalfa crop included in the model (Fig. 8.1). The photosynthetic input was corrected for respiration and for growth of plant parts not included in the model (e.g., in taproots and fibrous roots), but it represents a potential which may not be entirely used. Therefore, one of the flows out of MATS is for other uses, a mechanism of removing MATS in excess of use in top growth and storage.

MATS is primarily used to produce leaves (LEAF), stems (STEM), or total nonstructural carbohydrates accumulated in taproots (TNC). When a harvest is made, the leaves and stems are entirely removed and regrowth is initiated by the elongation of basal buds (BUDS) into new leaves and stems. The source of material for bud formation is accumulated TNC (Fig. 8.1). When BUDS is zero and the TNC supply is exhausted, harvesting the leaves and stems will kill the crop. The model predicts crop death when there is no photosynthesis (no input to MATS) and the supply of TNC is $\leq 5 \ \mathrm{g \ m^{-2}}$. Photosynthesis requires light interception by leaves so the condition of no photosynthesis occurs when all leaves have been removed and there are no buds to replace them. Flows between each component of the alfalfa plant model are determined by light intensity, day length, and the plant's physiologic status. Rates of photosynthesis, bud elongation, and leaf and stem growth all depend upon at least one of these variables.

The removal of leaves and stems can occur by three processes in ALSIM 1. They can die because of old age (senescence), they can be lost by a killing frost, or they can be harvested as hay. Insect defoliation is not included, but a fourth flow out of leaves for that factor would describe the situation. If the leaves and stems are harvested, the yields are retained in two state variables for harvested leaves (HLEAF) and harvested stems (HSTEM) which are summed to give the hay yield (HAY) since the start of simulation. The separate treatment of harvested leaves and stems makes it possible to calculate the proportion of leaves in the harvested hay; this is important in predicting hay quality. More information can be found in Fick (1975).

ALFALFA WEEVIL MODEL OF RUESINK

The cause–effect relationships among the components of the alfalfa weevil life system can be represented by the diagrammatic model in Figure 8.2. Each circle represents a component and corresponds to a life stage of the alfalfa weevil or of its parasite, *Bathyplectes curculionis*. For each of the components, equations have been written to describe the phenomena associated with the population dynamics of the two species. The approach to the mathematics of the model has been to write state and response equations in difference equation form using a 1-day time step. The following description is taken from Ruesink (1976).

As an example of the mathematics associated with each component, consider the detailed model for the first instar larva (Fig. 8.3). The vector $\vec{x}(k)$ is composed of n elements. The sum of the n elements is the total density of first instar larvae. As individuals age within the first instar, they move from the first through the nth element. The variable $y_1(k)$ represents the number of first instar larvae molting into second instars during day k, while $y_2(k)$ is the potential amount of food (in mg dry weight) consumed by the first instar larvae during day k, and $u(k)$ is the number of weevil eggs which hatch during day k. If $T_L(k)$ and $T_H(k)$ represent the low and high centigrade temperatures, respectively, on day k, then the instantaneous temperature, $T(t)$, can be approximated by the equation given in Fig. 8.3. Substituting this into the equation for developmental rate, $R(T)$, which is of the form proposed by Davidson (1944), and integrating over the interval of 1 day, will yield an average rate of development for day k, which is denoted by $R^*(k)$.

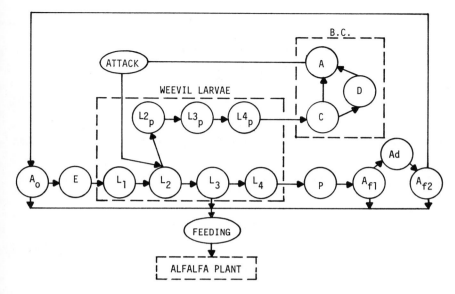

Figure 8.2. Diagrammatic model of the alfalfa weevil and *B. curculionis* life systems. Circles represent life stages of the insects (e.g., L2p is for parasitized second instar larvae), ovals represent processes that couple trophic levels, and arrows represent flows of information or materials. [From Ruesink, 1976. Copyright 1976, Board of Trustees, Michigan State University. Reprinted by permission.]

Feeding potential, $y_2(k)$, is modeled as directly proportional to average developmental rate, $R^*(k)$. The available data is presented as so many milligrams of food consumed per instar, and the equation for $y_2(k)$ in Figure 8.3 is one way to achieve a constant food intake per instar while allowing the daily consumption rate to vary with temperature.

The number of individuals molting out of the first instar, $y_1(k)$, is computed by the same process used to update the state vector, $x(\vec{k})$. The average developmental rate, $R^*(k)$, is multiplied by n, which is the length of vector $\vec{x}(k)$, to obtain a number representing the distance along $\vec{x}(k)$ that an individual should move during day k. If this number is not an integer, then the integer and fractional parts of this distance are separated into j and m, respectively, and the appropriate part of the individuals is moved along j elements and the remaining part is moved along $j + 1$ elements. Those individuals which, by this procedure, would move into elements labeled $n + 1$ or greater, are considered to have molted out of the first instar. In updating the state vector, each element must be reduced somewhat to account for mortality during day k.

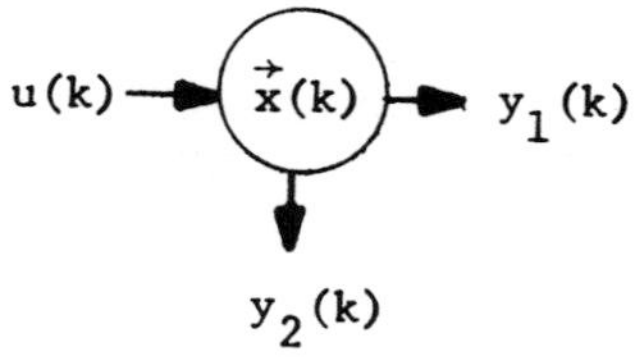

$$u(k) \rightarrow \vec{x}(k) \rightarrow y_1(k)$$

$$y_2(k)$$

$$T(t) = 1/2(T_H(k) + T_L(k)) + 1/2(T_H(k) - T_L(k)) \cos 2\pi t$$

$$R(T(t)) = \frac{.704}{1 + e^{4.04 - .166T(t)}} \quad ; \quad R^*(k) = \int_0^1 R(T(t))dt$$

$$j = \text{integer part of } \left[nR^*(k) \right]$$

$$m = \text{fractional part of } \left[nR^*(k) \right]$$

$$y_1(k) = s(k)\left[mx_{n-j}(k) + \sum_{i=n-j+1}^{n} x_i(k) \right]$$

$$y_2(k) = cR^*(k) \sum_{i=1}^{n} x_i(k)$$

$$x_i(k+1) = s(k)\left[(1-m)x_{i-j}(k) + mx_{i-j-1}(k) \right], \quad j+1 < i \leq n$$

$$x_i(k+1) = s(k)(1-m)x_1(k) + \frac{m}{j+m} u(k), \quad i = j+1$$

and if $j > 0$,

$$x_i(k+1) = \frac{u(k)}{j+m} \quad , \quad 1 \leq i < j+1$$

Figure 8.3. Submodel for the first instar larva. See text for definition of terms. [From Ruesink, 1976. Copyright 1976, Board of Trustees, Michigan State University. Reprinted by permission.]

SIMULATION STUDIES USING
THE COUPLED ALFALFA PLANT–ALFALFA WEEVIL MODEL

The two models described above have been coupled and additional logic has been added to describe the effects of insecticides and harvesting on the plant and on the alfalfa weevil and its parasite. To initialize the model and begin a simulation we must assign values to each of the state variables. Normally, we select September 1 as a starting point, since at that time essentially all of the alfalfa weevils are diapausing adults and essentially all of the *B. curculionis* are diapausing cocoons. Hence, the insect portion of the model can be initialized by assigning nonzero densities to these two state variables and setting the remaining state variables at zero. The plant portion of the model includes four state variables, namely biomass of leaves, stems, buds, and TNC, which is "total nonstructural carbohydrates."

In addition to the state variables, we must provide the latitude, so that day length can be computed, and we must indicate which biotype of alfalfa weevil is being simulated, since the eastern and western populations of this species do have a few differences that are not explainable by differences in climate.

Once the model is initialized, the simulation proceeds from day to day according to the environmental data. For each day we must provide maximum and minimum temperatures and solar radiation inputs. In those cases where we are trying to reproduce a historical situation (for example, when testing the accuracy of the model), these inputs consist of daily measurements of the actual maximum and minimum temperatures. When we are using the simulation in a predictive mode, we have the option of using (1) actual weather forecasts, (2) historical data from a previous year, or the average of several years, for the same period of the year, or (3) hypothetical data generated by a weather simulator. We have used all three options in our research program.

In this section we emphasize the benefits that have been realized from the modeling effort. There are three general areas of benefits that deserve attention: (1) it has helped improve our understanding of the alfalfa weevil life system, (2) it has guided our research efforts by showing us which areas need the most attention, and (3) it has helped us develop a field evaluation system based on a dynamic economic threshold. We will now discuss each of these benefits in some detail.

One aspect of a host-parasite system that is usually poorly understood is the group of processes that lead up to a given amount or percentage of parasitism. Our approach has been to study and model each process individually and then to combine the submodels into a description of the larger process.

The first process modeled is the production of eggs by the parasite, and the second is the deposition of those eggs into host larvae. At least three factors are important in determining the rates for those processes. Further, the number of eggs laid per female per day in laboratory studies is known to vary with the age of the individual female, reaching a peak near the mid-point of her adult life. In our model this is described by an equation fitting the line shown in Fig. 8.4A. Secondly, temperature is presumed to have some effect on oviposition rate, although our laboratory data have shown no difference over a range of temperatures from 15°C to 30°C. Hence our model tentatively

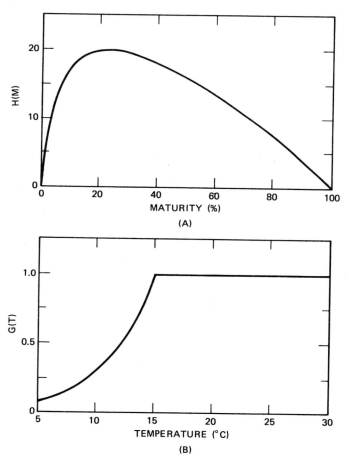

Figure 8.4. Oviposition rate of *B. curculionis* as affected by (A) age (maturity) of the adult female parasite, (B) temperature.

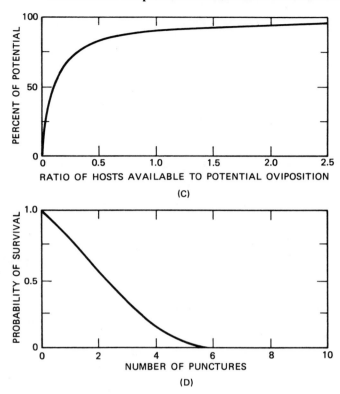

Figure 8.4. (Continued) (C) host density. Survival of host larvae (D) is reduced when the number of stings per host increases.

uses a function of the form shown in Figure 8.4B, which indicates a maximum of oviposition rate from 15°C to 30°C. Thirdly, the density of available hosts may affect the oviposition rate by *B. curculionis*. This parasite prefers third instar larvae; so in the model we assumed that second and third instars are equally preferred and that first and fourth instars are never attacked. Figure 8.4C illustrates the form of the function we use to compute the percentage of potential oviposition that occurs as host density varies. According to this function, about 90 percent of the potential eggs are laid when there are exactly enough hosts available for one egg per host. Preliminary laboratory data support the curve presented in Figure 8.4C, but as with the two previous figures, additional information is needed before the function can be accepted as being well understood.

The information contained in the three equations for the effects of parasite age, temperature, and host/parasite ratio have been combined into a single equation for computing the number of eggs laid per day (Y):

$$Y = \frac{WG(T) \sum\limits_{i=1}^{10} H(0.1iM)X(0.1iM)}{W + 0.1G(T) \sum\limits_{i=1}^{10} H(0.1iM)X(0.1iM)}$$

Where

W = host density

$X(.)$ = density of parasites of maturity (.)

$G(T)$ = effect of temperature

$H(.)$ = effect of *B.c.* maturity

The next step is to compute the distribution of parasite eggs among host larvae. There is no evidence that this species can recognize previously parasitized hosts. Laboratory data gathered by placing a single female in a small funnel with unparasitized larvae strongly suggest that the parasite eggs are distributed randomly among the available hosts; thus we use the Poisson distribution to compute how many hosts receive no eggs, one egg, two eggs, and so on. Furthermore, we assume that only one egg is laid per "sting" by the parasite. By sting, we mean the insertion of the ovipositor into a host larva.

Next we compute the mortality of host larvae that results from stings by the parasite independent of an egg being laid and the subsequent parasitism. We have used the data of Duodu and Davis (1974) to derive this function as shown in Figure 8.4D. They found, for example, that a single sting to a second-instar weevil larva caused about 23 percent mortality, while three stings caused about 72 percent mortality.

Finally, we consider the ability of a weevil larva to encapsulate eggs of *B. curculionis* and thereby effectively reduce the rate of parasitism. In the eastern United States, about 40 percent of the eggs are encapsulated (our unpublished data; Puttler, 1974) when only a single egg occurs per host. We have assumed that this percentage applies to each egg separately, in those larvae containing several parasite eggs. Thus if n eggs are inserted into a single weevil larva, the probability of all n being encapsulated is

$$(0.40)^n$$

and the probability of at least one remaining unencapsulated (and thus parasitizing the larva) is

$$1 - (0.40)^n$$

In the western United States the percent encapsulation is much lower (van den Bosch, 1964) and for such a population we use

$$(0.02)^n$$

as the probability of all n eggs being encapsulated.

This completes the description of the processes included in the model for parasitism of the alfalfa weevil by *B. curculionis*. We do not claim to have a perfect model, but it has helped tremendously in understanding how the natural system operates. We are thus convinced that our understanding and our ability to communicate our understanding have benefited immeasurably from this modeling effort.

A second benefit to research scientists is that areas needing attention can be identified and priorities can be assigned to them. For example, as our model of the alfalfa weevil system was being developed, we found that no quantitative information was available for some portions of the system even though a vast amount of research has been done on alfalfa, the alfalfa weevil, and its parasite. In formulating the prototype model we at first had to rely on "educated guesses" until the necessary data could be gathered. When the model containing such unsupported information was completed, we were able to evaluate the need for new research by performing on the model, what is known as sensitivity analysis. This consists of measuring how the response and behavior of a model varies as each of the parameter quantities in question is varied from its expected or average value. If the model is insensitive to a particular variable or constant, then that variable or constant does not deserve special attention in subsequent research. Analysis of the first version of our alfalfa weevil model revealed that the areas needing greatest research attention were related to diapause, oviposition, dispersal, and adult longevity of both the weevil and the parasite, *B. curculionis*, and to mortality during the weevil pupal stage and the parasite cocoon period. How the research was developed and the results obtained in response to these needs were dealt with in Chapter 7.

The third benefit of the modeling effort extends beyond the research community into applied pest management. Even though this model still requires much improvement it will be a good mimic of the full ecological context implied, we can and have used it to help derive a dynamic economic threshold for the weevil *H. postica*, that can be used to determine if an insecticide application is necessary. We call this threshold dynamic because it varies from 0.7 to 4 large larvae (third and fourth instars) per stem, depending on the height of the alfalfa and the time of the year. Figure 8.5 shows the two "Recommendation Charts" as they appear in a circular (Wedberg et al., 1977) available from the Illinois Cooperative Extension Service. In the past, growers have been making control decisions based on the severity of feeding

Alfalfa Weevil Pest Management Recommendation Chart 1

Number of larvae collected from a 30-stem sample

Total degree-days (dd)	\						Alfalfa height (inches)										
	2	3	4	5	6	7	8	9	10	11	12	13	14	15	16	17	18 or more
190-210																	
SPRAY	27	47	67	85	100	115	130										
Resample in 50 dd	0-26	0-46	0-66	0-84	0-99	0-114	0-129										
240-260																	
SPRAY	21	30	39	47	55	62	69	69	69								
Resample in 50 dd	0-20	0-29	0-38	0-46	0-54	0-61	0-68	0-68	0-68								
290-310																	
SPRAY		25	37	52	67	75	83	94	105	105	105						
Resample in 50 dd		0-24	0-36	0-51	0-66	0-74	0-82	0-93	0-104	0-104	0-104						
340-360																	
SPRAY					82	82	82	82	82	82	82	82	82	82	82		
Resample in 50 dd					14-81	14-81	14-81	14-81	14-81	14-81	17-81	17-81	17-81	17-81	17-81		
Resample in 100 dd					0-13	0-13	0-13	0-13	0-13	0-13	0-16	0-16	0-16	0-16	0-16		
390-510																	
SPRAY										52	52	58	64	68	72	76	80
Resample in 50 dd										8-51	8-51	8-57	14-63	14-67	14-71	18-75	18-79
Resample in 100 dd[a]										0-7	0-7	0-7	0-13	0-13	0-15	0-17	0-17

228

540 to harvest   (See Chart 2)

100 after harvest

SPRAY[b]	23	33	43	48	53	58	63
Resample in 50 dd	17-22	17-32	17-42	20-47	23-52	23-57	23-62
Resample in 100 dd[c]	0-16	0-16	0-16	0-19	0-22	0-22	0-22

150 or more after
harvest   (See Chart 2)

[a] If this field was sprayed more than 7 days ago, you can wait 200 degree-days to resample.
[b] See comment in text about windrow effects.
[c] If last preharvest sample had less than 30 larvae, the weevil season is over and you can quit sampling.

Alfalfa Weevil Pest Management Recommendation Chart 2

Total degree-days (dd)	Change in number of larvae since last sample		
	Decreased 10 or more	Within 10	Increased 10 or more
540 to harvest			
SPRAY or harvest	73	63	53
Resample in 50 dd	23-72	18-62	13-52
Resample in 100 dd[a]	0-22	0-17	0-12
150 or more after harvest			
SPRAY	78	58	48
Resample in 50 dd	28-77	18-57	0-47
Quit sampling	0-27	0-17	

[a] If sprayed more than 7 days ago, you can wait 200 degree-days to resample.

Figure 8.5. Recommendation charts derived from the coupled insect-plant simulation model. Note the influences of alfalfa height and accumulated degree-days on the recommended action.

on the alfalfa plant. In southern Illinois most growers were applying two insecticide treatments against alfalfa weevil each spring. Those growers who used this new system from 1975 through 1977 almost always achieved satisfactory control with a single treatment.

To use this program one has to do the following things:

1. Calculate degree–day accumulation by recording daily high and low temperatures from January 1 until the end of the alfalfa weevil season in late spring.
2. Count the number of larvae on a 30-stem sample.
3. Measure the height of 10 stems from the original 30.
4. Refer to Recommendation Charts 1 and 2. They provide directions for the entire weevil season. They tell when to resample, harvest early, or spray. They also tell when the weevil season is over and sampling may then cease.

A record of daily high and low temperatures should be kept from January 1 until the end of the alfalfa weevil season. These temperatures are converted to degree–days using a table provided in the Extension Circular. The degree–days used are based on a developmental threshold of 48°F. In areas where the alfalfa weevil lays eggs in the fall (e.g., southern Illinois), field sampling should begin when a total of 200 degree–days have accumulated since January 1. In more northerly areas where no significant fall or winter oviposition occurs, sampling need not begin until 400 degree–days have accumulated.

Each time a field is sampled, 30 stems must be collected from an area of the field large enough to be representative of the weevil infestation. The recommended procedure is to avoid field edges then walk across the entire field picking stems at evenly spaced intervals, without dislodging any larvae, and placing them into a 2- to 3-gallon container. Next, the stems are beat vigorously against the inside of the container for a few seconds. The dislodged larvae are counted, and 10 of the 30 stems are measured to determine alfalfa height.

Recommendation Chart 1 or 2 is then consulted to determine the proper course of action. For example, if during the first sampling of the season 44 larvae were found on alfalfa that averaged 3 inches tall, Chart 1 would say to resample in 50 degree–days (240 to 260). If 47 or more larvae were found, the recommendation would be to spray (see Chart 1).

When a field is sprayed it must be sampled again in about 100 degree-days to make sure the spray was effective and to detect any post-spray resurgence of the weevil population. A sample should also be taken 100 degree-days after first harvest to be sure that larval numbers are not high enough to retard growth of the new crop.

This system has been field tested in Illinois, Kentucky, Missouri, and Nebraska. On the whole the system has worked well, with the main complaints being that too many visits to each field are required and that it is bothersome to compute and accumulate degree-days.

OPTIMIZATION MODEL OF SHOEMAKER

The dynamic economic thresholds discussed in the preceding section give guidelines for the application of insecticide treatments. However, alfalfa weevil populations can also be suppressed by early harvests and by the parasite *Bathyplectes curculionis*. In this section we will discuss the use of an optimization method to determine the best combination of chemical, biological, and cultural control methods for alfalfa weevil.

Early harvests are an effective means of control because most of the alfalfa weevil eggs and small larvae are removed from the field with the hay. Harvests which take place later, when the weevils are larger and less susceptible to mortality factors caused by harvesting, are not able to cause such a significant reduction in the population.

The effectiveness of biological control depends in part upon synchrony between the parasite and host populations. The parasites attack primarily the second and third instar larvae. In warm years many of the weevils will have passed through the susceptible stages before the parasites become active.

Thus the decision of when to harvest and whether to apply insecticide depends upon the weather and the density of parasites as well as on the weevil density. The decision is further complicated by the fact that the total yield obtained in three harvests during the growing season is influenced by the timing of the first cut. Early harvests result in a lower yield for the first harvest, but this loss may be partly compensated by larger second and third harvests. The degree of compensation depends largely upon the weather.

The goal of the model discussed in this section is to incorporate the interactions among populations, weather and management practices in order to determine the insecticide and harvesting policies which are best. Developing the dynamic economic thresholds discussed in the previous section required a large number of computer simulations in order to calculate by trial and error the yield effect of insecticide treatments on yield for various densities of weevils and heights of alfalfa. To try to assess the impact of cutting time and parasite density as well as insecticide treatments on yield would dramatically increase the possible combinations of control alternatives. As a result, a very large number of computer simulations would be necessary—so many that the expense of computer calculations is beyond reasonable limits.

For this reason, the best combination of control methods has been calculated by a numerical algorithm called dynamic programming. Dynamic programming is one of a class of techniques known as optimization methods. These techniques determine the best or "optimal" management policy without simulating a model of a system for each possible alternative policy. Instead the best policy is calculated directly by manipulating the underlying mathematical structure of the model of the system.

The optimization model is based upon a description of the effect of weather and management policies on plant, parasite, and alfalfa weevil dynamics. The yield is determined by simulating Fick's ALSIM model with the appropriate weather data and cutting times. The weevil and parasite dynamics are predicted by a set of differential equations which incorporate descriptions of the temperature dependence of weevil development, possible asynchrony between parasite and host populations, and age-specific mortality during harvest of both alfalfa weevils and their parasites. The parameter values describing harvest and overwintering mortality are based upon field studies in Ithaca, New York, by R. G. Helgesen (Helgesen & Cooly, 1976; personal communication). The effect of feeding on the delay in alfalfa regrowth was estimated from the results of field studies by Fick (1975). Many of the remaining parameter values for rates of development and feeding were obtained from the alfalfa simulation model by Ruesink (1976). Further details of the mathematical structure of the model and the data base are given in Shoemaker (1977).

Three different weather patterns were considered. The patterns used were determined by a statistical analysis of past weather patterns. The values were obtained by calculating the mean and standard deviation of the physiologic time corresponding to six years of weather data for Ithaca, New York.

Based upon the equations describing the effect of cutting time and insecticide applications on yield, the model calculates the management policy which maximizes net economic returns. These returns are defined to be the value of the yields minus the costs of pest management. The insecticide and harvesting policies calculated to be best are a function of the densities of alfalfa weevil and parasites and of weather.

In Figure 8.6A is shown the optimal time of harvesting both for the case where there are no parasites and where there is a moderate density (10,000 per acre). The optimal time of cutting becomes earlier as the weevil density increases and there is more need for weevil control.

The results in Figure 8.6B are based on an average weather pattern while those in Figure 8.6A are for cool weather. A comparison of the two figures indicates that the optimal time of cutting becomes earlier as the weather becomes warmer because both the plant and weevil populations mature more quickly under warmer conditions. As a result the times when harvesting will

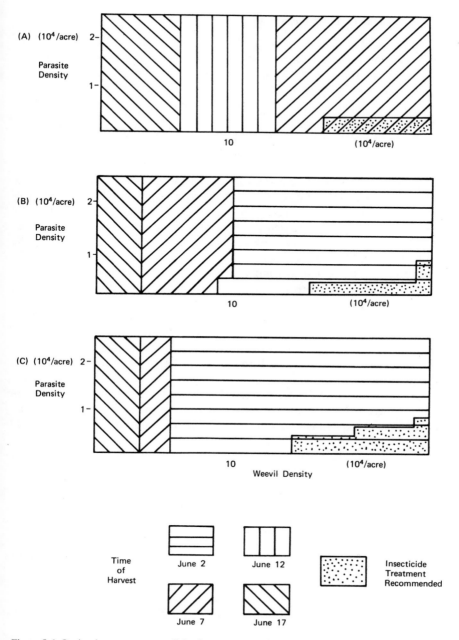

Figure 8.6 Optimal management policies for a one year planning horizon for (A) cool, (B) average, and (C) warm weather. Hatching represents the optimal time of harvest for a given parasite and weevil density. Superimposed dots note where insecticide treatments are recommended.

be an effective alfalfa weevil control agent occur earlier in the season. Since the crop matures more quickly, an early harvest causes less of a reduction in total yield. However, temperature affects each of the populations in a somewhat different way. Therefore, changes in weather alter the synchrony between the alfalfa crop and the alfalfa weevil population as well as the synchrony between the alfalfa weevil and its parasite.

The effect of parasites on control policies are estimated by comparing the two bars in each of the Figures 8.6 and 8.7. Notice that the times of cutting calculated to be optimal are in most cases the same whether or not parasites are present. The major impact is with the insecticide treatment policy. In the absence of parasites, an insecticide treatment is recommended for high densities of alfalfa weevils. However, an insecticide treatment is not economically justified when a moderate density of parasites are present. This occurs because an insecticide treatment reduces the number of parasites as well as the number of alfalfa weevils.

The results presented in Figures 8.6 and 8.7 assume that the economic value of alfalfa hay is proportional to the tonnage harvested. However, the value of the hay also depends upon its nutritive value, which decreases with late cutting. Research is currently underway to incorporate quality changes in the alfalfa hay into the alfalfa weevil management model. It is likely that when this additional factor is incorporated, the last cutting date, June 17, will be replaced as an optimal policy by an earlier cutting time in warm weather.

The optimal cutting times indicated in Figure 8.6C are even earlier than those in Figure 8.6B because of the warmer temperature. Insecticide applications are also recommended at lower densities of weevils than recommended for cooler weather.

The pest-management policies calculated to be optimal for a 3-year planning period and warm weather are presented in Figure 8.7. Because of

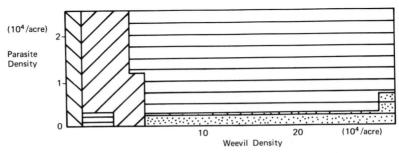

Figure 8.7. Optimal management policy for a three year planning horizon for warm weather. Hatching and dots are the same as in the previous figure.

the longer-term planning horizon, the effect of management decisions on population levels in subsequent years influences the choice of harvest date and insecticide treatment. Because of this effect, the optimal results presented in Figure 8.7 recommend a more intense control of the alfalfa weevil: harvest dates are generally earlier and insecticide treatments more frequent than in Figure 8.6.

The low cost of obtaining the results presented in Figures 8.6 and 8.7 illustrate the advantage of using an optimization method rather than exhaustive simulation to choose the best management program from among a large number of options. The results were obtained by one calculation of a computer model based upon a dynamic programming algorithm at a cost of $12. Given the number of different densities of weevils and parasites, the variety of pest management alternatives, and several weather patterns over each of several years considered, the number of possible combinations of these variables is over 1 million. Thus, to obtain the same information by exhaustive simulation as is presented in Figures 8.6 and 8.7, over 1 million simulations at a cost of thousands of dollars would be necessary. Of course, one would not make such a large number of simulations but would instead reduce the number of alternatives considered. Thus, for many situations in order to evaluate a full range of alternatives, it is necessary to use optimization methods.

ALFALFA AND THE EGYPTIAN ALFALFA WEEVIL MODELS OF GUTIERREZ

All of the work on insects discussed to this point pertains to the alfalfa weevil, *H. postica*. Similar efforts have been made for the Egyptian alfalfa weevil (EAW), *H. brunneipennis*, and the alfalfa plant under California conditions (Gutierrez et al., 1976) The alfalfa weevil is highly destructive, and is the key pest in alfalfa in California. Furthermore, disruption of the alfalfa ecosystem kills natural enemies and causes secondary pest outbreak not only in alfalfa, but also via migration to other crops (e.g., cotton, grape, etc.). Alfalfa is the prime insectary for beneficial species for the agroecosystem in much of California, hence its proper management is crucial.

The California Plant Model

During late fall, frosts kill the aerial parts of the plant, which begin regrowth when conditions become favorable during late winter or early spring. The crop grows approximately 500 to 555 D° ($>5°C$) at which time it is cut (approximately $\frac{1}{10}$ bloom). Regrowth of the crop begins from the root reserves, and continues until its photosynthetic machinery can provide all of the growth requirements. Alfalfa in California is cut several times during the season.

The amount of photosynthetic material (P) which is produced at time t can be described in general terms by

$$P = f(\text{solar radiation, leaf surface, age, } H_2O, \ldots)$$

In the simplest model the total plant weight (W) is composed of plant tissues (Q) and carbohydrate reserves (C)

$$W = Q + C = \int_0^\infty (L + S + R)\,da + C \tag{1}$$

where L, S, and R are leaf, stem, and root tissues, respectively. The growth rate of the crop is

$$dW/dt = dP/dt - \psi_1\,dQ/dt - \psi_2 \int_0^\infty W\,da - \psi_3\,dP/dt - \psi_4\,dC/dt \tag{2}$$

Note that $\Delta t = \Delta a$ in the model, but this is not a necessary condition. The various ψ_i values are respiratory loss rates associated with reconverting P to Q and C, maintenance respiration, and production of P, respectively. Other values could be assigned, but the above are the dominant ones. If we differentiate Equation (1) with respect to t and solve Equations (1) and (2) simultaneously for dC/dt, we get

$$dC/dt = \left[(1 - \psi_3)\,dP/dt - (1 + \psi_1) \cdot dZ/dt - \psi_2 \int_0^\infty W\,da\right]\Big/(1 + \psi_4) \tag{3}$$

or the rate of new carbohydrate accumulation in plant reserves. A similar mass flow model was presented by Jones et al. (1974) for cotton, but the model lacked age structure for the various components of Q.

In our model, the growth of the crop depends upon the supply (Z) of available carbohydrate $[P(t) + \alpha \cdot C]$, where α is the fraction of C available for the growth at time t. This supply is counter balanced against the demands (D) for maximum growth.

$$D_t = \left(\psi_1\,dQ/dt + \psi_2 \int_0^\infty W\,da + \psi_3\,dP/dt + \dot{L}(t) + \dot{R}(t) + \dot{S}(t)\right)\Delta t$$

$$= \hat{\psi}_1 + \hat{\psi}_2 + \hat{\psi}_3 + \Delta L + \Delta R + \Delta S)$$

If $Z \geq D$, $C_t = C_{t-1} + P_t - D_t$ and $r = 1$. But if $Z < D$ the $\hat{\psi}_i$ are satisfied first, and if $Z - \sum_j \hat{\psi} > 0$, then the plant grows only a fraction of its potential and $C_t = (1 - \alpha)C_{t-1}$. The scalar for the rate of actual growth (r_1), is

$$r_1 = \left(Z - \sum_{i=1}^{3} \hat{\psi}\right)\bigg/(\Delta L + \Delta S + \Delta R)$$

and

$$\frac{dW}{dt} = \left(\frac{dL}{dt} + \frac{dR}{dt} + \frac{dS}{dt}\right) \cdot r_1$$

Conceptually the population model for the alfalfa crop may be viewed as plant part models

$$\frac{\partial L}{\partial t} + \frac{\partial L}{\partial a} = -\mu_L(\cdot) \cdot L(t, a)$$

$$\frac{\partial R}{\partial t} + \frac{\partial R}{\partial a} = -\mu_R(\cdot) \cdot R(t, a)$$

$$\frac{\partial S}{\partial t} + \frac{\partial S}{\partial a} = -\mu_S(\cdot) \cdot S(t, a)$$

which are coupled via the metabolic pool model ($r = Z/D$) (see above and Wang et al., 1977). Both t and a are measured in physiologic time since the last cutting. The models are the continuous form of the well-known Leslie matrix (Leslie, 1945). In fact, $L(t, a)$, $R(t, a)$, and $S(t, a)$ are mass density functions and the various death rates (μ_i) are complex functions.

$$\mu_i(\cdot) = f(r, \text{herbivore damage, frost, cutting cycles} \ldots)$$

The models require that initial conditions be specified to assure a unique solution. These models are basically time-varying life tables (Hughes, 1963; Southwood, 1966; Gilbert et al., 1976). This mathematical model is conceptually similar to Fick's algorithm, with the only major discrepancy being that the Fick model lacks age structure, temperature-dependent metabolic costs are excluded, and net rather than gross photosynthesis is used.

The Egyptian Alfalfa Weevil Model (EAW)

One of the most important submodels developed by the California group is one for predicting the adult weevil migration during the fall period. Christensen et al. (1974) used a multivariate regression technique to predict weevil

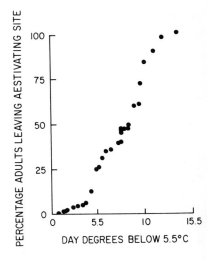

Figure 8.8. The percentage adult alfalfa weevils induced to leave aestivation. (N.B.: The curve, translated to a frequency histogram, is approximately normal.) [From Gutierrez et al., 1976.]

emergence from diapause. This regression model indicated that most of the variance in the emergence data was associated with day–night temperature differences. A reevaluation of those data indicated that the properties of weevils emerging from aestivation (θ) from a particular site could be more accurately predicted as a function of the physiologic time ($=$ day degrees, $D^\circ_{<5.5^\circ C}$) the population spent below its metabolic threshold (Fig. 8.8). The total $D^\circ_{<5.5}$ can be viewed as the induction time for termination of aestivial diapause, while the time above ($D^\circ_{<5.5}$) is the time available for migration provided the maximum temperature was above 13°C. The number (x) of adults entering the field per square foot were estimated, and plotted against the cumulative $D^\circ_{<5.5}$ since the beginning of migration. The relationship between θ and $D^\circ_{<5.5}$, and x on $D^\circ_{>5.5}$ are roughly linear, hence the following relationship holds:

$$\frac{\frac{1}{n}\sum_{i=1}^{i} X_{i,t}}{X_{max}} = \frac{\frac{1}{n}\sum \theta_{i,t}}{1}, \quad \text{or} \quad X_{max} = \frac{\bar{x}_t}{\bar{\theta}_t}$$

where, n samples were taken of adults in the field and from the diapause termination equation at corresponding times t. If this is so, then the relationship can be used to estimate X_{max}, or the maximum number which will be present in the field (see Fig. 8.9 for simulation results). For best results, several estimates should be made at each t.

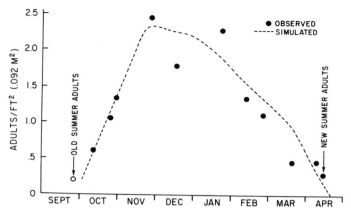

Figure 8.9. Simulated adult Egyptian alfalfa weevil populations at Davis, Calif., during 1973. [From Gutierrez et al., 1976.]

The model is further useful in that the immigration rate (β) can be computed, for example,

$$\beta = \frac{X_{max}}{(D^0_{tmax} - D^0_{t_0})}$$

and the age structure of the immigrating adult population (I_t) estimated on a daily basis

$$I_t = X_{max} \cdot (\theta_t - \theta_{t-\Delta t})$$

The population model for the EAW for one time step ($\tilde{\Delta}t$) can be written in discrete form as

$$\sum_{a_0}^{a_m} N_{i_{t+\Delta t}} = \left[\sum_{a_0}^{a_1} N_i + \sum_{a+1}^{a_2} (N_i 1 x_1 1 x_2) + \sum_{a+1}^{a_3} N_i + \sum_{a+1}^{a_5} (N_i 1 s_3) \right.$$

$$\left. + \phi_1 \phi_2 \sum_{a_4}^{a_5} \cdot N_i f_i \cdot \lambda_1 \cdot \lambda_2 \cdot \lambda_3 \cdot \xi \cdot \Delta t + I(a_3) \right]_t$$

$\text{TOTAL}_{t+\Delta} = [\text{EGGS} + \text{LARVAE} + \text{PUPAE} + \text{ADULTS}$

$+ \text{NEW EGGS} + \text{NEW IMMIGRANTS}]_t$

where

N_i = the number in age category i

$1x_1 = 1 - \mu_L$ = larval survivorship due to lack of food

$1x_2 = 1 - \mu_F$ is the frost survivorship on larvae of ages 118 to 210

$1x_3 = 1 - \mu_A$ = adult survivorship rate

I_t = immigration of adults into the field during t

F_i = age-dependent fecundity

$\lambda_1, \lambda_2, \lambda_3$ = weather effects on the feeding behavior of adults

ξ = larval feeding effects on adult fecundity (see below)

$a_0, a_1, \ldots, a_5$ = are the age boundaries for the weevils' life stages in day degrees

t = the current physiologic time step (daily)

ϕ_1, ϕ_2 = are corrections for adult sex ratio (= 0.5) and for egg viability (0.85)

The parasite larvae of this species of weevil are almost always encapsulated (killed), and hence not considered in the model. In addition, the parasitism rate is very low (<2 percent).

The continuous form (approximately) of this model is again the von Foerster model (1959),

$$\frac{\partial N}{\partial t} + \frac{\partial N}{\partial a} = -\mu_N(\cdot)N(t, a) + I$$

where $\mu_N(\cdot)$ is a complex mortality function, $N(t, a)$ is the number density function, and I is the net immigration rate which could be negative (see Chapter 5 for a more complete discussion). The EAW population model in reality is a submodel of the alfalfa plant model, and couples via the leaf population model.

Egyptian alfalfa weevil adults and larvae attack the growing tips of the alfalfa stems, and consume leaves (it prefers the younger aged leaves). This activity alters the age structure of leaves and the growth form of the plant both of which affect photosynthate production. In addition, its feeding causes the plant to divert photosynthate from growth to wound healing.

A complete model for the pest (see Gutierrez et al., 1976) would describe not only the feeding activities, but also the dynamics of the weevil population. Let us assume, for simplicity, that we have some time pattern and age structure of weevils which are attacking the crop. The mass of the crop at time t under a given weather regime will be influenced by the feeding it has

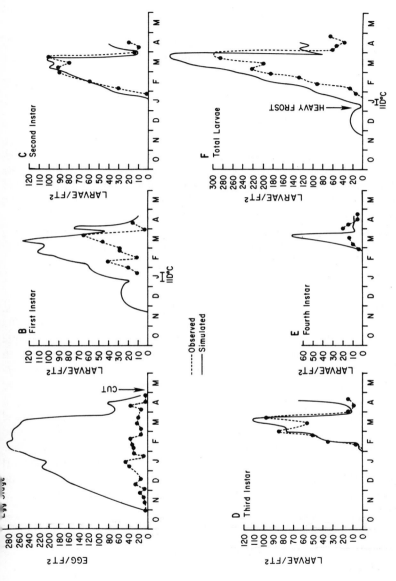

Figure 8.10. Observed and simulated populations of the Egyptian alfalfa weevil at Davis, Calif., during 1973. All counts are depicted per square foot (0.092 M^2): (A) eggs, (B) first instar, (C) second instar, (D) third instar, (E) fourth instar, and (F) total larvae. [From Gutierrez et al., 1976.]

sustained prior to that time. The dry matter required for wound healing (H = b · E) would be a further net loss potential (b) to the amount eaten [$E_t = \int g(a) \cdot N \, da$] by the population N, where g is some age-dependent consumption rate. Equation (2) then becomes

$$\frac{dW}{dt} = \frac{dP}{dt} - \psi_1 \cdot dQ/dt - \psi_2 \left[\int_0^\infty lx(a)L \, da + \int_0^\infty (S + R) \, da \right]$$

$$- \psi_3 \frac{dP}{dt} - \psi_4 \frac{dC}{dt} - \frac{dH}{dt}$$

and $lx(a)$ is the survivorship value from weevil feeding on leaves age $0 \le a \le m$. If this equation is substituted for Equation (2) and used to obtain Equation (3), we can see that H affects dC/dt directly. This fact becomes especially important if we consider that the crop's regrowth occur from the reserves. Hence we see that severe defoliation can affect not only the current crop, but also the next one via the reserves and/or the destruction of whole plants. The latter aspect has not been completely studied. Also note the fact that ψ_2 is a temperature-dependent rate, hence any factor which would reduce P under high-temperature conditions would greatly stress the plant.

A Field Simulation

A comparison of the observed and simulated EAW populations for Davis California for 1973 is shown in Figure 8.10. The simulated number of adult was extremely good; those for eggs and first instar larvae show considerable discrepancy. A careful examination of EAW egg distribution data in green and dried stems, and in the debris showed that the samples grossly under estimated the egg population. The discrepancies in first instar counts were in part due to the fact that they are difficult to find because they burrow into the growing tip. The simulations of the other life stages were quite adequate The rapid decline of the larval population during April occurred because food became limiting, and affected larval survivorship and adult fecundity (see Gutierrez et al., 1976). The simulations for the alfalfa crop were in general quite reasonable but uninteresting, hence they are not reported here. The use of these models in developing alfalfa pest management strategies is deferred to the next section.

MODELS BY REGEV AND GUTIERREZ
FOR OPTIMIZING EGYPTIAN ALFALFA WEEVIL MANAGEMENT

Regev, Gutierrez, and Feder (1976) describe a model for optimizing economic gain using Egyptian alfalfa weevil as an example. The model is a reduced description of the alfalfa–Egyptian alfalfa weevil simulation but put into an economic framework. The analysis considers the effects of secondary pest outbreaks and seeks to determine not only the optimal time to spray, but also the quantity. The economic problem is examined from both the single farmer's and society's point of view, and recognizes specific common property characteristics of the pest (i.e., the pest in any particular field comes from many fields). The analysis points to the gap between private and societal control benefits, offers estimates of shadow prices, and indicates a direction toward optimal pest-control policy from both the societal and farmer's point of view.

Individual farmers generally place major emphasis on single-season pest control, even though the system which gives him the best within-season return may cause trouble in later years. The optimal within-season solution obtained by Regev, Gutierrez, and Feder (1976) is contrasted with the current practices observed in California as shown in Table 8.1.

The optimal timing of weevil control applications is in the time period up to and including the peak adult weevil population (Fig. 8.9), but prior to the last frost when the new seasonal regrowth of the alfalfa begins. The control is directed against the adult stage rather than the current practice of spraying larvae. The resultant benefit is greater in spite of the fact that the insecticide is a more efficient larvicide than adulticide (e.g., given the parameters estimated here for insecticide application of 16 oz, about 20 percent of the adults, compared with 4 percent of the larvae, will survive). Insecticide applications made early in the season reduce not only adults but also the future eggs and larvae. The common current practice is to apply insecticides

Table 8.1. Current and Optimal Private Insecticide Application

Type of Solution	Application Time ($\Delta t = 60$ D°)									Total Amount of Pesticides Applied (oz)	Net Revenue per Acre
	1	2	3	4	5	6	7	8	9–21		
			(Oz/Acre)								
Current practices	0	0	0	0	0	0	0	16–32	0	16–32	$88–$101
Unrestricted private optimal solution		4.3	12.6	0	0	0	0	0	0	28.2	$116.84

approximately 180 D° *after* the last frost, only after a number of larvae can be found in the field. The amount of insecticide to be applied in practice varies greatly, ranging from less than 16 to 32 oz/acre, depending upon the timing of the first application and on the actual damage throughout the season. Comparison of the net gains from weevil control of the private optimal policy and current practices (Table 8.1) indicates an advantage for the first system. The most important practical conclusions of these results are the shift in the timing of pesticide applications and a concentration on the adults rather than the larvae.

While the sampling decision rules for common sampling methods are well in hand (Christensen et al., 1974), more sophisticated and less expensive sampling methods to help the farmers assess the level of adult pest infestation early in the season, and more effective insecticides which are not adversely influenced by winter weather must be developed before the results of this work can have practical utility.

From the societal viewpoint, the model recognizes specifically the fact that individual farmers have little regional control over either the within-season pest population levels or the interseason dynamics of the pest. Formation of a central decision agency or cooperatives (or pest consultants), on the other hand could enhance the farmers' income by considering the effects of pest control practices by all farmers in the region on the level of pest infestation in the following seasons. The model indicates that for an optimal regional policy, growers should *spray heavily during the first few seasons* to suppress the weevil populations regionally to a low level where they can be maintained using less insecticides. *On the surface, this would appear to be a good policy, but it ignores the very real threat of the development of pesticide resistance in the weevil population.*

The model was then modified to include a description of genetic selection under selective pressure by insecticides (Gutierrez et al., 1976). Some Russian work on *H. postica* indicates that nearly total resistance to heptachlor developed within a period of 6 years. They found that resistant individuals tended to have more than one generation per year and had higher fecundity. If the resistant individuals had the same "fitness" (progeny surviving per female in the absence of pesticides), the resistance genes would obviously be selected, once the trait appeared. Genetic theory, however, indicates that a new gene—one for resistance, for example—would not be well integrated into the gene pool immediately, and, as a result, total fitness would be initially lower. However, fitness of a resistance gene does increase with time and continued pesticide selection pressure. The model was then altered in the following way: a genetic mechanism for the selection of resistance genes due to pesticide pressures (the Hardy-Weinberg law) was introduced into the model, along with a low mutation rate, increasing fitness of the resistance gene(s) with time, higher fecundity of resistant females, and different geno-type pesticide caused mortality relationships for both larvae and adults (Gutierrez, Regev, & Shalit, 1979).

The above concepts were incorporated into the optimization model, and the following results emerged. Because of pest mobility, the pest resistance level in each field results from chemical pesticide application in the preceding season by *all* farmers. Since the individual farmer does not consider the external effect of his pest control policies on the development of pest resistance, it is only through a joint action by a central body (e.g., cooperatives, legislature, etc.) that the optimal pest control policy would be likely obtained. This policy calls for reduction in chemical pesticide use, and increases in biological methods of control so as not to increase pest resistance (Regev et al., 1976). Effective biological control of this pest is currently not available at this time. This analysis will however enable us to estimate the implicit costs of increasing pest resistance, which in turn could be used as an indicator of the amount of money that should be spent by society on research for such nonchemical control methods as host plant resistance, biological control, and cultural methods.

CONCLUSION

In the alfalfa subproject we have concentrated effort on two closely related pest species: the alfalfa weevil and the Egyptian alfalfa weevil. Modeling and analysis has progressed independently for the two, but many similarities can be noted. Each began by developing a descriptive simulation model that included the pest and its host. Each found that existing knowledge about the system was insufficient. Experiments were designed and performed to acquire the desired information, and this was used to refine the descriptive model. Analysis of the model provided insight into the behavior of the natural system.

In addition to descriptive simulation models, each group developed pest management models that included economic considerations. These models necessarily included additional simplifying assumptions, and in a sense, the biological portions are models of the simulation models. That is, the simulation models were developed first and were compared to the natural system, then the management models were developed and compared to the simulation models.

It is impossible in such a short report to cover all the interesting and practical findings achieved in 5 years, and in 5 years it is equally impossible to perform all the interesting and useful analyses that one would like to do. The possibilities appear endless, limited largely by one's imagination, ability, resources, and persistence. If the first five years are an indication of the future, and there is no reason to think otherwise, then pest management research in alfalfa and other crops can be expected to make solid additional progress.

LITERATURE CITED

Christensen, J. B., W. R. Cothran, C. E. Franti, and C. G. Summers. 1974. Physical factors affecting the fall migration of the Egyptian alfalfa weevil, *Hypera brunnei-pennis* (Coleoptera: Curculionidae): A regression analysis. *Environ. Entomol. 3*: 373–376.

Davidson, J. 1944. On the relationship between temperatures and rate of development of insects at constant temperatures. *J. Anim. Ecol. 13*:26–38.

Duodu, Y. A., and D. W. Davis. 1974. Selection of alfalfa weevil larval instars by, and mortality due to, the parasite *Bathyplectes curculionis* (Thompson). *Environ. Entomol. 3*:549–552.

Fick, G. W. 1975. ALSIM 1 (Level 1)—Users' manual. Agron. Mimeo, 75-20, Dept. of Agron., Cornell Univ., Ithaca, New York.

Gilbert, N., A. P. Gutierrez, B. D. Frazer, and R. E. Jones. 1976. *Ecological Relationships*. Freeman, Reading and San Francisco.

Gutierrez, A. P., J. B. Christensen, C. M. Merritt, W. B. Loew, C. G. Summers, and W. R. Cothran. 1976. Alfalfa and the Egyptian alfalfa weevil (Coleoptera: Curculionidae). *Can. Entomol. 108*:635–648.

Gutierrez, A. P., U. Regev, and H. Shalit. 1979. An economic optimization model of pesticide resistance: alfalfa and Egyptian alfalfa weevil—an example. *Environ. Entomol. 8*:101–107.

Gutierrez, A. P., U. Regev, and C. G. Summers. 1976. Computer model aids in weevil control. *Calif. Agric. 30*(4):8–9.

Helgesen, R. G., and N. Cooley. 1976. Overwintering survival of the adult alfalfa weevil. *Environ. Entomol. 5*:180–182.

Hughes, R. D. 1963. Population dynamics of the cabbage aphid, *Brevicoryne brassicae* (L.). *J. Anim. Ecol. 32*:393–424.

Jones, J. W., A. C. Thompson, and J. D. Hesketh. 1974. Analysis of SIMCOT: Nitrogen and growth. *Proc. Beltwide Cotton Prod. Res. Conf.*, Memphis, pp. 111–117.

Leslie, P. H. 1945. On the use of matrices in certain population mathematics. *Biometrika 35*:213–245.

Puttler, B. 1974. *Hypera postica* and *Bathyplectes curculionis*: Encapsulation of parasite eggs by host larvae in Missouri and Arkansas. *Environ. Entomol. 3*:881–882.

Regev, U., A. P. Gutierrez, and G. Feder. 1976. Pests as a common property resource: A case study of alfalfa weevil control. *Am. J. Agric. Econ. 58*:186–198.

Regev, J., H. Shalit, and A. P. Gutierrez. 1976. Economic conflicts in plant protection: the problem of pesticide resistance; theory and application to the Egyptian alfalfa weevil, in G. A. Norton and C. S. Holling (eds.): Proceedings of a Conference on Pest Management, 25–29 Oct. 1976. Int. Inst. Appl. Systm. Anal. (IIASA), Laxenburg, Austria, pp. 281–299.

Ruesink, W. G. 1976. Modeling of pest populations in the alfalfa ecosystem with special reference to the alfalfa weevil, in R. L. Tummala, D. L. Haynes, and B. A. Croft (eds.), *Modeling for Pest Management*. Michigan State University Press, East Lansing, pp. 80–89.

Shoemaker, C. A. 1977. Pest management models of crop ecosystems, in C. A. Hall and J. Day (eds.), *Ecosystem Modeling in Theory and Practice*. Wiley-Interscience, New York, pp. 537–574.

Southwood, T. R. E. 1966. *Ecological Methods*. Methuen, London.

van den Bosch, R. 1964. Encapsulation of the eggs of *Bathyplectes curculionis* (Thomson) (Hymenoptera: Ichneumonidae) in larvae of *Hypera brunneipennis* (Boheman) and *Hypera postica* (Gyllenhal) (Coleoptera: Curculionidae). *J. Insect Pathol. 6*:343–367.

Von Foerster, H. 1959. Some remarks on changing populations, in Frederick Stohlman, Jr. (ed.), *The Kinetics of Cellular Proliferation*. Grune & Stratton, New York.

Wang, Y., A. P. Gutierrez, G. Oster, and R. Daxl. 1977. A general model for plant growth and development: Coupling cotton–herbivore interactions. *Can. Entomol. 109*: 1359–1374.

Wedberg, J. L., W. G. Ruesink, E. J. Armbrust, and D. P. Bartell. 1977. Alfalfa weevil pest management program. Illinois Cooperative Extension Service Circ. 1136.

9

THE SYSTEMS APPROACH AND GENERAL ACCOMPLISHMENTS TOWARD BETTER INSECT CONTROL IN POME AND STONE FRUITS

D. Asquith

Pennsylvania State University
Fruit Research Laboratory, Biglerville, Pennsylvania

B. A. Croft

Pesticide Research Center
Michigan State University, East Lansing, Michigan

S. C. Hoyt

Tree Fruit Research Center
Washington State University, Wenatchee, Washington

E. H. Glass

Department of Entomology
New York State Agricultural Experiment Station, Geneva, New York

R. E. Rice

San Joaquin Valley Agricultural Research and Extension Center
University of California, Parlier, California

INTRODUCTION

The Integrated Pest Management Subproject for Pome and Stone Fruits, part of a broader multicrop project entitled "The Principals, Strategies, and Tactics of Pest Population Regulation and Control in Major Crop Ecosystem," can boast, after 5 years of research effort, of many significant accomplishments while humbly admitting that only the tip of the iceberg has been chipped.

The goals of the multi-crop, multidisciplinary project and of the pome and stone fruits project have been to (1) develop ecologically based and structured systems of management of pest populations at noneconomic densities so as to optimize economic returns on a continuing basis consistent with minimal environmental damage and (2) demonstrate that agricultural research can be done in a more productive way than in the past through unified, interdisciplinary approaches utilizing systems analysis.

To attain the above goals the pome and stone fruits group has pursued the following objectives: (1) develop new and improved methods for monitoring the dynamics of pest populations and determining the need to take action; (2) develop computer simulation programs which will forecast pest population dynamics with sufficient lead time for making appropriate pest management decisions; (3) develop selective and nonchemical control tactics or systems for the key direct fruit pests which occur without effective natural control and which normally are present at low levels; (4) evaluate the effects of other orchard management practices (e.g., disease and weed control, irrigation, etc.) on arthropod pest management systems and interpret their interactive relationships within the context of total pome and stone fruits production programs; and (5) evaluate the impact of insecticides and other fruit production practices on the stability of selected pest complexes (e.g., phytophagous mites) which are normally regulated by biological control agents, and evaluate whether increasing diversity of pests, alternate prey, and natural enemies will enhance short or long-term arthropod pest control, and if so, determine the cost and efforts required to maintain such diversity.

Since inception of the NSF/EPA Pome and Stone Fruits Project 5 years ago, accomplishments have been many and varied. Space does not permit discussion of all the research involved. In the following categories, examples that illustrate these research accomplishments toward the achievement of the objectives are presented: (1) sampling and monitoring methods; (2) economic injury levels; (3) biological control methods; (4) chemical control methods and selection of chemicals tolerated by predators; (5) basic biological studies; (6) modeling and orchard ecosystem analysis; (7) implementation of integrated pest management systems; and (8) economic evaluation. The economic results from application of the research findings by orchardists are evaluated, and the benefits environmentally may be inferred by the reductions in use of disrupting and hazardous insecticides. Examples of research will be presented from all states associated with the NSF/EPA Pome and Stone Fruits Project whether funded or not.

SAMPLING, MONITORING, AND FORECASTING

The development of sound sampling and monitoring methods is essential to the collection of data for development of economic thresholds, economic injury levels, simulation models of faunal interactions, and pest management decisions.

Navel Orangeworm—Artificial Egg Trap

The navel orangeworm, *Paramyelois transitella* (Walker), is not a pest of pome and stone fruits, but it is a pest of almonds, a related nut crop, which is grown in orchards. The artificial egg trap developed for this insect is an excellent example of development of a modern, efficient sampling and monitoring technique. Various techniques for monitoring populations of the navel orangeworm were compared in almond orchards in the San Joaquin Valley of California. Adult populations were monitored with pheromone traps containing virgin female moths and with black light traps. Oviposition was monitored by use of an egg collection device baited with an oviposition attractant and by periodic examinations of "mummy" nuts collected from the trees. Results of these comparisons showed that the artificial egg trap accurately reflected seasonal population trends of the moths and the egg laying patterns by females until "hull-split" of the new crop began and competition for oviposition was created between traps and nuts (Rice, 1976).

After the efficiency of the egg trap was validated, it was used to study populations of the navel orangeworm in 1974 and 1975. Seasonal variations in moth emergence and flight as indicated by egg laying patterns were readily identified in both years. Egg deposition curves closely followed male moth flight patterns during the early season but did not correlate as well from mid-summer through fall. Differences in moth and egg populations could be detected with egg traps in orchards that received varying levels of post-harvest sanitation. Sticky traps baited with the egg trap attractant showed a sex ratio of responding navel orangeworm moths of approximately 11 females: 1 male. Attractant odors given off by the egg trap bait are not specific for navel orangeworm but also attract the raisin moth, *Cadra figulilella* (Gregson). However, only eggs of navel orangeworm were found on the egg traps in the field (Rice et al., 1976).

Forecasting Codling Moth Phenology

At present in the United States, the potential for biological control of the codling moth, *Laspeyresia pomonella* (L.), is limited. Improved control and a reduction in the number of spray applications can be accomplished by better timing of chemical control measures in relationship to the moth's phenology. In the past, a major limitation to this approach has been the lack of effective monitoring methods. Recently, sex attractant or pheromone traps have been developed which have proven to be valuable tools for following the abundance and activity of adult male moth populations during a season. However, since control measures should be timed not only in relation to adult abundance but primarily to prevent larval damage, the ability to detect egg development and predict egg hatch are also critical requirements.

Riedl, Croft, and Howitt (1976) defined the relationship of pheromone trap catch to emergence and oviposition of the codling moth for the two-generation climate of Michigan. Male moth catch anticipated emergence and oviposition during spring flight when the trap displayed greatest efficiency, but lagged behind emergence and closely followed oviposition during the second generation. Trapping efficiency declined toward the end of the first generation and was generally lower during the summer flight period. Factors which possibly relate to this efficiency loss are discussed. First catch and the catch peaks in both generations were evaluated as reference points for the prediction of phenological events (particularly egg hatch) in both generations. Of the four forecasting methods evaluated, degree day and developmental unit summations, starting from first catch, for the preoviposition and incubation period were most reliable in predicting the beginning of spring brood egg hatch.

In California, Falcon, Pickel, and White (1976) developed a computer model known as " BUGOFF " in an effort to provide more accurate methods for determining when to apply insecticides for controlling the codling moth. The model is based on the assumption that temperature regulates activity, growth, and development of the codling moth. As in the Michigan system, determining early events in the seasonal development of the codling moth and coupling them with day-degree accumulations has proved to be the best way developed so far for predicting later events in the seasonal history of this pest.

Tufted Apple Budmoth

Currently, the tufted apple budmoth [*Platynota idaeusalis* (Walker)] is the most serious pest of apple in Pennsylvania. Determination of the components of the sex attractant produced by female moths as *trans*-11-tetradecenyl alcohol and *trans*-11-tetradecenyl acetate at Geneva, New York, by Hill et al. (1974) has helped in timing control measures for this pest. Field trials in Pennsylvania to determine the best lure to use in "sticky traps" demonstrated that the combination of these two compounds in a 1:1 ratio was very effective. The acetate itself is not an attractant but when added to the alcohol it increases or synergizes attractancy.

Trap Selection

Bode et al. (1973) conducted field trials to evaluate the 1C and the 1CP Pherotrap® for catching male tufted apple budmoths. The 1C design caught about twice as many moths as the 1CP; hence the 1C Pherotrap was chosen for use in monitoring programs.

Sampling of the Mite Predator *Stethorus punctum*

The predatory coccinellid, *Stethorus punctum* (LeConte), is the most important predator of the European red mite and two-spotted spider mite in Pennsylvania apple orchards. Being able to count this predator accurately is important in predicting the biological control of pest mites. The usual method of counting the predator in Pennsylvania is for a trained observer to make a 3-minute observation around the periphery of a tree, recording on a hand counter all *S. punctum* adults and larvae that he sees. An experiment was conducted to relate the number of *S. punctum* counted in a 3-minute observation with the number present per 1000 leaves. Each of 10 trees was sampled using the 3-minute observation, and then estimates were made of the number of predators/1000 leaves, using a "beating sample" method. Employing regression analysis, the functional relationship between the number of predators/1000 leaves and the number/3-minute observation was estimated for both adult and larval populations. The predictive equations and R-squared values are given below. For modeling purposes, these predictions are combined with an estimation of the number of leaves/tree to estimate the absolute predator population/tree.

ADULTS $Y = 14.3 + 0.0047x^2$ R-SQUARED $= 63$ percent
LARVAE $Y = 0.862x$ R-SQUARED $= 89$ percent

where Y equals the number of $S. punctum/1000$ leaves and x equals the number of $S. punctum/3$-minute observation. Expectedly, there was a closer conformity between the two methods for the larvae because of their slower mobility in comparison to the adults.

A further experiment was designed to test the accuracy of the 3-minute observation. Two hypotheses were tested: (1) there is no real difference between observers in the number of $S. punctum/3$-minute observation when the same tree is counted by different observers; and (2) the number of $S. punctum/$ 3-minute observation is directly proportional to the number per tree. The first hypothesis investigates the possibility that the 3-minute observation may be biased by a tendency for different observers to systematically count more or less predators. Establishing the validity of the second hypothesis is important because it is very desirable to use a population index in which, for example, twice as many predators/3-minute observation means twice as many predators/tree.

The experiment was conducted by counting predators on each of four trees consecutively by three different trained observers using the 3-minute observation. Predators on each tree were then counted using a more accurate 8-minute section count. The results were analyzed using an analysis of covariance model in which the covariate was the 8-minute count and the AOV factor was the observer. No statistically significant difference was found between observers. A straight line through the origin accurately described the relationship between the number of $S. punctum$ adults and larvae/3-minute observation and the number/8-minute count, suggesting the validity of the second hypothesis.

Sampling for
Tufted Apple Budmoth Egg Masses
in Pennsylvania

Knowing the distribution of tufted apple budmoth (TABM) egg masses in apple trees can be helpful in adjusting sprayers for control operations. Five trees were sampled for TABM egg masses on at least 18 dates throughout a season. Each tree was divided into top and bottom levels and north, east, south, and west quadrants. The number of hatched plus unhatched egg masses on leaf clusters were counted in each of the eight sections on each date. Throughout the season, 148 first brood and 296 second brood egg masses were counted. A statistically significant difference between top and bottom levels and among directional quadrants was found for first brood egg masses. In contrast, no statistically significant differences between levels or among quadrants were found for second brood egg masses. The average percentage of egg masses found in each of the eight tree sections is given in Table 9.1.

Table 9.1.   Percent of Total Tufted Apple Budmoth (TABM) Egg Masses/Trees—Arendtsville, 1974—Percentages computed from cumulative egg mass counts, averaged over five trees (W. M. Bode, unpublished).

	North	East	South	West	Total
	TABM—First Brood				
Top	27%	14%	20%	11%	72%
Bottom	8%	11%	6%	3%	28%
Total	35%	25%	26%	14%	100%
	TABM—Second Brood				
Top	18%	10%	14%	11%	53%
Bottom	11%	12%	13%	11%	47%
Total	29%	22%	27%	22%	100%

Dispersion Pattern and Foliage Injury of *T. mcdanieli*

Tanigoshi, Browne and Hoyt (1975) studied population growth and dispersion of the McDaniel spider mite, *Tetranychus mcdanieli* McGregor, on Red Delicious apple trees grown in a light and temperature controlled walk-in room. Lloyd's index of mean coding ($\overset{*}{m}$) was used to interpret aggregation patterns; however, mean colony size ($\overset{*}{C}$) proved to be a more generally useful population parameter ($\overset{*}{C} = \overset{*}{m} + 1$). The regression of $\overset{*}{C}$ on mean crowding ($\overset{*}{m}$) and the ratio of mean crowding to $\overset{*}{C}$ were used to determine and characterize the patterns of dispersion obtained under the experimental conditions of the study.

The ratio of mean crowding to $\overset{*}{C}$ was the proportion ($\overset{*}{p}$) of sampling units (i.e., individual leaves) that would have a colony size of $\overset{*}{C}$ individuals per unit, all other units being unoccupied. Characterization by the concomitant parameters $\overset{*}{C}$ and $\overset{*}{P}$ (= 100 p) provides a more sentitive and meaningful method of describing populations than does the mean density.

The data indicated that adult female populations of *T. mcdanieli* tended to be distributed as clumped colonies on apple trees. A rating system for damage by plant feeding mites was devised to evaluate the reliability of gross observational estimates as indicators of spider mite levels.

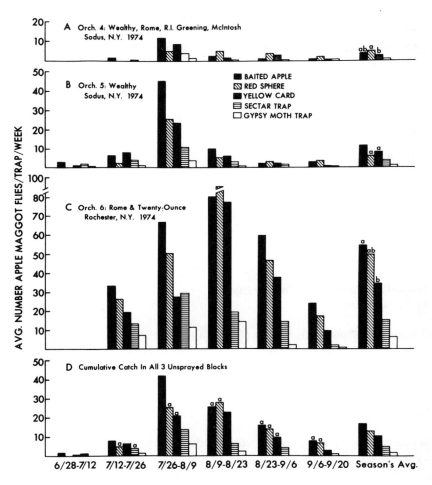

Figure 9.1. Average number of apple maggot flies/trap/week in three unsprayed apple orchards. [From Reissig, 1975.]

Trapping Apple Maggot Flies

In New York and some other northeastern states, the apple maggot may be crucial to the success or failure of integrated control systems. The ability to eliminate sprays for this pest or to time them to have only minimal deleterious effects on natural enemies depends on the efficiency of trapping apple maggot adults. Results of a test in New York in 1974 with five different types of traps in three unsprayed apple orchards revealed differences in their efficiency (Fig. 9.1).

Apple Scab and Other Apple Diseases

It is difficult to time chemical sprays to control the apple scab fungus, *Venturia inaequalis* [Cke.] Wint., because fungal development and its ability to infect apple foliage and fruit are closely tied to the combined occurrence of moisture and favorable temperatures. In Michigan, A. L. Jones and co-workers have developed the following sophisticated systems for predicting infection periods of this fungus: (1) A forecast system for primary infection (e.g., infection of new apple foliage by the overwintered fungus); (2) VISIM: An overall simulation model for the whole growing season; and (3) EVE-PLOT Plotting Package: A plotting package for producing computer drawn graphs of environmental and biological measurements.

Lewis (1974) compiled a list of interacting factors (Table 9.2) which determine the level of control of fruit diseases obtained per pound of fungicide. Careful attention to these nine factors enables the grower to assess the efficiency of various fungicides for controlling scab and other apple diseases.

Table 9.2.   Major Interacting Factors Affecting Fruit Disease Control (Lewis, 1974)

1. Plant susceptibility	6. Its concentration
2. Fungus population	7. Spray timing
3. Rainfall—especially duration	8. Spray equipment
4. Tree growth	9. Application method
5. The chemical	

ECONOMIC INJURY LEVELS

Economic thresholds and injury levels for apple pests are difficult to determine. There are great differences among apple cultivars in their attractiveness to pests and their susceptibility to pest injury. Type of planting, orchard site, soil type, and weather are among a few of the factors that confuse attempts to determine pest densities associated with given economic injury levels. Nevertheless, economic injury levels are important in assessing the need for control actions and in deciding on control strategy and tactics.

McDaniel Spider Mite

Tanigoshi and Hoyt (unpublished) hypothesized that one destructive effect of mite feeding on apple foliage is a reduction in fruit growth rate, with the reduction in fruit size at harvest depending upon the length of time that the growth rate is reduced. Preliminary analyses support this hypothesis. However, as the season progresses, growth rate per se becomes progressively less important, since a reduction in growth rate just prior to harvest could have little effect on fruit size. Conversely, early and continuing reductions in growth rate would have correspondingly cumulative effects.

Preliminary data suggested that a healthy tree has a relatively high tolerance for mite feeding from mid-July until harvest. Also, no significant effect on fruit growth was found until the density of the mites multiplied by the time they are present (mite days) was substantial. An exact threshold value for this cannot be given because of variation between trees, cultivars, and other variables.

Data analysis (i.e., ANOVA, regression) of the phenological relationship of apple growth and McDaniel spider mite feeding revealed a significant

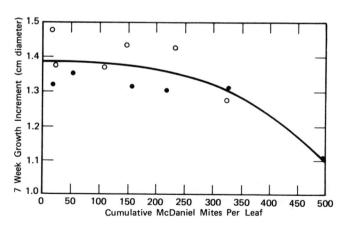

Figure 9.2. Curvilinear regression model demonstrating a McDaniel spider mite effect on fruit growth increment ($r = -0.846$, significant at 0.01 level). [From L. Tanigoshi & S. C. Hoyt, unpublished.]

mite effect, reducing fruit growth increment over a 7-week period (89th–130th day post-bloom) by 0.29 cm (Fig. 9.2). This reduction represents an economic loss of over one box size at harvest. The data indicated that the relationships for all functions were nonlinear and multivariable. By coupling the fruit growth model, Growth Rate of Increment = [(time of year) + (size)] − (*T. mcdanieli*) − [*Panonychus ulmi* (Koch)] − [*Aculus schlechtendali* (Nalepa)] to SIMBUG, a stochastic model projecting fruit size is planned. The functions describing mite effects include an interaction term for time of year. These functions may have limits imposed upon them by interactions between various phytophagous species. Moreover, each mite function has two components (e.g., the instantaneous effect of mite density and the cumulative effect of mite damage). As this simulation model is refined, horticultural and climatic variables of the ecosystem can be and will be included.

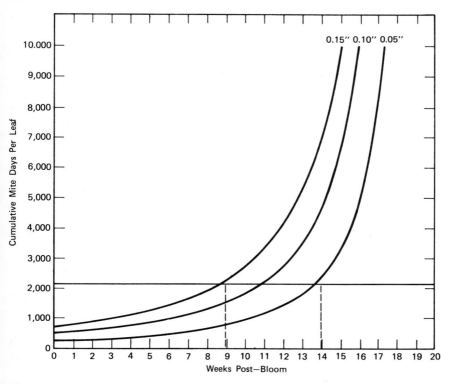

Figure 9.3. Prototype decision-making economic injury level index. [From L. Tanigoshi & S. C. Hoyt, unpublished.]

Decision-Making Economic Injury Level Index

In Fig. 9.3 the allowable limits of mite days of feeding by the McDaniel spider mite to prevent economic loss from exceeding the specific levels are depicted. The levels chosen are arbitrary, but the loss of 0.10 inch (0.25 cm) approximates a loss of one northwest apple box size. The projection curves were obtained by estimating the maximum reduction in size that would occur if the cumulative mite-days of feeding were constant from any given point in time to harvest. This is analagous to a pesticide application where the population drops suddenly to zero. Once feeding damage has been sustained, its effect is subsequently manifested as a continuing cumulative reduction of fruit size at harvest. These material flows (Fig. 9.4, mite management flow chart) directly influence the flow of information upon which "control decisions" by the orchardist will ultimately be made.

The ultimate goal of this study is thus to utilize Washington's computer simulation model as a research tool for performing designed experiments to answer specific questions about the sensitivity of various components of an apple ecosystem. It will allow simulation in minutes of relations which would possibly require years of effort to research and/or analyze the consequences, while great risk to real resources is indulged. Based upon specific assumptions, X-variables, and field data, higher resolving indices of economic injury for decision-making than the prototype shown in Figure 9.3 will be synthesized for commercial use. The utility and efficacy of these indices will be predicated upon periodic mite sampling and processing by a mite counting service, using the standard mite brushing machine (Tanigoshi & Hoyt, unpublished).

Pear Psylla

The greatest injury to pear trees and fruits by the pear psylla, *Psylla pyricola* Foerster, is indirect rather than direct. Psylla nymphs and adults suck plant juices from the tree and secrete honeydew copiously. The honeydew itself causes russeting of pear fruits, but it is also an excellent growing medium for unsightly sooty mold fungus.

Investigation of the economic injury level of pear psylla on fresh market d'Anjou pears indicated the level to be between 34 and 100 nymphs/sample. A sample consisted of 10 spurs for the first generation and five spurs plus five terminals for succeeding generations (ca. 50 leaves/sample). Psylla nymph levels greater than 200/sample for only a few weeks resulted in 65 percent cullage. A sugar analysis technique was employed to determine foliar honeydew concentrations in pear orchards. Foliar honeydew concentrations appeared to be a better predictor of psylla-caused russeting than psylla

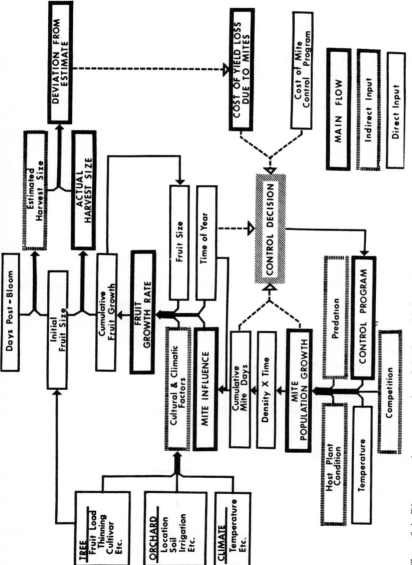

Figure 9.4. Diagrammatic representation of the flow of information through WSU's modeling effort of the pome and stone fruit subproject. [From L. Tanigoshi & S. C. Hoyt, unpublished.]

nymph counts. The removal of psylla honeydew from pear foliage and fruit with washes utilizing ground spraying equipment, effectively reduced psylla russet. Washes with a wetting agent added were superior to washes of water only (Brunner, 1975).

BIOLOGICAL CONTROL STUDIES

One of the purposes of the project has been to learn how to take maximum advantage of biological control and other alternatives to pesticides, and thus reduce the amount of pesticides applied to our crops. Biological control methods are frequently complicated because of the numerous factors involved in the interaction between a natural enemy, its prey, the weather, and various other crop production features. Research presented in this section reveals various features regarding the interaction of a natural enemy and its prey which are useful in management of the prey (pest).

Woolly Apple Aphid

The woolly apple aphid, *Eriosoma lanigerum* (Hausmann), developed large populations in many New York orchards in 1974. Collections late in the season showed it to be heavily parasitized by *Aphelinus mali* (Haldeman). In 1975, the aphid and parasite populations were monitored and the data indicated the woolly apple aphid had been held in check by the parasite.

Discrete Time Model of a
Pear Psylla-Predator Interaction

The development of psylla and immature stages of a predatory anthocorid, *Anthocoris nemorallis* (F.) and the fecundity and longevity of winter-form pear psylla were investigated at constant temperature, with the objective to develop a model of the predator-prey interaction. Pear psylla developed from the egg to adult in 104, 39, and 25 days at 50°F, 60°F, 72°F, and 80°F, respectively. The fecundity of winter-form psylla was greatest at 70°F, averaging 271 eggs/female. Following eclosion, nymphs required 26 and 15 days to molt to the adult stage at 60°F and 80°F, respectively. The form of the functional response curves for three pear psylla predators was determined.

A simple psylla-predator model was developed from information presented in 1976 (Vol. 2, Pome and Stone Fruits Subproject report to NSF and EPA, and published material). Simulation experiments suggested that a single dormant spray application would be the poorest program if regulation of pear psylla by natural enemies is to be encouraged. Suggestions were made for future research on specific components required for refinement of the model.

European Red Mite vs Predatory Mites in New York

Biological control of the European red mite by natural enemies may become important to the integrated control system in New York. *Amblyseius fallacis* Garman and *Typhlodromus pyri* Scheuten are the two predatory mite species commonly found in commercial apple orchards. The latter species over-winters under the bark of the trees as well as in the groundcover. It is present when the first European red mite eggs hatch. Although the life cycle of *T. pyri* is somewhat longer than that of *A. fallacis*, it has been observed to successfully check European red mite populations, and is a principal regulating factor for red mites on apples in parts of Europe (Van de Vrie et al., 1972). These two predatory species are commonly found in New York where six to eight applications/season of organophosphorus and other orchard pesticides are used. Results from laboratory bioassays indicate that both *A. fallacis* and *T. pyri* are tolerant to azinphosmethyl. *A. fallacis* is extremely susceptible to carbaryl and phosalone whereas *T. pyri* is highly tolerant to both materials. *A. fallacis* is also extremely susceptible to demeton and dimethoate, but highly tolerant to Imidan® and moderately so to parathion, whereas *T. pyri* tolerance to Imidan is low (Watve & Lienk, 1976).

Granulosis Viruses of
Codling Moth and Redbanded Leafroller

In Pennsylvania an evaluation was made of the ability of granulosis viruses of codling moth, *L. pomonella*, and redbanded leafroller, *Argyrotaenia velutinana* (Walker), to suppress populations of those pests on apple (W. M. Bode, unpublished). In 1973, the two viruses were applied at the rate of 50 larval equivalents (LE) per tree, and compared to a standard chemical treatment (azinphosmethyl plus lead arsenate). A molasses-based adjuvant and powdered charcoal were used with the viruses. Treatments were applied to single-tree plots with a back-pack mist-blower. Redbanded leafroller virus was applied 11 times between May 16 and September 4. Codling moth virus was applied eight times beginning June 6. At harvest codling moth entries were found in 12.25 percent of the apples from check trees, contrasted to only 5.25 percent from trees receiving the codling moth virus (a 57 percent reduction). On the other hand, less than 1 percent of the apples on trees receiving the standard chemical treatment had codling moth entries. The redbanded leafroller populations were too low to permit an evaluation of the treatments.

Nuclear Polyhedrosis Virus of Tufted Apple Budmoth

A nuclear polyhedrosis virus (NPV) was isolated from diseased larvae found in an orchard near Idaville, Pennsylvania, in 1974. It is transmitted by ingestion of contaminated food, and has been propagated in larvae reared on artificial diet in the laboratory. Polyhedral inclusion bodies (PIBs) for study and bioassay were purified by sucrose density gradient centrifugation. The average diameter of PIBs was 0.94 μm (range 0.32 to 2.08 μm). Electron micrographs of thin-sectioned PIBs revealed typical structure, although the high frequency with which nucleocapsids occurred in pairs seemed unusual.

Bioassays were performed in the laboratory to determine the insecticidal activity of the virus. Third stage TABM larvae were fed small pellets of food inoculated with a range of dosages of purified PIBs. The slope of the log dosage versus probit mortality regression line was 1.176; the LD_{50} was about 31 PIBs per third-stage larva; and the LD_{95} was about 340 PIBs. When third-stage larvae were fed a dosage of 1×10^3 PIBs, the median lethal time (LT^{50}) was about 16.5 days at 80°F. This virus disease develops rather slowly and does not eliminate larvae rapidly, but most larvae which become infected do eventually die. For field use this virus should be applied to populations consisting primarily of young larvae which would become infected with smaller dosages and die in shorter times.

In 1975 the NPV was sprayed onto apple trees at the rate of approximately 2.3×10^{10} PIBs per tree. At weekly intervals five applications were made for first-brood TABM larvae (June 13 to July 16) and another five for second-brood larvae (August 4 to September 6). The population of first-brood larvae on virus-treated trees was reduced 97 percent from that on check trees, and the second-brood population was reduced 93 percent. At harvest 16 percent of the fruits from check trees were injured by TABM larvae, while only 3 percent of the apples from virus treated trees were damaged.

In 1976 a bioassay of residual virus activity was performed. Tufted apple budmoth nuclear polyhedrosis virus was applied to apple trees with a tractor-mounted mist blower at the rate of about 3.8×10^{10} PIBs per tree. Leaf samples were collected from sprayed trees at 24-hour intervals and in the laboratory 50 neonate TABM larvae were put onto leaves from each exposure period. These test larvae were examined daily and mortality was recorded. Results demonstrated that the initial virus deposit on leaves produced lethal

infections for nearly all test larvae. Virus activity declined fairly rapidly out-
doors, although about 50 percent of the original activity remained about 4
days after exposure.

Distribution of *Stethoris punctum* in Relation to Densities of European Red Mite

Distributional patterns of *S. punctum* in relation to densities of the European
red mite are an important factor in this predator's ability to control the pest.
These factors were investigated within an apple orchard and within an apple
tree in Pennsylvania. *S. punctum* adults were able to respond to increasing
mite populations on specific trees in the orchard. A prey population averaging
only two to three motile mites/leaf is sufficient to enable the predator population to reproduce and increase in size. Average daily temperatures above
65°F were found to influence favorably the ability of *S. punctum* to find its
prey. Within-tree distribution showed that *S. punctum* adults search out areas

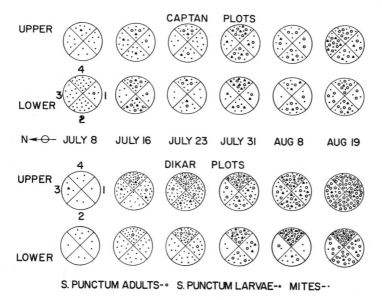

S. PUNCTUM ADULTS-∘ S. PUNCTUM LARVAE-• MITES-·

Figure 9.5. Distributional patterns of *Stethorus punctum* adults and larvae with respect to its
prey, *Panonychus ulmi*, within apple trees. Each circle represents the number of *S. punctum*
adults and larvae (per 1-min. observation) and *P. ulmi* (per leaf) on each section of a Stayman
apple tree. [From Hull et al., 1976]

within the tree where mite populations are highest. *S. punctum* larvae were also found feeding in higher numbers in these areas. Mite populations increased more rapidly within the tree row where the leaves were not as heavily covered with pesticide material. As the season progressed, however, the distribution of this red mite population became fairly uniform around the tree, partially through the concentration of *S. punctum* and its effects in the areas of the tree containing the higher mite populations (Fig. 9.5) (Hull et al., 1976).

Aphid Predator Survey in Apple Orchards

In Pennsylvania, there are numerous predators of the species of aphids that attack apple. To date a species of predator with tolerance for the common insecticides used in apple orchards has not been found. In the hope of finding

Order and Family	Genus and Species	Date Collected
Diptera	*Toxomerus marginatus* (Say)	July
Syrphidae	*Toxomerus politus* (Say)	July
	Toxomerus germinatus (Say)	May–July
	Allograpta obliqua (Say)	June–July
	Metasyrphus americanus (Wied.)	May–August
	Metasyrphus latifasciatus (Macq.)	July
	Syrphus knabi Shannon	August
	Syrphus rectus Osten Sacken	June–July
	Syrphus torvus Osten Sacken	March–April
	Sphaerophoria contigua Macq.	June–July
	Sphaerophoria philanthus (Meigen)	May–July
	Sphaerophoria sp.	July–August
	Platycheirus immarginatus (Zett.)	June
	Platycheirus sp.	May
	Melanostoma sp.	July–August
	Carposcalis sp.	May
	Eristalis transversa Wiedmann	April–May
	Eristalis arbustorum (L.)	April–May
	Helophilus fasciatus Walker	May
Neuroptera	*Chrysopa nigricornis* Burm.	July
Chrysopidae	*Chrysopa quadripunctata* Burm.	June–July
	Chrysopa oculata Say	June–August
	Chrysopa carnea Stephens	June–August
	Chrysopa rufilabris Burm.	July–September
Hemerobiidae	*Hemerobius stigmaterus* Fitch	August
	Hemerobius humulinus Linn.	June–August
	Micromus posticus (Walker)	August–Novembe

a predator(s) that offered some promise of biological control of aphids in commercial apple orchards, James W. Travis made collections of predators in the spring, summer, and fall of 1975 and 1976 (data unpublished). Several species of Syrphidae, Chrysopidae, and Hemerobiidae were taken. The species of Syrphidae were determined by F. C. Thompson of the Systematic Entomology Laboratory, USDA; D. J. Shetlar of Pennsylvania State University identified the Chrysopidae and the Hemerobiidae.

Tufted Apple Budmoth Parasite Survey

In Adams County, Pennsylvania, a survey was conducted by Robert W. Koethe to determine the species of parasites attacking the tufted apple budmoth, *P. idaeusalis*, the most serious lepidopterous pest of apples in this area. Collections of first and second brood larvae were made in selected orchards from June 1975 through October 1975. Species of Tachinidae were identified by C. W. Sabrosky, of Braconidae by Paul M. Marsh, and Ichneumonidae by R. W. Carlson, all three specialists with the Systematic Entomology Laboratory, USDA. The survey is being continued and made more extensive.

Order and Family	Genus and Species	Date Collected
Diptera		
Tachinidae	*Actia interrupta* (Curran)	July–August
Hymenoptera		
Braconidae	*Agathis* sp.	July–August
	Meteorus sp.	July
	Apanteles conanchetorum Vier.	October
Hymenoptera		
Ichneumonidae	*Triclistus* sp.	August
	Itoplectis conquisitor (Say)	August
	Phytodietus sp.	October
	Enytus sp.	October
	Hercus pleuralis (Prov.)	October

THE NEED FOR CHEMICALS AND THEIR SELECTIVE USE —RESISTANCE OF PREDATORS TO PESTICIDES*

Pesticides remain the primary tool in the management of pest populations in orchard ecosystems because alternative methods that can hold most pest populations below economic injury levels have not been found or developed.

*Material presented in this section was essentially taken from an article by Croft (1977).

Evidence that resistance to pesticides has developed among arthropod predators and parasitoids is meager indeed compared to that for pest species. Brown (1976) estimated that at least 268 species of pests have developed resistant strains, yet less than a dozen resistant natural enemies have been reported (Croft & Brown, 1975). There are probably several reasons for this disproportionality, including lack of study, the less dramatic effects of such a development, and probably an inherent lower frequency of appearance of resistance among natural enemies as compared to pests (Croft & Brown, 1975). It is the purpose of this discussion to review the cases of developed resistance among phytoseiid mite predators. What little is known about cross-resistance, the toxicology and mechanisms of resistance, the genetics of resistance, and the general principles governing selection of resistant strains of these mites will be discussed. In some cases the impact of these developments on practical pest control programs is noted.

Resistant Phytoseiid Mites

Seven phytoseiid species have been reported resistant to one or more insecticides in widely scattered areas of the world. These species include *Amblyseius chilenensis*, *Amblyseius fallacis*, *Amblyseius hibisci* (Chant), *Phytoseiulus persimilis* (Athias-Henriot), *Typhlodromus caudiglans* Schuster, *Typhlodromus occidentalis* (Nesbitt), and *T. pyri* (Table 9.3).

Resistance to azinphosmethyl and phosmet has developed in *A. chilenensis* populations from Uruguayan apple orchards. As yet the process of selection is still in progress since normal field rates applied for codling moth or leafroller (*Argyrotaenia sphaleropa* Meyrick) control give kills of predators, but do not cause so much mortality that a practical program of integrated control of the European red mite is not possible (Dover et al., 1979).

Among *A. fallacis* populations, strains resistant to DDT, carbaryl, and several organophosphorus-related compounds (e.g., azinphosmethyl, parathion, phosmet, diazinon, stirofos) have been found (Table 9.3). Cross-resistance to these OP compounds extends to at least 11 other insecticides commonly applied to control deciduous fruit pests (Croft & Nelson, 1972; Croft, Brown & Hoying, 1976). Only in localized populations is resistance to DDT and carbaryl known; however, to OP compounds high levels have been developed and resistance is widespread throughout the deciduous fruit growing regions of mid-western and eastern North America (Croft & Brown, 1975). Some success in developing an OP × carbaryl bi-resistant strain by hybridization and selection was reported by Croft and Meyer (1973). However, practical attempts to establish, maintain and manage these resistant features in field populations have met with only limited success (Meyer,

Table 9.3.   Cases and Levels of Resistance to Insecticides Among 7 Phytoseiid Mite Species (Croft, 1977)

Species	Insecticide	Resistance Fold Level	Method Used[a]	Reference
Amblyseius chilenensis	phosmet	10	SD	Croft et al., 1977
	azinphosmethyl		SD, R	Croft et al, 1977
Amblyseius fallacis	azinphosmethyl	100	SD	Motoyama et al., 1970
		83	SD	Rock & Yeargan, 1971
		117	SD	Ahlstrom & Rock, 1973
		944	R	Croft & Meyer, 1973
		74	SD	Croft et al., 1976
	carbaryl	25, 77	R, SD	Croft & Meyer, 1973
		8	R	Croft & Hoying, 1975
	DDT		R	Smith et al., 1963
	diazinon	119	SD	Croft et al., 1976
	Gardona	25	SD	Rock & Yeargan, 1971
		28	R	Croft & Meyer, 1973
	parathion	103	R	Motoyama et al., 1970
		59	SD	Rock & Yeargan, 1976
		152	SD	Croft et al., 1976
	phosmet	45	SD	Rock & Yeargan, 1971
Amblyseius hibisci	parathion		R	Kennett, 1970
Phytoseiulus persimilis	parathion		NA	Helle & Van de Vrie, 1974
	demeton		NA	Helle & Van de Vrie, 1974
Typhlodromus caudiglans	DDT		F	Herne & Putman, 1966
Typhlodromus occidentalis	azinphosmethyl	101	SS	Croft & Jeppson, 1970
		104	SD	Ahlstrom & Rock, 1973
Typhlodromus pyri	azinphosmethyl	20	SD	Watve & Lienk, 1976
		10	SD	Hoyt, 1972
		14	SD	Collyer & Geldermalsen, 1975

[a]SD = slide-dip; R = residue application; SS = slide spray; F = field application; NA = no information available.

1975; Croft & Hoying, 1975). Study of resistance in *A. fallacis* populations has indicated that while OP resistance is easily developed and is stable in this species, carbaryl resistance is only developed under conditions of intensive selection and is much more unstable. Croft and Hoying (1975) concluded that if widespread resistance to organophosphates among fruit pests were to develop and chemical control programs were to become dependent on intensive applications of carbaryl, then resistance would develop among *A. fallacis* to a sufficient level to allow for integrated mite control.

For the species *A. hibisci* occurring on citrus in California, *P. persimilis* in European greenhouses, and *T. caudiglans* in Canadian peach orchards, only limited and isolated cases of resistance have been shown (Table 9.3). Other than the use of *Phytoseiulus* in controlling spider mite pests in greenhouses, no widespread exploitation or practical utilization of these resistant predatory mite populations has been developed.

Of the phytoseiids which have been studied, probably no species is more inherently tolerant or resistance-adapted than is *T. occidentalis*. As early as 1952, its tolerance to parathion was noted (Huffaker & Kennett, 1953). Since then, populations highly resistant or tolerant to several organophosphorus-related compounds (e.g., azinphosmethyl, phosmet, parathion, TEPP, phosalone) have been reported in the semi-arid fruit-growing regions of western North America. Based on these developments, practical programs of integrated mite control over widespread acreages have been developed on such major crops as apples, peaches, plums, and grapes. It is likely that resistance and cross-resistance to other chemical groups (e.g., carbaryl) occur in populations of this species, but as yet they have not been documented.

The resistance patterns in populations of *T. pyri* are considerably different from those observed in *A. fallacis* and *T. occidentalis*. Even though OP compounds have been widely applied where this species occurs in the fruit-growing regions of northwestern and northeastern North America, western and eastern Europe, and New Zealand, high levels of resistance to this class of insecticides have not been developed. In New Zealand, Hoyt (1972) first reported 10-fold resistance to azinphosmethyl, but he noted that this was too low for practical use in integrated mite control programs. Collyer and Geldermalsen (1975) thereafter found a population in New Zealand to have a 14-fold resistance to this same compound. More recently, Watve and Lienk (1976) reported an LC_{50} level which would give a nearly 20-fold resistance to azinphosmethyl when compared with Hoyt's (1972) data for a susceptible strain; at this level predators were able to better tolerate field applications and provide appreciable biological control of *P. ulmi* in New York apple orchards. Although the potential for resistance to OP compounds appears to be limited in this species, several studies have indicated that *T. pyri* may either be tolerant to carbaryl or it has developed high levels of resistance to this compound (van de Vrie, 1962; Watve & Lienk, 1976; Collyer & Geldermalsen, 1975).

Cross-Resistance

In Table 9.4 the cross-resistance and toxicity relationships between seven broad-spectrum insecticides and the four phytoseiid mites, *A. chilenensis, A.*

Table 9.4.   Relative Toxicity of Field Application Rates and Cross-Resistance Relationships of 7 Broad Spectrum Insecticides Among 4 Azinphosmethyl-Resistant Phytoseiid Mite Species (Croft, 1977)

	Species			
Chemical Compound	Amblyseius chilenensis	Amblyseius fallacis	Typhlodromus occidentalis	Typhlodromus pyri
Azinphosmethyl	M[a]	L,	L,	M,
Phosmet	M, (+)[b]	L, (+)	L, (+)	M, (+ ?)
Parathion	H, (−)	M, (+)	L, (−)	M, (+ ?)
Stirophos	L, (+)	M, (+)	M, (+)	NI[c]
Phosalone	NI	H, (−)	L, (+ ?)	L, NI
TEPP	NI	H, (−)	L, (+ ?)	NI
Carbaryl	L, (−)	H, (−)	H, (−)	L, (− ?)

[a]L = lowly; M = moderately; H = highly toxic.
[b](−) = no evidence of cross-resistance to azinphosmethyl. (+) = positive evidence; ? = inferred from indirect data.
[c]NI = no information available.

fallacis, *T. occidentalis*, and *T. pyri* are summarized. As noted previously, azinphosmethyl resistance has reached a level sufficient to render field rates virtually innocuous to both *A. fallacis* and *T. occidentalis* but not to *A. chilenensis* or *T. pyri*. Phosmet (Imidan®) has not shown a strong cross-resistance relationship with azinphosmethyl and similar toxicity ratings also exist between these two compounds for each of the four species. The response of the four mite species to parathion is extremely variable; whereas it is very toxic to *A. chilenensis* and is not cross-related to azinphosmethyl; for *A. fallacis* it is only moderately toxic and closely linked with azinphosmethyl (Motoyama et al., 1970, Rock & Yeargan, 1971; Croft, Brown & Hoying, 1976). For *T. pyri*, parathion is moderately toxic and any required tolerance may also extend to azinphosmethyl (Watve & Lienk, 1975, 1976). For *T. occidentalis*, parathion is virtually innocuous, but this seems to be due to inherent tolerance rather than to the acquisition of any resistance or cross-resistance. This was first indicated by relative immunity shown by *T. occidentalis* populations almost immediately after parathion was registered for use (Huffaker & Kennett, 1953). Further evidence was provided by Croft and Jeppson (1970), who reported similar LC_{50} levels to parathion among two intensively selected strains (Washington and Utah) versus two strains originating from untreated habitats (California). More recently its inherent tolerance to parathion has been indicated by an indigenous Australian population which initially tolerated field applications of parathion but was highly susceptible to azinphosmethyl (Field, 1974).

Stirofos resistance is cross-linked with azinphosmethyl in each of the four species, but its relative innocuousness to *A. chilenensis* is in contrast to *A. fallacis* and *T. occidentalis* (Table 9.4). Phosalone and TEPP show high toxicity and no cross-resistance with azinphosmethyl and possibly just the opposite relationship in *T. occidentalis* populations. Lastly, carbaryl is innocuous to *A. chilenensis* and *T. pyri* and is extremely toxic to *A. fallacis* and *T. occidentalis*; there is no cross-resistance between this carbamate insecticide and azinphosmethyl.

In summary, the data in Table 9.4 would suggest that the four phytoseiid species, *A. chilenensis*, *A. fallacis*, *T. occidentalis*, and *T. pyri*, present response differences with respect to insecticide toxicity and cross-resistance relationships which are extremely varied. One would conclude from this small sampling that the interspecific differences within the family are even in excess of those of their spider mite prey of the Tetranychidae which are considered to be highly variable (compare data for the four species, *Tetranychus urticae*, *T. pacificus*, *Panonychus ulmi*, and *P. citri* in Helle, 1965; Helle & van de Vrie, 1974).

Toxicology and Mechanism of Resistance

The toxicology and mechanisms of resistance among a natural enemy species has been investigated in only two populations of *A. fallacis* resistant to azinphosmethyl. Motoyama et al. (1971) studied the possible resistance mechanism in vivo and reported that an OP resistant strain from North Carolina degraded azinphosmethyl faster than a susceptible strain and that it suffered less inhibition of its cholinesterase activity. No differences were detected between S and R strains in bi-molecular rate constants and they indicated that the resistance mechanism was attributable not to a modified cholinesterase but rather to a higher nonspecific esterase activity. In the resistant strain, they reported a sixfold increase in desalkylation of azinphosmethyl to the monodesmethyl derivative (methylphosphorothiolothionate). This also was accompanied by a threefold increase in A-esterase activity.

In light of this activity Croft, Brown, and Hoying (1976) evaluated the relationship between resistance levels and the molecular structure of 16 organophosphate related compounds in a Michigan strain of *A. fallacis* which had been exposed mainly to azinphosmethyl and to limited applications of diazinon. They especially paid attention to differential responses to methyl and ethyl esters in the cross-resistance spectrum. Their results were somewhat inconclusive (Table 9.5); significantly high resistance ratios to the

Table 9.5.   Resistance Spectrum of the Belding Strain of A. fallacis to OP Compounds (Croft, Brown, & Hoying, 1976)

	Methyl		Ethyl	
Phosphorodithioates $\diagup P \overset{\diagup S}{\underset{}{\diagdown}} S -$				
Azinphosmethyl	74	Azinphosethyl	35
Malathion	33	Carbophenothion	8
Phosmet	21	Phosalone	9
Dimethoate[a]	135	Ethion (Str.)	5
Phosphorothioates $\diagup P \overset{\diagup S}{\underset{}{\diagdown}} O -$				
Methyl parathion	28	Parathion	152
Fenitrothion	17	Diazinon	119
Fenthion	21	Demeton (Str.)	20
Phosphates[a] $\diagup P \overset{\diagup O}{\underset{}{\diagdown}} O -$				
Phosphamidon	29	TEPP	4

[a]Straight-chain compound.

methyl phosphorodithioates, dimethoate, malathion, and phosmet were observed, while significantly low resistance ratios to ethyl phosphorodithioates, carbophenothion, phosalone, and ethion were found. However, cross-resistance to azinphosmethyl was relatively high. With respect to the ethyl versus ethyl phosphorodithioates, resistance levels were appreciable to either group, although the ethyl derivatives, parathion and diazinon, gave particularly high resistance ratios. Only two straight-chain phosphates, phosphamidon (methyl) and TEPP (ethyl), were evaluated; cross-resistance was moderate to the former insecticide and extremely low for the latter.

Genetics of Resistance

In this program, the genetics of resistance had been evaluated only for
single Michigan strain of *A. fallacis* (Croft, Brown, & Hoying, 1976)
azinphosmethyl. Using standard crossing and backcrossing methods wit
an R and an S strain, it was demonstrated that reciprocal cross offsprir
showed almost identical dosage-mortality lines and a partial dominance wa
indicated (Fig. 9.6). When F_1 females were backcrossed with S males, of
spring showed a d-m line with long inflections at the 40 to 48 percent lev
(Fig. 9.7). These d-m lines were reasonably congruent with theoretical lin
calculated for a 1:1 proportion of hybrid (R_+) and S($++$) genotype an
thus reflected a clear segregation and that the resistance factor was principal
due to a single gene allele.

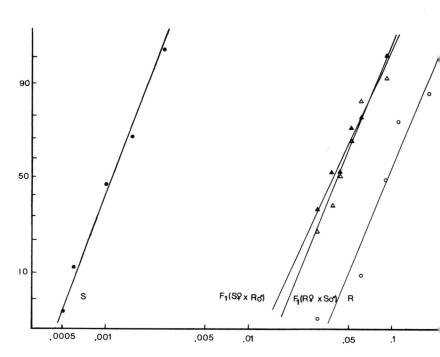

Figure 9.6. Dosage-mortality lines for females of the Rose Lake S strain and the Belding
strain and for the female hybrid offspring of the S♀ × R♂ (solid triangles) and R♀ × S♂ (ope
triangles) crosses. [From Croft, Brown & Hoying, 1976.]

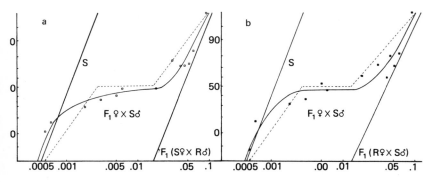

Figure 9.7. Dosage-mortality lines for the female offspring of the back-crosses with S-strain males of (a) the R♀ × S♂ hybrid females, and (b) the S♀ and R♂ hybrid females, as compared to the lines expected for a 50–50 ratio (dotted); d–m lines for the S and either type of F_1 females taken from Figure 9.6. [From Croft, Brown & Hoying, 1976.]

Tolerance of Pesticides by the Coccinellid Predator *Stethorus punctum*

In Pennsylvania, Asquith et al. (Asquith & Hull, 1973; Asquith, Hull & Travis, 1976; Asquith et al., 1976a,b) tested numerous insecticides in the field for tolerance by *S. punctum* to satisfy the need for using compatible materials in the Pennsylvania integrated pest management program for apples. The list of pesticides found to be tolerated by *S. punctum* within certain dosage limits is presented in Table 9.6.

The Pennsylvania Integrated Pest Management System Without Dinocap

The original Pennsylvania system of integrating biological control of the European red mite, *P. ulmi* (Koch), by the ladybird beetle, *S. punctum*, with chemical control of other pests relied on the use of dinocap to suppress mite populations until *S. punctum* numbers became large enough to subdue the mite infestations. With the possible introduction of new fungicides for mildew that do not suppress mite populations, new methods of manipulating mite populations had to be found that would take advantage of the natural control potential of *S. punctum*. In one experiment, five insecticide treatments were applied to apples in a season-long program with replicated tree plots of which two replicates received the fungicides, captan plus sulfur, and two replicates received Dikar®. The mite (Fig. 9.8) populations in all captan

Table 9.6.   Pesticides[a] Compatible with the Pennsylvania Integrated Pest Management Program for Apples (D. Asquith, unpublished)

Fungicides	Insecticides	Acaricides
Benlate	Guthion	Carzol
Captan	Imidan	Plictran
Cyprex	Zolone	Vendex
Polyram	Penncap M	Omite
Dikar	Cygon	
Karathane	Systox	
Sulfur	Lannate	
	Thiodan	
	Superior Oil	

Note: EC formulations less satisfactory than wettable powders.
[a]It is necessary with many pesticides to use specifically selected dosages.

plots began to increase earlier than in the Dikar® plots, and this encouraged an earlier increase in the *S. punctum* populations in the captan plots. With this movement of *S. punctum* into the captan plots and one timely application of 0.094 lb Plictran® AI/acre, the mite populations remained at low level throughout the season. The mite populations in the Dikar® plots increased more slowly due to the suppressing effect of dinocap, but eventually attained even higher numbers. *S. punctum* populations subsequently moved in to those plots too, supplemented again by one timely application of 0.125 lb Plictran® AI/acre, reduced the mite populations to low levels where they were eventually eliminated by *S. punctum* predation.

Now, with proper timing of low dosages of acaricides, the Pennsylvania Pest Management System for apples can be operated successfully without the use of dinocap with its mite suppressing effect.

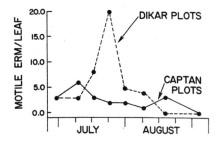

Figure 9.8. The Pennsylvania system without dinocap. [From D. Asquith, L. A. Hull & P. D. Mowery, unpublished.]

European Red Mite—
Replacing Dinocap with an Acaricide in
the Pennsylvania IPM Program for Apples

An experiment was conducted to investigate new principles of European red mite control in Pennsylvania apple orchards. Current control practices utilize oil to prevent hatching of the overwintered eggs, dinocap (a mite suppressant and the active ingredient in Karathane®), and Dikar® applied throughout the season, and the mite predator *Stethorus punctum*. This experiment was designed to develop a mite control program that does not rely upon either oil or dinocap.

The experiment was conducted in a 2-acre mixed variety block of apples in Arendtsville, Pennsylvania, during 1974. No oil or dinocap was used. The *S. punctum* population increased throughout the season and exerted continuous pressure on the mite population. During July and early August the *Amblyseius fallacis* [a predatory mite (above)] population achieved high numbers and contributed to control of the European red mite. Plictran® WP 50 was applied three times for a total of 8 oz./acre for the season. The European red mite population was kept below an average of 3.6 motile stages/leaf throughout the season (Fig. 9.9). The experiment demonstrated that

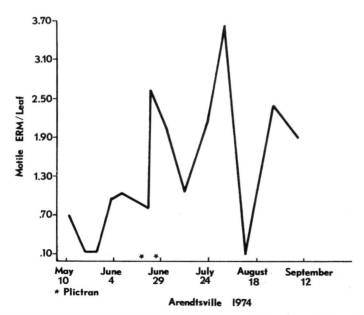

Figure 9.9. The seasonal history of the European red mite population. [From D. Asquith, L. A. Hull & P. D. Mowery, unpublished.]

integrated biological and chemical control of the European red mite in Penn-sylvania apple orchards can be successful without the use of oil and dinocap. This is an important conclusion because dinocap, now widely used for powdery mildew control, may soon be replaced by a material(s) which has little or no effect on the European red mite. The experiment also demonstrated the utility of using low dosages of an acaricide in manipulating the European red mite population in order to make more effective use of mite predators.

BASIC BIOLOGICAL AND ECOLOGICAL STUDIES

To aid in making decisions regarding control strategies and tactics, it is essential to understand the basic biology and ecology of both the pests and natural enemies present in a crop ecosystem. One cannot make intelligent decisions regarding the timing of sprays, elimination of sprays, and attempts to conserve important natural enemies without an understanding of the details of the biology and ecology of the organisms involved. A few examples of basic studies conducted by the Pome and Stone Fruits Group follow.

Field Studies with
a Sex Pheromone of the San Jose Scale

The presence of sex pheromones in diaspid scales has been demonstrated in the California red scale, *Aonidiella aurantii* (Maskell) and yellow scale, *A citrina* (Coquillett) (Tashiro & Chambers, 1967; Moreno et al., 1972). The related San Jose scale, *Quadraspidiotus perniciosus* (Comstock), does not reproduce in the absence of males (Gentile & Summers, 1958) and this suggests the presence of a sex pheromone in this species also.

Field studies in California have shown the occurrence of a pheromone produced by virgin females of San Jose scale which attracts the males. Several host fruits were tested in the laboratory for scale acceptance and longevity. These included plum, nectarine, quince, pear, apple, pomegranate, potato, and a wild gourd. Of these prospective hosts, only the plum cultivar Empress and the gourd *Cucurbita foetidissima* H.B.K., proved adequate for extended field use. Although the gourd is not a natural host, scale crawlers settle readily on gourds and mature within 1 to 2 days of those maturing on plums.

Four replicates each of traps containing 0, 100, 300, and 500 females 22 days old on plums were placed in scale-infested nectarine trees on July 23, 1973. By August 20, the mean numbers of male San Jose scales collected per treatment were 19.0, 126.8, 231.5, and 228.5, respectively. Data from treatments with females were statistically different from the traps with no females, indicating that the female scales were producing a male attractant. Although there was also a trend toward greater numbers of males being attracted to increasing numbers of females, this trend was not statistically supported in this test.

A second test using blank traps, 60 females/trap on plums and 60 females/trap on gourds result in collections of 12, 429, and 278 males, respectively, over a 3-week period. In a similar test using blank traps and 300 females each on plums and gourds/trap, the males collected were 1, 104, and 129, respectively. In addition to confirming the presence of a sex pheromone in virgin female scales, these latter tests showed that females on gourds, as well as those on plums, could produce the pheromone, and that it was attractive when released from gourds.

Two tests using plums as host fruits were conducted to determine the optimum female age for maximum pheromone production. The results of these tests (Fig. 9.10) showed that San Jose scale females began to attract males when 22 to 23 days old and reached maximum attractiveness between 28 and 31 days. With few exceptions, females stopped pheromone production on plums at 50 to 55 days of age.

The presence of a sex pheromone in virgin San Jose scale females suggests that, following further development, a synthetic attractant for this worldwide pest, which occurs on a wide spectrum of Rosaceous fruit trees, could be used

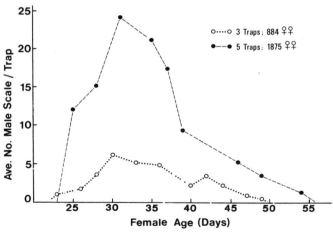

Figure 9.10. Collections of male San Jose scale on sticky traps containing virgin females of increasing age. [From Rice, 1974.]

in pest management programs similar to the use of pheromones of many other crop pests. Even without a synthetic pheromone mimic, Shaw et al. (1971) have shown the feasibility of using California red scale females to detect infestations in a scale eradication district in southern California (Rice, 1974).

The Role of *Amblyseius fallacis* in Michigan Apple Orchards

For the past 5 years, detailed studies have been conducted to assess the basic interactions of within-tree and ground cover populations of the principal apple pest species, *P. ulmi*, *T. urticae*, and *A. schlechtendali*, and the most important predator of these prey, *A. fallacis*. In summary data from the interaction studies between plant-feeding mites and their natural enemies in both the ground cover and tree plots have provided an understanding of how biological control of these pests is accomplished. Also this information has led to the development of certain manipulative strategies based on the timing and concentration of selective pesticides or habitat modifications in the orchard such as ground cover management which can greatly enhance biological control success and minimize a reliance on chemical control measures (Croft & McGroarty, 1977). A discussion of various factors affecting *A. fallacis* populations follows.

The Significance of *Amblyseius fallacis* Populations in the Ground Cover

The population of *A. fallacis* (Garman) was studied in the ground cover of nine commercial apple orchards in Michigan from 1972 to 1974. Densities were observed to fluctuate as *A. fallacis* moved into the tree during July and then back to the ground after reducing the European red mite population in the tree. *A. fallacis* was found most abundant in the ground cover directly under the tree canopy, where it was observed feeding on the twospotted spider mite, searching for other food sources, or preparing to enter diapause. In most cases, *A. fallacis* was found to have a slightly aggregated distribution on the orchard floor.

The study showed that the ground cover is important in the biological control of the European red mite because it is the principal habitat of *A. fallacis*. In all orchards where successful biological control of the European red mite was observed, high numbers of *A. fallacis* were found in the ground cover preceding their movement into the trees and the successful interaction with the pest mite in the trees. In orchards where the ground cover did not support a population of *A. fallacis*, biological control of the European red mite was not observed and chemical control was required (McGroarty & Croft, 1976).

Dispersal Behavior of Amblyseius fallacis in the Laboratory

A specific behavior described in this publication was suspected to be involved in the dispersal of the phytoseiid mite, *A. fallacis*, in nature. The behavior consisted of a series of events which were stimulated in a receptive mite by an air speed exceeding 1 mile per hour: (1) initially receptive mites altered their behavior from a random search movement to a directional movement toward the edge of an arena where (2) they terminated all forward motion, (3) began to orientate to the air flow, and (4) eventually assumed an anteriorly raised stance downwind from which they (5) frequently dispersed via air currents. It was found that the preovipositing adult females, ovipositing adult females, and to a lesser extent the adult males were the principal life stages that exhibited this behavior and actively dispersed. Starvation and to a less extent temperature affected the dispersal behavior of the ovipositing adult females and to a limited extent of the adult males and the preovipositing adult females (Johnson & Croft, 1976).

Dispersal Behavior of Amblyseius fallacis in the Field

In the course of developing the population model for *A. fallacis*, certain critical types of biological research were needed. One was related to the dispersal rates of *A. fallacis* into and out of apple trees. This relation was particularly critical since the rate of migration in early season determines the initial predator-prey ratios which are established and ultimately determine if, or at what level, biological control will occur. Although studies to determine specific rates of predator migration into the tree are still in progress, the general relationships between within-tree populations of prey and predators, and with predators on the ground and dispersal out of the tree have been ascertained (Johnson & Croft, 1976).

Some Aspects of Functional and Numerical Responses of Amblyseius fallacis to Tetranychus urticae

Adult females of the predator *A. fallacis* were fed different densities of *T. urticae* eggs. Various numerical and functional responses were studied. Additional work was done on the effects of starvation of the predator. This research supplied input to the pest control model for apple orchards in Michigan (Blyth & Croft, 1976).

The Role of the Apple Rust Mite

One interspecific interaction between two phytophagous mite species, which appears to have significance insofar as mite pest management is concerned,

is that between *P. ulmi* and *A. schlectendali*. In addition to the benefits provided by *A. schlectendali* as an alternate food source of *A. fallacis* (see discussion and figure in Hoying & Croft, 1976), there is evidence that *A.
schlectendali* may inhibit the development of *P. ulmi* by what is believed to be
a foliage conditioning mechanism. We have observed conditions in commercial apple orchards where insecticides were used intensively, apple rust
mites were common, no major predator populations were present, and where
spider mites were virtually absent. The following is our interpretation of the
factors which seemed to create and maintain this situation which was studied
intensively for a 4-year period (Croft, unpublished data).

Spray records indicated that the pesticides which were being applied in the
orchard each year were maximally favorable to the development of apple
rust mites. Rust mites became abundant in early season each year in these
blocks. However, only a moderate amount of "silvering" and "bronzing"
occurred and this damage was not readily apparent until late season. Apparently, the rust mite conditioning of the leaves in early season in advance of the
development of substantial *P. ulmi* populations affected the subsequent
reproduction of this spider mite which is known to be extremely sensitive to
small changes in leaf nutrition (Rodri guez, 1958). It was hypothesized that
this effect influenced *P. ulmi* to the extent that over a protracted period of
time this spider mite was virtually eliminated as a pest from the orchard.
Subsequent laboratory experiments measuring the rate of reproduction by
P. ulmi were made on leaves which had sustained previous rust mite feeding.
Results showed a substantial reduction in *P. ulmi* oviposition even on leaves
which had had only moderate exposure to rust mites. In addition, release
studies of *P. ulmi* onto trees having received different amounts of previous
feeding by *A. schlectendali* in the field showed a similar effect. Studies conducted subsequently have indicated that it is possible, by spraying insecticides
which favor *P. ulmi* as opposed to *A. schlectendali*, to change a block of
apples which previously had had only a rust mite problem into one in which
P. ulmi became the principal mite pest, and vice versa (Hoying & Croft, 1976).

Life History and Development of
the Tufted Apple Budmoth
in Relation to Temperature

It was determined that there are two generations of tufted apple budmoth
annually in southcentral Pennsylvania. Larvae of the second brood overwinter in leaf litter or in other sheltered places in the orchard. The larvae may
overwinter in any larval instar. Diapause apparently does not occur. In the
spring the larvae complete development by feeding on any of a variety of

herbaceous plants on the ground. Flight periods of male moths were monitored with pheromone traps. Moths were caught from May through October, but the highest weekly catches occurred near the middle of June and the beginning of September. Tree foliage was examined at frequent intervals to determine when eggs were being deposited. First brood eggs were laid from about June 1 to July 10 in 1974 and 1975. Second brood eggs were laid from August 1 to September 15. Caged female moths in an insectary laid an average of about 7.5 egg masses each. The average number of eggs per egg mass collected from an orchard was about 60. Therefore, the average female may lay at least 450 eggs (Bode, 1975).

A laboratory study to estimate the relationship between temperature and larval development of the tufted apple budmoth was conducted during 1974 to 1975. Newly hatched larvae were placed singly into plastic cups and raised on artificial diet at constant temperatures of 60°F, 65°F, 70°F, 75°F, 80°F, 85°F, and 90°F. The larvae were examined daily, and the length of each instar was recorded. Sixty or more larvae were started at each temperature. The proportion of larvae reaching the adult stage increased from 32 percent at 60°F to 80 percent at 80°F. At higher temperature, survival decreased. A logistic model was used to describe the functional relationship between rate of development per day, egg hatch to adult emergence, and temperature (Fig. 9.11). The parameters of this model were statistically estimated using nonlinear regression techniques. The model fit the data well, as indicated by the R-squared value of 90 percent. In comparison, fitting a linear, degree-day summation model to the data resulted in an R-squared value of 82 percent. No statistical difference was found between the rates of development for male and female larvae. Based on Figure 9.11, the estimated maximum developmental rate is approximately 4 percent/day at 90°F (Berkett et al., 1976).

Biological Components of Phenology for " Parameterizing " the PETE System

Modeling the phenology of a biological event (e.g., insect, disease, plant), is essential for incorporation into the PETE system (predictive extension timing estimation system) of sequential phases. At present approximately 17 orchard-inhabiting species have been brought to various stages of development in this system (Table 9.7). The primary effort in the sequence of development is to gather the necessary biological information by a thorough search of existing literature. All available information pertaining to life history, duration of stages relative to temperature and other factors, oviposition rates, and stage-specific mortalities is gathered and organized.

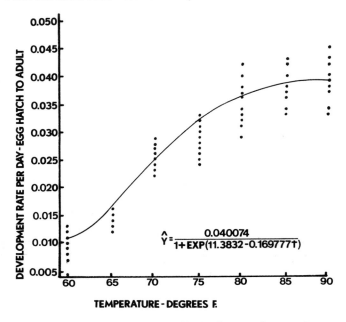

Figure 9.11. Estimated relationship between daily developmental rate and temperature for tufted apple budmoth for period from egg hatch to adult emergence. [From Berkett et al., 1976.]

The second phase, referred to as "parameterization" in Table 9.7, is necessary for inserting the insect of concern into the generalized insect model already developed and referred to elsewhere as the set of computer software routines called PETE. When possible, the biological information is arranged such that for each state (1) minimum and maximum thresholds for accumulation of degree days, (2) mean length of state (in units of degree days), and (3) an estimate of variance (in degree days squared) for each life stage within the population are obtained. PETE can also use information based on average length of stage (in calendar days) for specific temperatures and their associated variances. These parameters are then organized in a form acceptable to the PETE model.

The model is generated with these parameters, and phase three, validation, is then begun. Validation essentially is the comparison of model output with real-world data obtained from sources independent of those from which the model parameters were initially derived. This phase can never be conducted with 100 percent assurance but it is a very impressive tool. Perhaps its best use is in accurately predicting events such as first and/or peak emergence before they actually occur in the field. If the model and parameters prove to be accurate to an acceptable error or tolerance range then implementation, phase four, can be undertaken.

Table 9.7.   Developmental Phase in Phenological Modeling of a Deciduous Fruit Complex for Inclusion in a Predictive Extension Timing Estimation (PETE) System (Adapted from Welch et al., 1978)

Organism	Literature Search	Parameterization	Validation	Implementation
Apple tree	×	× [a]		
Apple scab	×	×	×	×
Codling moth	×	×	×	×
Redbanded leafroller	×	×	×	×
Lesser appleworm				
Oriental fruit moth	×	×	×	×
Apple maggot	×	×	× [b]	
Plum curculio	×			
Obliquebanded leafroller				
Fruit tree leafroller				
Green fruitworm	× [d]			
Tentiform leafminer	×	× [c]		
Rosy apple aphid				
Green apple aphid	×			
Woolly apple aphid				
White apple leafroller	×	×	×	
San Jose scale	×	×		
Tarnished plant bug	× [d]			
European red mite	×	×	×	×
Twospotted mite	×			
Apple rust mite	×			
Grape berry moth	×			
Peach tree borer				
Lesser peach tree borer				
Cherry fruit fly	×			

[a]Parameterization incomplete, need additional data.
[b]Validation currently being conducted.
[c]Parameterization currently being performed.
[d]No useful data found to date from literature.

If, however, the model is shown to be inaccurate in the validation phase, new biological information must be assembled. A procedure called sensitivity analysis is a useful tool to direct efforts to improve inaccurate parameters. This procedure reveals the parameters to which the model is most and least sensitive. Those parameters to which model accuracy is most sensitive are the most important ones and they must be the most accurately determined. More error can be tolerated in parameters that do not greatly affect model accuracy.

In this way the modeling approach is shown to be important in that even i
the major goal of accurate prediction fails to be reached, new and importan
research is indicated by the weaknesses in the information already know
(Croft, Howes, & Welch, 1976).

Influences of Temperature on Pest and Predator Mites

Influence of Constant Temperature
on Population Increase of Tetranychus mcdanieli

Constant temperature studies on the biological development of the McDanie
spider mite provided the data necessary to construct life tables and to analyz
developmental curves for this species. Optimum growth and reproductiv
rates were attained at $35 \pm 2°C$. At this temperature, indices of intrinsi
rate of natural increase, and finite rate of increase were maximum whil
minimum generation time was attained. Evaluation of various methods fo
fitting the developmental curves indicated that the more general method o
polynomial regression analysis provided the flexibility needed to determin
the best curve for each life stage of the mite. In Figure 9.12 curves for the egg
and active life stages were quartic in form, while those for the quiescen
stages and preoviposition period were cubic (Tanigoshi et al., 1975a).

Influence of Temperature
on Population Increase of Metaseiulus occidentalis

Constant temperature studies on the biological development of this phytoseiid predator also provided the data necessary to construct life tables an
analyze its developmental curves. Polynomial regression curves for the egg
and larval life stages were quadratic in form; those for the protonymphal an
deutonymphal stages were cubic; and for the preoviposition period wa
quartic (Fig. 9.13). From the life history and life table data, M. occidentali
was found to attain its optimum developmental and reproductive rates a
32°C. The intrinsic rate of natural increase, r_m, was 0.279 individuals/female,
day and the population possessed the capacity to double every 2.48 days. A
curvilinear regression model relating r_m to developmental period wher
developmental period is a function of temperature was derived. From thi
model, the temperature for optimum population development was 33.4°C
with a maximum r_m of 0.305 (Tanigoshi et al., 1975b).

Empirical Analysis of Variable Temperature Regimes
on Life Stage Development and Population Growth
of Tetranychus mcdanieli

Life stage developmental studies of the McDaniel spider mite, T. mcdanieli
were conducted under both alternating temperature regimes and naturall

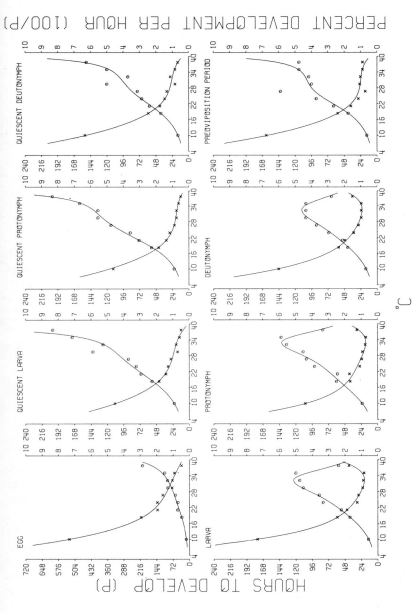

Figure 9.12. Developmental curves of immature female *T. mcdanieli*. [From Tanigoshi et al., 1975a.]

289

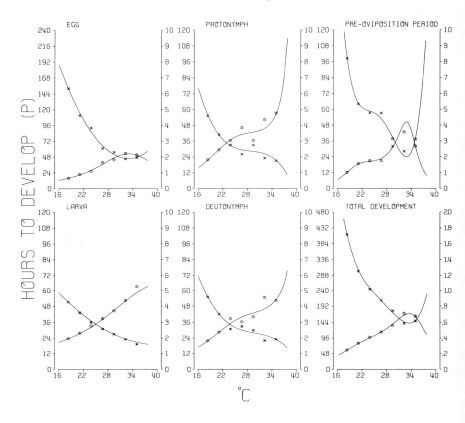

Figure 9.13. Developmental curves of *M. occidentalis*. [From Tanigoshi et al., 1975b.]

varying diurnal temperatures. A model, describing *T. mcdanieli* development under variable temperature patterns, was derived from life stage regression equations previously reported for constant temperature regimes. When rounded to the nearest quarter day, maximum deviation of model estimates from observed values for female egg to egg development was ± 0.25 day at the temperatures utilized in this study. Furthermore, the reported close correlation between r_m and developmental time, under static temperatures favorable for *T. mcdanieli* survival, was extended to a model relating r_m to variable temperature patterns (Tanigoshi et al., 1976).

Influence of Temperature on
Population Development of European Red Mite

The rate of development of *P. ulmi* from egg to adult was studied over a range of temperatures from 13°C to 32°C. Total fecundity and rate of egg production was studied over this same range of temperatures, but abnormally low fecundity was obtained by the methods used. Fecundity was reduced at 30°C compared to 20°C or 25°C. Temperatures of 32°C appear to be close to the upper limit for both development and egg production. A constant temperature of 35°C caused total mortality of eggs and females. This species appears to have a lower optimum temperature, and to be less heat tolerant than *T. mcdanieli*.

Effect of " Preconditioning " of the Foliage by Rust Mite
on the Population Development of European Red Mite

Comparisons were made of the development and survival of European red mite (groups of 200 larvae) on Red Delicious leaves very heavily damaged by rust mite (1500–2000 mites/leaf) and on undamaged leaves. Developmental times and percentage survival were similar on damaged and undamaged leaves, but 60 percent of the young female adults moved from damaged leaves onto undamaged leaves lower down the shoot, as opposed to only 3 percent moving when development was on undamaged leaves. The total fecundity was unaffected by these conditions of development when movement was permitted, but was reduced by 90 percent when adult females were forced to remain on the heavily rust-mite-damaged leaves. In additional tests with field-collected leaves having lower rust mite densities and rather mild damage, there was no evidence of reduced fecundity.

Searching Behavior and Growth Rates
of *Anthocoris nemoralis*, a Predator of the Pear Psylla

The behavior and growth rates of this predator were investigated using pear psylla as prey. Stadia duration and prey consumption on instars and adults at three prey densities were determined. The minimum number of prey per day required for completion of development to the adult stage was found to be between 5 and 10 eggs or small nymphs. Reproductive capacity and longevity were determined for females reared and maintained at two prey

densities. Searching patterns of *A. nemoralis* correlated well with the distri
bution of pear psylla on pear leaves, that is, there was a concentration o
searching along the midrib and leaf periphery. When prey was located a
thorough search of the area was then made (Brunner & Burts, 1975).

MODELING AND ECOSYSTEM ANALYSIS

To paraphrase Huffaker and Croft (1976), modeling, as broadly defined
includes the conceptualization of a process by mental, pictorial, flow chart o
mathematical means, and it has as its objective to understand, manage, o
predict some feature or totality of the process in question. For prediction
there is of course a trade-off between monitoring the development of ar
insect, natural enemy, or crop population and forecasting its expected
development based on simulation or methematical models. The more one
can predict the less he needs to monitor, and vice versa. Models, through
sensitivity testing and validation, can also play an important role in eluci-
dating control elements and critical parameters in a pest management system
to which research can be directed. (See further Chapters 2, 4, 6, 8, 10, and 13.

As many individual components of a pest management system are developed as subunits or submodels, they must be integrated or coupled into a
total crop management system. The Pome and Stone Fruits Group has
developed subunits or submodels, several of which are described below, and
is entering this phase of integrating the submodels (a complex of variables
so that each component can be managed in a way to give optimal benefits

The development of forecasting systems for the codling moth in both
California and Michigan was discussed in the section "Sampling, Monitoring, and Forecasting," as was also Michigan's apple scab forecasting and
predictive systems. Modeling was an integral part of the development of
these systems.

Models for Predicting
Control of the European Red Mite
by *Amblyseius fallacis*

In Michigan, models of varying degrees of sophistication have been developed for predicting control of populations of the European red mite in
apple orchards by the predatory mite *A. fallacis*. The simplest system employs a decision-making index for estimating the probability of biological
control of spider mites by *A. fallacis*. The index is divided into six regions
based on the spider mite and *A. fallacis* populations (Fig. 9.14). To categorize an orchard in one of the regions at a given time, weekly counts of prey

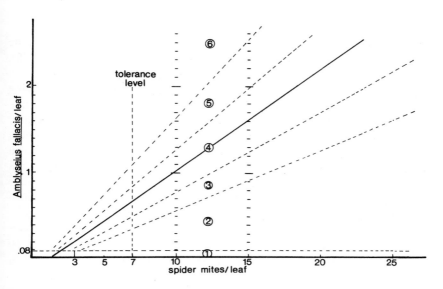

Figure 9.14. A decision-making index for estimating biological control of spider mites by *Amblyseius fallacis*. [From Croft, 1975.]

Table 9.8 Spray Recommendations for Pest/Predator Ratio Regions in Figure 9.14 (B. A. Croft, 1975)

Region	Suggested Recommendation	Probability for Biological Control
1	As bronzing appears, spray Plictran miticide 50 WP at full rate (4–6 oz/100 gal or 16–20 oz/A)	Very low
2	If bronzing appears, spray Plictran 50 WP at 2 oz/100 gal or 8 oz/A	Equal to or less than 10%
3	If bronzing appears, spray Omite 30 WP at $1\frac{1}{4}$ lb/100 gal or 5 lb/A or Plictran 50 WP at $1\frac{1}{2}$ oz/100 gal or 6 oz/A (see Table 9.4 for use limitations)	Greater than 10% but less than 50%
4	Wait 1 week, biological control should occur soon; if not, spray Omite at $1\frac{1}{4}$ lb or Plictran $1\frac{1}{2}$ oz/100 gal (see Table 9.4 for use limitations)	Approximately 50%
5	Same as region 4	Greater than 50% but less than 90%
6	Wait 1 week—biological control is almost certain.	

293

and predator are needed. Spray recommendations are made according to the region, based on pest/predator ratios as shown in Table 9.8 (Croft, 1975). A more sophisticated model simulates the population dynamics interaction of the European red mite and this predator, *A. fallacis*, for a growing season. Model inputs needed to use this system are (1) current or real time weather received daily from a 27-site network in the Michigan Agricultural Reporting System operated by the National Oceanographic and Atmosphere Administration (NOAA) and located throughout the Michigan fruit belt and (2) the stage distribution and density counts of spider mites and predators taken early in the season from a 10-acre block of apples where pest mites are just beginning to increase. Model outputs include predictions of the dynamics of both species on a time/density scale (Fig. 9.15), statements about the probability of biological control occurring without the aid of selective acaricides and/or rate recommendations for applying a selective material to establish a more favorable predator/prey ratio (Croft, Howes, & Welsh, 1976).

Modeling Temperature-Dependent Rate of Development of the Tufted Apple Budmoth

A logistic equation was derived from the experimentally determined relationship of tufted apple budmoth development to temperature. The developmental rate is responsive to temperature change between about 15.6°C (60°F) and 32.2°C (90°F). At 15.6°C development from egg hatch to adult emergence requires about 100 days; at 23.9°C it requires about 35 days; and at 32.2°C about 25 days. The equation was written into a computer program to simulate TABM larval and pupal development. Input to the program consists of estimates of the proportion of the total population in each larval instar and the daily mean temperature. Using the predicted developmental rates, population development is simulated through the larval and pupal stages. Output includes the daily status of the developing immature population and the percentage of the population emerging as adults each day.

Computer Simulation for Predicting the Number of *Stethorus punctum* Needed to Control the European Red Mite in Pennsylvania Apples

A computer model of the European red mite and its apple orchard environment was developed to simulate a mite infestation, complete with predation and the relevant environmental conditions, at the rate of a growing season a

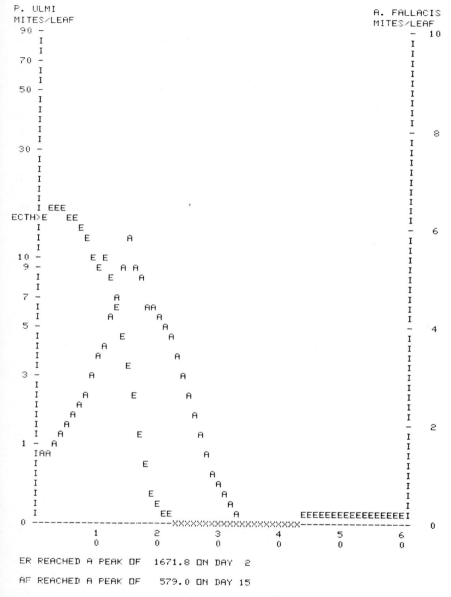

Figure 9.15. Predictions of the dynamics of the European red mite and *Amblyseius fallacis* on a time/density scale. [From Croft, Howes & Welch, 1976.]

minute. The model was developed for two purposes: (1) to conduct simulation experiments designed to evaluate various pest management tactics or strategies and (2) as the basis for a predictive system to be used in pest management decisions. European red mite developmental equations used in the model are based on the work of Cagle (1946). Feeding rates for the predator *Stethorus punctum* were estimated from results of insectary experiments (Hull et al., 1977). Other quantitative information required on mite and predator populations, the orchard spray program, and tree size and growth are based on Pennsylvania experience. The model is programmed in Fortran IV and runs on an IBM 370/168 computer in *batch environment* in contrast with "time sharing." Batch environment refers to the method of putting data into the computer program. Under this system all the data for a given period, such as a week, are saved and inserted into the computer program at one time, whereas with time sharing small amounts of data may be fed into the computer program at any time. Core requirement is approximately 7000 words (Mowery et al., 1975).

An Analytic Model for Description of Temperature-Dependent Rate Phenomena in Arthropods

Logan et al. (1976) deduced a new descriptive model of temperature-dependent rate phenomena to describe developmental time and ovipositional data for the McDaniel spider mite. The derived equation accounted for asymmetry about the optimum temperature and was of particular utility for description of events at or above optimum temperatures. Ovipositional and developmental rate functions were used in a temperature-driven, discrete-time, simulation model describing McDaniel spider mite population dynamics excluding predation. Temperature dependence of instantaneous population growth rate was determined by fitting the derived rate-temperature function to data generated through simulation at various fixed temperatures. The functional relationship of important population parameters to temperature provided the mechanism for inclusion of phenological effects on mite populations in a synoptic apple pest management model. Predation and its effect are included in a mite management model.

Two derived functions were fit to several published rate-temperature data sets. Adequacy of description (as indicated by R^2 values) indicated general applicability of both functions for description of such temperature-controlled, biological processes. It was also concluded that the singular perturbation method of matched asymptotes has potentially wide application in ecology, and an appendix detailing the application of this method is included in the paper.

IMPLEMENTATION OF PEST MANAGEMENT SYSTEMS

Again paraphrasing Huffaker and Croft (1976), the systems implementation phase is the culmination effort which concentrates or simplifies the best system of control methodologies and integrates monitoring, modeling, and management tactics and strategies into a system of delivery to the pest manager. Although effective systems can be based on traditional methods (i.e., off-line mode, without mathematical models) use of real-time weather acquisition systems which interface with biological monitoring systems provide for rapid delivery of better decision-making information to the pest manager, including feedback data which provide updates on the state of the entire crop-pest system.

A Computer-Based Extension Pest Management Delivery System

Extension entomologists traditionally have faced the problem of collecting, integrating, and reporting data relevant to agriculture with only crude information processing capabilities at their disposal. Croft, Howes, and Welch (1976) discussed the use of computer-based extension delivery systems to augment the flows of information in a modern pest management implementation program. They identified the needs and constraints associated with an integrated pest management (IPM) data acquisition/delivery system developed for pests of apples and other crops in Michigan (Fig. 9.16). They emphasized the fact that these needs impel a reassessment of customary methods in extension entomology. They give examples of information application types including biological data, weather data, pest population and management models, memoranda, reference data, system documentation, and educational or performance evaluation programs. PMEX, a computer-based Pest Management Executive System, is presented. The system consists of a central computer with appropriate softwear and telecommunications network linking extension offices scattered over a wide area. At these sites, remote data terminals are used to update and interrogate a large data-base associated with on-line pest management. The authors discuss the implications of prototype IPM extension systems in this class as opposed to extension delivery systems in general.

A Mobile Van Prototype for Biological Monitoring of Deciduous Fruit Pests

To evaluate the utility of a mobile monitoring and educational unit, a prototype unit was tried in the field in Michigan. The unit consisted of a van containing a programmable calculator (for models and statistical analyses), a

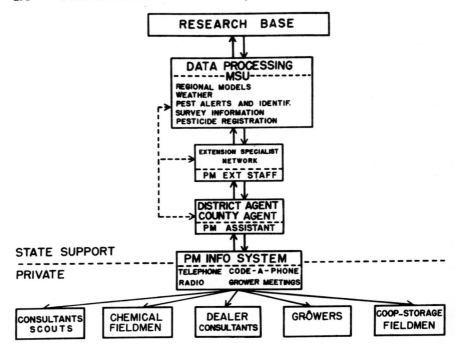

Figure 9.16. Personnel and principal information flows associated with IPM in Michigan. [From Croft, Howes & Welsh, 1976.]

data terminal (for calculator output and interaction with remote computers), a slide-tape projector (for audio visual educational material), and a battery power supply (to permit operation in the orchard itself).

Although the hardware is flexible enough to support a varity of pest programs, the van was operated originally as a mobile mite counting service. On grower requests, the van was dispatched to potentially mite-infested orchards. Leaf samples were then taken and counted using a sequential sampling alogrithm programmed on the calculator to optimize manpower usage. Both predatory (beneficial) and phytophagous (pest) mites were counted. When the calculator had sufficient data, it generated a control recommendation based on a regression management model described previously. This recommendation was printed out at the data terminal.

Concurrently with these operations, the grower (if present) was viewing a 30-minute slide tape presentation on integrated mite control. Pictures of pests and predatory mites were shown, along with descriptions of their life histories. The lethality to predators of various pesticides was discussed and the grower was told how an integrated pest management program could be devised. The importance of obtaining the mite counts was also developed and all this was related to the presence of the van in his orchard. Specific grower questions were answered by the mite van personnel.

The van was equipped to contact the Michigan State University central computer via a long distance telephone call at the end of the day. The staff of the van could then interact with the University's data base using the data terminal. At this time the results of the day's counts (stored on cassette tape) could be sent to the computer. More sophisticated models could then be used to validate any borderline regression model recommendations.

MITESIM Validation

In 1976, a computer-predictive system for European red mite management was validated in Pennsylvania. The MITESIM computer model considers the mite population on a tree to be a system. Components of this system represent mite life stages. The first component represents summer egg incubation. Newly hatched larvae are represented in a component that simulates development through the immature stages. And there is a component for adult development and egg laying. The environment of this mite system is made up of orchard temperatures, the *S. punctum* population, and the orchard spray program.

The specific validation discussed here was carried out in a university-owned apple orchard, which was managed as a commercial orchard in full production. European red mite and *S. punctum* counts were made at 7 to 12 day intervals. Daily maximum and minimum temperatures were recorded in a weather station located approximately $\frac{1}{2}$ mile from the orchard. The MITESIM system accurately predicted changes in the mite population (Table 9.9). On July 9, the computer predicted the need for an acaricide treatment due to the low number of predators and the size of the pest mite population (Table 9.10, recommendation 4). Plictran® 50 W, 2 oz/acre, was applied July 15. Because only a low dosage of acaricide was used, the mite population again began to increase. This time, however, the computer predicted biological control by *S. punctum* and recommended no further treatment. Additional acaricide was not used and subsequent counts showed the prediction to be reliable (Table 9.9). The mite population reached a peak of 7.1 motile mites/leaf on July 26 and was kept at a low level by *S. punctum* throughout the remainder of the season. The seasonal history of the mite population in this orchard illustrates how the MITESIM system can be used to time an acaricide application so as to maintain the mite population below the economic injury level until the predator population reaches an effective controlling density (Mowery, Asquith, & Hull, 1977).

Table 9.9. Counts of Motile European Red Mites and *Stethorus punctum* Larvae and Adults (Mowery et al., 1977)

| | Motile ERM/Leaf | | *S. punctum* per 3 Minutes |
	Actual	Predicted[a]	
July 2	0.7		1.8
July 9	9.8	4.9(2)[b]	3.3
July 16[c]		12.4(4)	
	3.2	1.7(2)	22.0
July 26	7.1	10.9(3)	38.0
Aug. 3	2.1	0.2(1)	57.8
Aug. 11	1.1	0.0(1)	27.8
Aug. 23	0.8	0.4(1)	11.5

[a]Predicted on the date of the previous count.
[b]Numbers in parentheses are MITESIM recommendations (see Table 9.10).
[c]The first prediction given for July 16 (12.4 mites/leaf) assumed no acaricide, the second (1.7 mites/leaf) assumed an acaricide would be applied. The acaricide was actually applied on July 15, resulting in the actual mite population shown of 3.2 mites/leaf.

Table 9.10. MITESIM Recommendations (Mowery et al., 1977)

1. No acaricide recommended. Continue regular counts of mites and *S. punctum*.

2. No acaricide recommended. Make next count of *S. punctum* in 7 days.

   a. Some patches of light bronzing may appear.

3. No acaricide recommended as prospect of biological control by *S. punctum* is likely.

   a. Some patches of light bronzing may appear.

4. Apply a low dosage of acaricide, such as Plictran® 50 W, 2 to 3 oz/acre/half spray, in order to suppress mite population. Other acaricides may be used to accomplish the same purpose. Do not attempt to eliminate the entire mite population. Make a count of mites and *S. punctum* 3 to 7 days after application of acaricide.

5. Tree bronzing possible. Check trees. Acaricide application may be necessary if bronzing is extensive and the mite population is increasing.

ECONOMIC EVALUATION OF INTEGRATED PEST MANAGEMENT*

Sound economic evaluations of the cost/benefits or cost/effectiveness of plant-protection practices in the context of the total crop production system are essential tasks in the development of a sound integrated pest management (IPM) program on any crop. While new methods and tactics of IPM beyond the conventional approach (i.e., strict chemical control) have been available for deciduous fruit pests for more than two decades, precise documentation of the economic and environmental benefits of these programs were mostly lacking prior to the last few years. While it has been relatively easy to compute the benefits accruing from IPM by comparing data from one season to another and using such basic indicators as reductions in pesticide use or the associated cost-savings, too often concurrent check plot treatments were not used for comparative purposes. Economic evaluations tended to be more of a post-implementation analysis rather than for determining of a feasibility of control. Furthermore, assessments were usually static in nature and were not monitored at frequent intervals. In all cases they were greatly limited by our inability to document the more complex effects of factors such as the cost of resistance development, the economics and long-term ecological consequences of insecticide applications, and other lag-time phenomena such as the beneficial but delayed action of arthropod natural enemies.

In recent years, more effort has been made to remedy these deficiencies and to improve the economic evaluation of deciduous fruit IPM methods. Because of the way that individual IPM practices (e.g., integrated mite control systems) have been integrated into deciduous fruit crop protection systems (i.e., those largely based on chemical control measures) there has been an initial tendency to make rather fragmented component cost assessments. For example, the economic value of a particular tactic against a particular pest was first evaluated in relation to the traditional means of control at an individual pest and individual grower level and seldom in relation to the total crop management system. Only limited effort was made to look beyond the farm unit to a group of growers in proximity to each other, or in a larger region subject in common to a problem or control effect. The emphasis in economic assessment of deciduous fruit IPM practices has progressed in recent years from simple evaluations of costs and pesticide reductions to more complex assessments where multiple and variable components are included in the evaluation. In the remaining portions of this presentation, specific examples of new innovations in tree fruit pest management (IPM) and a more detailed discussion of the economics of these developments are reviewed.

*Materials for this section essentially were taken from an article by B. A. Croft (1977).

Economic analysis has been done mainly for single control tactics. Economic studies of new methodologies for deciduous fruit pest control have principally been directed toward (1) integrated pest management control programs for plant-feeding mites, (2) the sterile male approach to codling moth (*L. pomonella*) eradication or control, (3) pheromone control systems, (4) improved spray timing and spray application methods, and (5) the value of scouting or intensive pest monitoring in a supervised IPM program for entire deciduous fruit pest complexes.

Integrated Mite Control Programs

Several reviews (Hoyt & Burts, 1974; Croft & Brown, 1975) have described programs of integrated mite control which have been implemented throughout the deciduous fruit growing region in North America. The essence of these programs is the combined utilization of insecticide-resistant or tolerant predators of plant feeding mites (e.g., phytoseiid mites, *Stethorus* sp., predator complexes) together with chemical controls selectively applied for other fruit pests and physiologically selective acaricides (tricyclohexyl-hydroxytin, Omite®, Vendex®) to readjust predator: prey ratios when the predators are not sufficiently abundant to provide effective biological control.

In Washington, a program utilizing organophosphate-resistant strains of *T. occidentalis* for biological control of *T. mcdanieli*, *P. ulmi*, and *A. schlectendali* has been used commercially since 1965. This program was a much needed replacement for a chemical mite control program which had become increasingly more costly due to pesticide resistance problems. Furthermore, mite control was becoming increasingly more difficult to obtain. In Figure 9.17, the costs/acre of materials for insect and mite control for a 1000-acre block of apples using a standard spray program (1957–1966) versus an integrated mite control program (1967 to 1969) is graphed (from Hoyt & Caltagirone, 1971). Data for the 1970 to 1975 period represents the average trend for control costs and does not show yearly fluctuations (S. C. Hoyt, personal communication). As can be seen in Figure 9.17, total control costs have declined from an average of $60 to near $20 to $30/acre; of this total reduction in pesticide use, the principal cost saving has come from the nearly complete elimination of summer acaricide applications used for phytophagous mite control.

Similar successes in reducing mite control costs have been realized in Pennsylvania and Michigan, although the total reduction in pesticide use (30%–40%) has not been so dramatic as in Washington due to the greater complex of insect and disease pests other than mites and the smaller proportion represented by mite control costs in relation to the total spray bill.

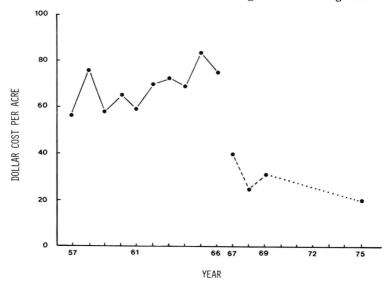

Figure 9.17. Costs/acre for mite control in a 1000 acre block of apples in Washington during the period 1957 to 1975. [From Croft, in press, as adapted from Hoyt & Caltagirone, 1971.]

Acaricide use, however, has been reduced by 75 percent in Pennsylvania where an integrated control program for the European red mite, *P. ulmi*, based on conservation of *S. punctum* populations has been utilized during the past 8 years (Croft et al., 1975). In Michigan the principal mite predator controlling *P. ulmi* is an O-P-resistant phytoseiid mite, *A. fallacis*. In that state, a 90 percent reduction in mite control costs to about $2.50/acre has been achieved in experimental plots where integrated mite control has been practiced for 5 years and a 69 percent decline to about $8.00/acre has been reached over the same period in some commercial orchards. Research sponsored by the NSF/EPA grant project has enabled entomologists in Michigan and Pennsylvania to keep their respective systems of integrated control of pest mites active and viable. Differing weather patterns from year to year place a strain on integrated systems. Also the necessity to add new insecticides for the control of important pests makes research on the effect of insecticides on the integrated system a continuing necessity. Continuing support of research on the biology of pests and predators, on the effects of pesticides on these organisms, and on systems analysis has laid the foundation for continuing success with integrated pest management systems in apple orchards.

The Sterile Male Technique for Codling Moth Control

A second IPM tactic for fruit pest control which has received cost analysis is the genetic sterility method which has been experimentally evaluated in Washington and British Columbia, Canada. Studies of codling moth control by the sterile male technique were begun in British Columbia in 1956. Research beginning in 1966 and continuing up to the present time has demonstrated it to be an effective means for controlling this pest, but reported estimates for control costs were $56.25/acre in 1973 (Hoyt & Burts, 1974) which is about twice the cost for chemical control of the codling moth. This figure includes costs for labor, rearing media, release of moths, and pheromone traps for monitoring populations, but does not include the cost of a rearing facility. A more recent figure of $42.49/acre (M. D. Proverbs, personal communication) indicates that improvements in rearing and releasing methods are reducing the cost for implementation. However, the most recent conclusion of those working with the project is that while the sterile male method for codling moth control *is* technically feasible, it is not economically acceptable (Hoyt & Burts, 1974). For this reason only minimal research is continuing on this program.

Pheromone Control Systems

Research on the use of pheromones to disrupt insect communication and thus provide commercial pest control is at the experimental stage of development for several important fruit insect pests. Recently, a feasibility study to evaluate the economic costs for developing and registering the pheromone of the codling moth for *direct control* purposes was conducted by Zoecon Corporation of Palo Alto, California (Siddall & Olson, 1976). In Table 9.11, the basic process development, toxicology, and registration and formulation costs, in relation to the sale and income potential of such a program, were analyzed for feasibility assessment. Expected sales returns were projected at $20/acre ($30/acre costs to the grower) for approximately 500,000 acres in South Africa, Australia, and western North America (California, Oregon, Washington, British Columbia, and Colorado) where the codling moth is the only or primary fruit insect pest. Their analysis assumed a penetration or coverage of approximately 5.5 percent of the above total, which is a relatively small fraction, but one of their basic assumptions was that in the early stages of development it would be difficult to persuade growers to change completely to this new and comparatively untried method of pest control.

Table 9.11. Calculation of Internal Rate of Return on Investment—Codling Moth Control by Pheromonal Disruption of Mating—Years 1 to 7[a]

	1	2	3	4	5	6	7
A. DEVELOPMENT COSTS							
1. Process development (target $200/lb)							
a.i. formulated	$(125)	(125)					
2. Toxicology & registration (tolerance exemption)	$(50)	(130)	(150)		($000)		
3. Formulation (select and develop 1 formulation)	$(40)			(20)			
	$(215)	(255)	(150)	(20)			
4. G&A, overhead (50%)	$(108)	(127)	(75)	(10)			
Total Develop. & Regist.	$(323)	(382)	(225)	(30)			
B. SALES, INCOME & EXPENDITURE							
1. Market penetration			(Total potential = 473,000A)				
0.5% (Exp. permit)		2,360A					
1.0% (Exp. permit)			4,730A				
3.0% (Full registration)				14,190A			
5.0% (Full registration)					23,650A		
5.5% (Full registration)						26,015A	
5.0% (Full registration)							23,650A
2. Sales ($000) at $20/acre		24	95	284	473	520	473
3. Cost of goods at 10/g/A/season 44c/g a.i. formulated		(10)	(21)	(62)	(104)	(114)	(104)
4. Sales expense (15%)		(4)	(14)	(43)	(71)	(78)	(71)
5. Net cash flow	$(323)	(372)	(165)	149	298	328	298
6. Discount factor (6% rate)	×0.94	×0.89	×0.84	×0.79	×0.75	×0.71	×0.67
7. Discount cash flow	$(305)	(331)	(139)	118	224	233	200
8. Discount cash flow, cumulated	$(305)	(636)	(775)	(657)	(433)	(200)	0

[a]From Siddal and Olson, 1976. Pheromones in agriculture—from chemical synthesis to commercial use, in *Pest Management with Insect Sex Attractants*, Martin Beroza (ed.) *Am. Chem. Soc. Symp. Ser. 23*:88–90. Reproduced with permission.

The conclusions were as follows: an investment of approximately $1 million would be required for development of a codling moth control program on a commercial basis. Only if significant improvements in developmental costs or expectations of substantially greater market penetration were realized to the extent of providing a 15 percent investment return, would the project receive major attention by private industry. They pointed out that before such an effort might be undertaken, considerable research is needed in the field to assure the effectiveness of a pheromone control system that is comparable to that achieved by conventional chemical control means.

Improved Spray Timing and Spray Application Methods

It has long been known that the economy of scale for the number of spray applications applied per season versus the percentage of clean fruits produced in a commercial apple orchard takes on a curvilinear function of the form of Figure 9.18. That is, we know that under average infestation conditions we may reduce the number of broad-spectrum insecticide applications by approximately 50 percent, and only reduce our harvested clean fruit percentage from approximately 99 to 95. To date, entomologists in the United

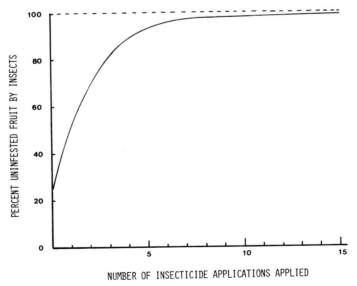

Figure 9.18. Economy of scale relationship between uninfested fruit and number of spray applications applied per season. [From Croft, in press.]

Table 9.12.   Cost of Spray Programs per Acre for Seasonal Insect Control in Michigan Apple Orchards (Thompson et al., 1973)

	Spray Method				
	Complete Program	Alternate Middle Row	Reduced Rate	Extended Interval	Early Termination
Orchards sprayed	117	14	28	21	10
Cost/acre	$29.13	$25.72	$13.68	$13.59	$11.63
Percent uninjured fruit	99	99	98	99	97

States have had very little impact in the market place or with establishment of laws governing quality standards to lower the economic threshold for clean fruit to the 95 percent level listed above. They have, however, been quite successful in developing better systems of spray timing and spray application which have allowed us to reduce either the amount of chemicals applied/treatment or the number of sprays applied/season without sacrificing any significant loss in quality of fruit.

Tables 9.12, 9.13, and 9.14 confirm the general economy of scale function shown in Figure 9.18. These figures summarize research conducted in Michigan, Pennsylvania, and New York, respectively, on the economic benefits of utilizing different spray timing and application methods in obtaining IPM of apple pests, mainly insects. In each case, various methods of

Table 9.13.   Use of Pesticides on Apples in Pennsylvania Under IPM Versus non-IPM Programs (Asquith, 1972)

	Cost of All Pesticides		Rate of Guthion® Recommended	
Program	$/acre	Reduction versus full schedule	Lb 50 WP/ acre/ season	Reduction versus full schedule
Full schedule	85.15		6.0	
Alternate middle spray schedule	60.23	29%	4.0	33%
Integrated pest management	46.12	46%	2.0	67%

Table 9.14.   Summary of Results with Minimal Spray Programs for Insect Control in New York Apple Orchards—1971–1973[a]

	Post-Bloom Insecticide Sprays			
	Average No. complete sprays per season	Percent reduction from full program	Percent reduction from grower standard	Average Percent Uninjured Fruit
Alternate middle	3.7	47	25	97+
Extended interval	4.6	34	6	97+
Cooperating grower standard	4.9	30		94+
Full chemical	7.0			98+

[a]From Trammel, 1974, reprinted by permission Dept. Entomology, New York State Agric. Expt. Station.

reducing application (e.g., treating alternate middle rows, extended intervals between treatments, etc.) have been developed which give reduction either in cost or pesticide use in the range of 20 to 60 percent, in comparison to traditional calendar date spraying methods, without causing significant loss in damaged fruit.

Although specific cost figures are not readily available now, similar kinds of reductions in pesticide use have resulted by using ultra-low volume methods for application of insecticides (Howitt et al., 1966).

Also, improved spray timing and more precise evaluation of the need for such treatments, using pheromone baited traps for a variety of tree-fruit insect pests, have had considerable impact on reducing spray applications in commercial orchards throughout the fruit growing areas of North America (Batiste et al., 1973; Madsen & Vakenti, 1973; Riedl et al., 1976).

**Scouting Systems for
Supervised Integrated Pest Management**

During the past 5 years there has been a significant emphasis on development and implementation of intensive monitoring systems for IPM of entire complexes of deciduous fruit pests, including insects, disease pathogens, rodents, nematodes, weeds, and so on (also referred to as supervised control, scouting programs, or use of applied insect ecologists in pest control advising). For apple pests, California, Michigan, Mew York, Pennsylvania, and Washington have been especially involved. These states produce more than 70 percent of the deciduous fruit crop in the United States. In these programs, the principal economic comparison is whether the cost savings gained from more precise monitoring of pests and the consequent reductions in pesticide use (with side benefits environmentally) are equal to or greater than the increased costs of equipment and labor resulting from the more detailed field monitoring of pest conditions that is required.

Comparative data are still a bit fragmentary, but several preliminary assessments are available. In Washington, private consultants providing monitoring services for IPM of all pests as well as for fertilizer and cultural management advice, charge approximately $20/acre for apple programs. Cost reduction figures for pesticides alone for the 1974 season showed a savings of $35.25/acre, from $64.25/A for a check group, and as low as $29.00/A for a program under a more intensive IPM monitoring system (Eves et al., 1975). In Michigan, costs for extension service scouting of apples for insect and disease pests alone in 1973 and 1974 were reported to be $6.57 and $9.02/acre, respectively. Estimated cost savings provided by these programs were $26.40/A in 1973 and only $8.00/A during 1974, which was an abnormally bad year for apple scab (*V. inaequalis*) development (Thompson et al., 1974).

New Innovations in Orchard IPM

The emphasis given to IPM in the last decade has created a need among pest managers for better information on the growth and development status of their crop, the pests, natural enemies, and other factors. This has prompted development of several new innovative systems of pest modeling, other biological monitoring, and extension delivery of the required information. Most of these innovative developments employ computer science technologies particularly suited to handling large data sets and allow relatively rapid turnaround cycles, from the input of field data to the computer and back to the pest manager. Prototype simulation models for several apple pests have been developed and are currently undergoing feasibility study and use in operational management systems—that is, codling moth (Croft, 1976a; Falcon et al., 1976), plant-feeding mites (Mowery, Asquith & Bode, 1975; Dover et al., 1979), and apple scab (Jones, 1974). A prototype mobile facility with computer-linked sender-receiver terminal has been used for pest survey and extension education purposes (Croft, 1976a; Croft, Howes & Welch, 1976;

described above). Also an extensive computer-based extension system for on-line management of multiple crop ecosystems, including deciduous fruits, is under development and feasibility study (Croft, Howes & Welch, 1976).

Each such technology must be evaluated with respect to its initial costs of development and maintenance in relation to its potential benefits and efficiency including comparison with traditional means of carrying out these functions. Not only is there a need for feasibility studies of this nature at the initial implementation stage of a new technology, but there is an even greater need to conduct long-term economic appraisal of the results of such technologies relative to the various crop ecosystem components which themselves change through time. Such appraisals and the data secured will then provide concrete proof of the long-term benefits possible from such innovative technologies with environmental and other societal costs and benefits not ignored.

Future Economic Analysis of IPM

Only in the past 3 to 5 years have economists and entomologists begun to work closely together in addressing the more long-term and more complex questions dealing with IPM. Unfortunately, very little research of this type has been done on deciduous fruit problems. Recently, studies have been reported for other crops wherein researchers are considering a multiplicity of factors, such as crop damage and compensation, pest densities in time, and pesticide timing and quantity to define the term "economic threshold" more precisely (Headley, 1972; Hall & Norgaard, 1973; Talpaz & Borosch, 1974; see Chapters 6 & 8). Tanagoshi (unpublished data) has more precisely evaluated several of these parameters for plant-feeding mites on apples in Washington. The economics related to increasing pesticide resistance of pests over time has been modeled by Heuth and Regev (1974) (and see Chapter 8). Several workers have presented assessments of optimal ways to utilize component control strategies, including use of natural enemies for IPM (Shoemaker, 1973; Feder & Regev, 1975; Regev et al., 1975). Research to develop optimal solutions for a group of growers or a whole region within a range of common circumstance has received preliminary evaluation for a major alfalfa pest (Regev et al., 1975; see Chapter 8) and for *Heliothis* attacking cotton (Chapter 5). Eventually, it may be possible to assess the impact of IPM activities via input-output analysis as is often done for major economic factors (i.e., for the automobile or tourist industries in Michigan) in the context of a regional or national cropping industry (i.e., deciduous fruit production).

LITERATURE CITED

Abdelrahman, I. 1973. Toxicity of malathion to the natural enemies of California red scale, *Aonidiella aurantii* (Mask.) (Hemiptera: Diaspidae). *Aust. J. Agric. Res. 24*: 119–133.

Adams, C. H., and W. H. Cross. 1967. Insecticide resistance in *Bracon mellitor*, a parasite of the boll weevil. *J. Econ. Entomol. 60*: 1016–1020.

Ahlstrom, K. R., and G. C. Rock. 1973. Comparative studies on *Neoseiulus fallacis* and *Metaseiulus occidentalis* for azinphosmethyl toxicity and effects of prey and pollen on growth. *Ann. Entomol. Soc. Am. 66*: 1109–1113.

Asquith, Dean. 1972. The economics of integrated pest management. *Pa. Fruit News 51*: 27–31.

Asquith, D., and L. A. Hull. 1973. *Stethorus punctum* and pest–population responses to pesticide treatments on apple trees. *J. Econ. Entomol. 66*: 1197–1203.

Asquith, D., L. A. Hull, and J. W. Travis. 1976. Test of insecticides, 1973. *Insecticide and Acaricide Tests 1*: 13–14.

Asquith, D., L. A. Hull, J. W. Travis, and P. W. Mowery. 1976a. Apple, insecticide test, 1974. *Insecticide and Acaricide Tests 1*: 15–16.

Asquith, D., L. A. Hull, J. W. Travis, and P. W. Mowery. 1976b. Apple, test of insecticides, 1975. *Insecticide and Acaricide Tests 1*: 17–19.

Atallah, Y. H., and L. D. Newsom. 1966. The effect of DDT, toxaphene and endrin on the reproductive and survival potentials of *Coleomegilla maculata*. *J. Econ. Entomol. 59*: 1181–1187.

Batiste, W. C., A. Berlowitz, W. H. Olson, J. E. Detar, and J. L. Joos. 1973. Codling moth: estimating time of first egg hatch in the field—a supplement to sex attractant traps in integrated control. *Environ. Entomol. 2*: 387–391.

Berkett, L. P., P. D. Mowery, and W. M. Bode. 1976. Rate of development of *Platynota idaeusalis* at constant temperatures. *Ann. Entomol. Soc. Am. 69*: 1091–1094.

Blyth, E. J., and B. A. Croft. 1976. Some aspects of the functional and numerical response of *Amblyseius fallacis* (Garman) to *Tetranychus urticae*. *Abst. Proc. N. Cent. Br., Entomol. Soc. Am. 30*: 89.

Bode, W. M. 1975. The tufted apple budmoth in Pennsylvania, 1974. *Pa. Fruit News. 54*: 57–58.

Bode, W. M., D. Asquith, and J. P. Tette. 1973. Sex attractants and traps for tufted apple budmoth and redbanded leafroller males. *J. Econ. Entomol. 66*: 1129–1130.

Brown, A. W. A. 1976. The progression of resistance mechanisms developed against insecticides, in Am. Chem. Soc. Symp. *Pesticide Chemistry in the Twentieth Century*, American Chemical Society, Washington, D.C., Vol. 37, pp. 21–34.

Brunner, J. F. 1975. Economic injury level of the pear psylla, *Psylla pyricola* Foerster, and a discrete time model of a year psylla–predator interaction. Ph.D. thesis, Washington State University.

Brunner, J. F., and E. C. Burts. 1975. Searching behavior and growth rates of *Anthocoris nemoralis* (Hemiptera: Anthocoridae), a predator of the pear psylla, *Psylla pyricola*. *Ann. Entomol. Soc. Am. 68*: 311–315.

Cagle, L. R. 1946. Life history of the European red mite. *Va. Agric. Exp. Stn. Tech. Bull.* 98.

Chambers, H. W. 1973. Comparative tolerance of selected beneficial insects to methyl parathion. Communication to *Annu. Meet. Entomol. Soc. Am.*, November 28, p. 68.

Chestnut, T. L., and W. H. Cross. 1971. Arthropod parasites of the boll weevil *Anthonomus grandis*: 2. Comparisons of their importance in the United States over a period of thirty-eight years. *Ann. Entomol. Soc. Am. 64*: 549–557.

Collyer, Elsie, and M. van Geldermalsen. 1975. Integrated control of apple pests in New Zealand. 1. Outline of experiments and general results. *N.Z. J. Zool. 2*: 101–134.

Croft, B. A. 1972. Resistant natural enemies in pest management systems. *Span 15*: 19–21.

Croft, B. A. 1975. Integrated control of apple mites. Mich. State Univ. Coop. Ext. Serv. Bull. E-825.

Croft, B. A. 1976a. Pest management systems for phytophagous mites and the codling moth, in Proc. 3rd USSR/USA symp. on integrated pest management (Lubbock, Texas, Sept. 1975). Texas A&M University Publ., pp. 99–131.

Croft, B. A. 1976b. Establishing insecticide-resistant phytoseiid mite predators in deciduous tree-fruit orchards. *Entomophaga 21*: 383–394.

Croft, B. A. 1977. Resistance in arthropod predators and parasites, in D. L. W. Watson and A. W. A. Brown (eds.), *Pesticide Management and Insecticide Resistance*. Academic, New York, pp. 337–393.

Croft, B. A. In press. The economics of integrated pest management in orchards. Proc. 4th USSR/USA symp. on integrated pest management (Yalta, USSR. July 25–Aug. 1, 1976). (In press Russ.)

Croft, B. A., J. Briozzo, and J. B. Carbonell. 1977. Resistance to organophosphorus insecticides in the predaceous mite, *Amblyseius chilenensis*. *J. Econ. Entomol. 69*: 563–565.

Croft, B. A., and A. W. A. Brown. 1975. Responses of arthropod natural enemies to insecticides, *Annu. Rev. Entomol. 20*: 285–335.

Croft, B. A., A. W. A. Brown, and S. A. Hoying. 1976. Organophosphorus—resistance and its inheritance in the predaceous mite *Amblyseius fallacis*. *J. Econ. Entomol. 69*: 64–68.

Croft, B. A., J. A. Howes, and S. M. Welch. 1976. A computer-based, extension pest management delivery system. *Environ. Entomol. 5*: 20–34.

Croft, B. A., and S. A. Hoying. 1975. Carbaryl resistance in native and released populations of *Amblyseius fallacis*. *Annu. Rev. Entomol. 4*: 895–898.

Croft, B. A., S. C. Hoyt, Dean Asquith, R. E. Rice, and E. H. Glass. 1975. The principles strategies and tactics of pest population regulation and control in the stone and pome fruits subproject, in C. B. Huffaker and Ray F. Smith (eds.), *The Principles*,

Strategies and Tactics of Pest Population Regulation and Control in Major Crop Ecosystems, Vol. I. Int. Cent. Integr. and Biol. Control, University Calif., Berkeley, pp. 98–105.

Croft, B. A., and L. R. Jeppson. 1970. Comparative studies on four strains of *Typhlodromus occidentalis*. II. Laboratory toxicity of ten compounds common to apply pest control. *J. Econ. Entomol. 63*:1528–1531.

Croft, B. A., and D. L. McGroarty. 1977. The role of *Amblyseius fallacis* in Michigan apple orchards. *Res. Rep. 333, Mich. Agric. Exp. Stn.*

Croft, B. A., and R. H. Meyer. 1973. Carbamate and organophosphorus resistance patterns in populations of *Amblyseius fallacis*. *Environ. Entomol. 2*: 691–695.

Croft, B. A., and E. E. Nelson. 1972. Toxicity of apple orchard pesticides to Michigan populations of *Ambylseius fallacis*. *Span 1*: 576–579.

Dover, M. J., B. A. Croft, S. M. Welch, and R. L. Tummala. 1979. Biological control of *Panonychus ulmi* (Acarina: Tetranychidae) by *Amblyseius fallacis* (Acarina: Phytoseiidae) on apple: a prey-predator model. *Environ. Entomol. 8*: 282–292.

Eves, J. D., J. D. Chandler, R. L. Britt, and R. F. Harwood. 1975. Third annual report—Washington extension deciduous tree fruits pest management project, in National Extension Fruit Pest Management Workshop (Yakima, Wash. Mar. 1975). Washington Sta. University Publ., pp. 24–28.

Falcon, L. A., C. Pickel, and J. White. 1976. Computerizing codling moth. *Fruit Grower 96*:8–14.

Feder, G., and U. Regev. 1975. Biological interactions and environmental effects in the economics of pest control. *Calif. Agric. Exp. Stn. Publ.*

Field, R. P. 1974. Occurrence of an Australian strain of *Typhlodromus occidentalis* (Acarina: Phytoseiidae) tolerant to parathion. *J. Aust. Entomol. Soc. 13*:255–256.

Gentile, A. G., and F. M. Summers. 1958. The biology of San Jose scale on peaches with special reference to the behavior of males and juveniles. *Hilgardia 27*:269–285.

Georghiou, G. P. 1967. Differential susceptibility and resistance to insecticides of co-existing populations of *Musca domestica, Faunia canicularis, F. femoralis* and *Ophyra leucostoma. J. Econ. Entomol. 60*:1338–44.

Georghiou, G. P. 1972. The evolution of resistance to pesticides. *Annu. Rev. Ecol. Syst. 3*:133–168.

Hall, D. C., and R. B. Norgaard. 1973. On the timing and application of pesticides. *Am. J. Agric. Econ. 55*:198–201.

Headley, J. C. 1972. Defining the economic threshold, in *Pest Control Strategies for the Future*. National Academy of Sciences, Washington, D.C., pp. 100–108.

Helle, W. 1965. Resistance in Acarina: Mites. *Adv. Acarol. 3*:71–93.

Helle, W., and M. van de Vrie. 1974. Problems with spider mites. *Outlook on Agric. 8*: 119–125.

Herne, D. H. C., and A. W. A. Brown. 1969. Inheritance and biochemistry of OP-resistance in a New York strain of the two-spotted spider mite. *J. Econ. Entomol. 62*:205–209.

Herne, D. H. C., and W. L. Putman. 1966. Toxicity of some pesticides to predaceous arthropods in Ontario peach orchards. *Can. Entomol. 98*:936–942.

Heuth, D. and U. Regev. 1974. Optimal agricultural pest management with increasing pest resistance. *Am. J. Agric. Econ. 56* : 543–552.

Hill, A. R. Cardé, A. Comeau, W. Bode, and W. Roelofs. 1974. Sex pheromones of the tufted apple budmoth (*Platynota idaeusalis*). *Environ. Entomol. 3* : 249–252.

Howitt, A. J., E. J. Klos, P. Corbett, and A. P. Shea. 1966. Aerial and ground ultra low volume applications in the control of diseases and pests attacking deciduous fruits. *Mich. Agric. Exp. Stn. Q. Bull. 49* : 90–102.

Hoying, S. A., and B. A. Croft. 1976. The role of the apple rust mite *Aculus schlectendali* (Nalepa) in Michigan apple orchards. *Abst. Proc. N. Cent. Br. Entomol. Soc. Am. 30* : 89.

Hoyt, S. C. 1972. Resistance to azinphosmethyl of *Typhlodromus pyri* from New Zealand. *N.Z. J. Sci. 15* : 16–21.

Hoyt, S. C., and E. C. Burts. 1974. Integrated control of fruit pests. *Annu. Rev. Entomol. 19* : 231–252.

Hoyt, S. C., and L. E. Caltagirone. 1971. The developing programs of integrated control of pests of apples in Washington and peaches in California, in C. B. Huffaker (ed.), *Biological Control*. Plenum, New York, pp. 395–421.

Huffaker, C. B., and B. A. Croft. 1976. Integrated pest management in the U.S. : Progress and promise. *Environ. Health Perspect. 14* : 167–183.

Huffaker, C. B., and C. E. Kennett. 1953. Differential tolerance to parathion in two *Typhlodromus* predatory on cyclamen mite. *J. Econ. Entomol. 46* : 707–708.

Hull, L. A., D. Asquith, and P. D. Mowery. 1976. Distribution of *Stethorus punctum* in relation to densities of the European red mite. *Environ. Entomol. 5* : 337–342.

Hull, L. A., D. Asquith, and P. D. Mowery. 1977. The functional responses of *Stethorus punctum* to densities of the European red mite. *Environ. Entomol. 6* : 85–90.

Johnson, Donn T., and B. A. Croft. 1976. A study of the " dispersal " behavior of *Amblyseius fallacis* (Garman) in the laboratory (Acarina: Phytoseiidae). *Proc. N. Cent. Br. Entomol. Soc. Am. 30* : 49–52.

Johnson, Donn T., and B. A. Croft. 1977. Laboratory study of the " dispersal " behavior of *Amblyseius fallacis* (Garman) (Acarina: Phytoseiidae). *Ann. Entomol. Soc. Am. 69* : 1019–1023.

Jones, A. L. 1974. Monitoring and controlling fruit diseases in Michigan in 1973. *Mich. Sta. Hort. Soc. Proc. 103* : 100–109.

Kennett, C. E. 1970. Resistance to parathion in the phytoseiid mite *Amblyseius hibisci*. *Mich. Sta. Hort. Soc. Proc. 63* : 1999–2000.

Kot, J. T., T. Plewka, and I. Krukierek. 1971. Relationship in parallel development of host and parasite resistance to a common toxicant—Final technical report PL-480, E21-Ent-19, F6-Po-203. *Inst. Ecol. Polish Acad. Sci., Wyk Dz. Inf. PAN Zam. 25/71.*

Lewis, F. H. 1974. Partial analysis of the problem of minimizing fungicide usage on apples. *Pa. Fruit News. 53* : 63–66.

Logan, J. A., D. J. Wollkind, S. C. Hoyt, and L. K. Tanigoshi. 1976. An analytic model for description of temperature dependent rate phenomenon in arthropods. *Environ. Entomol.* 5:1133–1140.

Madsen, H. F., and J. M. Vakenti. 1973. Codling moth: use of codlemone-baited traps and visual detection of entries to determine need of sprays. *Environ. Entomol.* 2: 677–679.

McGoarty, D. L., and B. A. Croft. 1976. Sampling of populations of *Amblyseius fallacis* (Acarina: Phytoseiidae) in the ground cover of Michigan commercial apple orchards. *Proc. N. Cent. Br. Entomol. Soc. Am.* 30:49–52.

McMurty, J. A., C. B. Huffaker, and M. van de Vrie. 1970. Ecology of tetranychid mites and their natural enemies: A review. I. Tetranychid enemies: Their biological characters and the impact of spray practices. *Hilgardia* 40:331–390.

Meyer, R. H. 1975. Release of carbaryl-resistant predatory mites in apple orchards. *Environ. Entomol.* 4:49–51.

Moreno, D. S., G. E. Carman, R. E. Rice, J. G. Shore, and N. S. Bain. 1972. Demonstration of a sex pheromone of the yellow scale, *Aonidiella citrina*. *Ann. Entomol. Soc. Am.* 65:443–446.

Motoyama, N., G. C. Rock, and W. C. Dauterman. 1970. Organophosphorus resistance in an apple orchard population of *Typhlodromus (Amblyseius) fallacis*. *J. Econ. Entomol.* 63:1439–1442.

Motoyama, N., G. C. Rock, and W. C. Dauterman. 1971. Studies on the mechanism of azinphosmethyl resistance in the predaceous mite *Neoseiulus (T.) fallacis*. *Biochem. Physiol.* 1:205–215.

Mowery, P. D., D. Asquith, and W. M. Bode. 1975. Computer simulation for predicting the number of *Stethorus punctum* needed to control the European red mite in Pennsylvania apple trees. *J. Econ. Entomol.* 68:250–254.

Mowery, Paul D., Dean Asquith, and L. A. Hull. 1977. MITESIM, computer predictive system for European red mite management. *Pa. Fruit News* 56(4):64–67.

Naqvi, S. M. Z. 1970. Comparative pesticide tolerance of selected freshwater invertebrates in Mississippi. Ph.D. thesis, Dept. of Zoology, Mississippi State University.

Newsom, L. D. 1974. Predator insecticide relationships. *Entomophaga Mem. Ser. 7*.

Pielou, D. P., and R. F. Glasser. 1951. Selection for DDT tolerance in a beneficial parasite, *Macrocentrus ancylivorus*. 1. Some survival characteristics and the DDT resistance of the original laboratory strain. *Can. J. Zool.* 29:90–101.

Redmond, K. R., and J. R. Brazzel. 1968. Response of the striped lynx spider, *Oxyopes salticus* to two commonly used pesticides. *J. Econ. Entomol.* 61:327–328.

Regev, U., A. P. Gutierrez, and G. Feder. 1975. Pest as a common property resource: a case study in the control of the alfalfa weevil. *Calif. Agric. Expt. Stn. Publ.*

Reidl, Helmut, B. A. Croft, and A. J. Howitt. 1976. Forecasting codling moth phenology based on pheromone trap catches and physiological-time models. *Can. Entomol.* 108:449–460.

Reissig, W. H. 1975. Evaluation of traps for apple maggot in unsprayed and commercial apple orchards. *J. Econ. Entomol.* 68:445–448.

Rice, R. E. 1974. San Jose scale: Field studies with a sex pheromone. *J. Econ. Entomol.* *67*:561–562.

Rice, R. E. 1976. A comparison of monitoring techniques for the navel orangeworm. *J. Econ. Entomol. 69*:25–28.

Rice, R. E., L. L. Sadler, M. L. Hoffmann, and R. A. Jones. 1976. Egg traps for the navel orangeworm, *Paramelois transitella* (Walker). *Environ. Entomol. 5*:697–700.

Rock, G. C., and D. R. Yeargen. 1971. Relative toxicity of pesticides to organophosphorus-resistant orchard populations of *Neoseiulus fallacis* and its prey. *J. Econ. Entomol. 64*:350–352.

Robertson, J. G. 1958. Changes in resistance to DDT in *Marcocentrus ancylivorus* Rohw. (Hymenoptera: Braconidae). *Can. J. Zool. 35*:629–33.

Rodriguez, J. G. 1958. The comparative NPK nutrition of *Panonychus ulmi* (Koch) and *Tetranychus telarius* (L.) on apple trees. *J. Econ. Entomol. 51*:369–373.

Shaw, J. G., D. S. Moreno, and J. Fargerlund. 1971. Virgin female California red scales used to detect infestations. *J. Econ. Entomol. 64*:1305–1306.

Siddall, J. B., and C. M. Olson. 1976. Pheromones in agriculture—from chemical synthesis to commercial use, in Morton Beroza (ed.), *Pest Management with Insect Sex Attractants. Am. Chem. Soc. Symp. Ser. 23*:88–90. American Chemical Society, Washington, D.C.

Shoemaker, Christine. 1973. Optimization of agricultural pest management. III. Results and extensions of a model. *Math. Biosci. 18*:1–22.

Spiller, D. 1958. Resistance of insects to insecticides. *Entomologist 2*:1–18.

Talpaz, H., and I. Borosch. 1974. Strategies for pesticide use: Frequency and applications. *Ann. J. Agric. Econ. 56*:769–775.

Tanigoshi, L. K., L. W. Browne, and S. C. Hoyt. 1975. A study on the dispersion pattern and foliage injury by *Tetranychus mcdanieli* (Acarina: Tetranychidae) in simple apple ecosystems. *Can. Entomol. 107*:439–446.

Tanigoshi, L. K., R. W. Browne, S. C. Hoyt, and R. F. Lagier. 1976. Empirical analysis of variable temperature regimes on life stage development and population growth of *Tetranychus mcdanieli* (Acarina: Tetranychidae). *Ann. Entomol. 69*:712–716.

Tanigoshi, L. K., S. C. Hoyt, R. W. Browne, and J. A. Logan. 1975a. Influence of temperature on population increase of *Tetranychus mcdanieli* (Acarina: Tetranychidae). *Ann. Entomol. Soc. Am. 68*:972–978.

Tanigoshi, L. K., S. C. Hoyt, R. W. Browne, and J. A. Logan. 1975b. Influence of temperature on population increase of *Metaseiulus occidentalis* (Acarina: Phytoseiidae). *Ann. Entomol. Soc. Am. 68*:979–986.

Tashiro, H., and D. L. Chambers. 1967. Reproduction in the California red scale, *Aonidiella aurantii* (Homoptera: Diaspididae). 1. Discovery and extraction of a female sex pheromone. *Ann. Entomol. Soc. Am. 60*:1166–1170.

Thompson, W. W., C. F. Stephens, L. G. Olson, J. E. Nugent, and T. B. Sutton. 1973. Michigan apple pest management annual report: 1972. Coop. Ext. Serv. Mich. Sta. Univ.

Thompson, W. W., C. F. Stephens, L. G. Olson, J. E. Nugent, and T. B. Sutton. 1974. Michigan apple pest management annual report: 1973. Coop. Ext. Serv. Mich. Sta. Univ.

Trammel, K. 1974. Apple pest management potential in New York. *N.Y. State Hort. Soc. Proc. 119*:114–125.

Van ve Vrie, M. 1962. The influence of spray chemicals on predatory and phytophagous mites on apple trees in laboratory and field trials in the Netherlands. *Entomophaga. 7*:243–250.

Van de Vrie, M., J. A. McMurtry, and C. B. Huffaker. 1972. Ecology of tetranychid mites and their natural enemies. III. Biology, ecology and pest status and host–plant relations of tetranychids. *Hilgardia 41*:343–432.

Watve, C. M., and S. E. Lienk. 1975. Responses of two phytoseiid mites to pesticides used in New York apple orchards. *Environ. Entomol. 4*:797–800.

Watve, C. M., and S. E. Lienk. 1976. Resistance to carbaryl and six organophosphorus insecticides of *Amblyseius fallacis* and *Typhlodromus pyri* from New York apple orchards. *Environ. Entomol. 5*:368–370.

Welch, S. M., B. A. Croft, J. F. Bruner, and M. F. Michels. 1978. PETE: An extension phenology modeling system for management of a multi-species pest complex. *Environ. Entomol. 7*:487–494.

10

ACCOMPLISHMENTS TOWARD IMPROVING INTEGRATED PEST MANAGEMENT FOR CITRUS

L. A. Riehl

Department of Entomology
University of California, Riverside, California

R. F. Brooks
C. W. McCoy

Agricultural Research and Education Center
University of Florida, Lake Alfred, Florida

T. W. Fisher

Department of Entomology
University of California, Riverside, California

H. A. Dean

Texas A&M University Agricultural Research and Extension Center
Weslaco, Texas

319

INTRODUCTION

The general objectives, organization, and pattern of research of the International Biological Program's (IBP) Integrated Pest Management Project, supported by NSF and EPA mainly and dealt with in this book, were covered in some detail in Chapter 1, and the modeling aspects in Chapter 2. The salient features relating to the citrus subproject concerning background and justification, its specific objectives, plan of the program, and conduct of research are treated here, although this entails some duplication of, in a general but not the specific sense, material described in Chapters 1 and 2 and in other chapters dealing with other subprojects. The modeling effort for citrus is covered in Chapter 11.

Citrus is a major United States food crop, constituting in 1968 (just prior to this program's being planned) about 3 percent of all food consumed in the country, with a consumption of 41.7 lbs. per capita. The national citrus crop for the 1968–69 year was valued at $664,682,000. In Florida in 1969 citrus comprised 34 percent of the total value of that state's agricultural commodities. Citrus comprises about 33 percent of all fruits consumed in the United States (fresh and processed), and the United States produces about 36 percent of the world's oranges, 30 percent of its lemons, and 75 percent of its grapefruit. Citrus is rich in its natural vitamin C and other nutritional qualities. Its general appeal to the consumer is obvious.

Pest control practices for citrus present many opportunities for improvements in economic gain to the grower while substantially lessening the adverse environmental effects which the heavy use of pesticides for citrus pest control have occasioned, and continuingly so. During 1966 some 10 million lbs. (active ingredients) of insecticides and 18 million lbs. of fungicides were used in United States citrus.

Citrus was thought to offer unusual opportunities as a model of what can be done in scientific pest management. We have already established a very strong base of bio-ecological knowledge relating to the population dynamics and management of citrus pests and their natural enemies. The first great success in biological control was achieved against the cottony cushion scale on citrus. This pest, of devastating potential, remains under excellent biological control after some 85 years, and this fact must be considered in any decisions made relative to treatments for other pests on this crop. Integrated control or pest management has been practiced in scattered instances for many years in California citrus groves, and elsewhere. Insectaries for the mass production and field colonization of natural enemies were established in the 1920s in California. These methods, which have reduced the need for insecticides, were based partly on basic biology and ecology, partly on intuition and empirical developments. None has yet achieved generally widespread utilization but they indicate definite possibilities.

The Ecological Society of America has concluded (as stated in the management component of this proposal) that "much basic information already known to science is not finding its way into the decision-making process" (BioScience, 1970, 20(24):1285). To a considerable extent this is where we stand today in citrus pest management ecology. We already have a large part of the scientific information required for a successful solution. On the other hand we still lack some essential knowledge of certain key pests, natural enemies, or processes, and we thus far have not achieved industry or grower acceptance and application of what is already known to be practical for much improved pest management.

From extensive (but still insufficient) foreign exploration and study in the native home areas of many citrus pests we know that many, if not most, are held under natural control by parasites or predators in one place or another. We know of, and have imported and established some of these natural enemies but many more await discovery, testing, and use. Taxonomic experts estimate that only 10 to 25 percent of the parasitic Hymenoptera are known. Nonetheless we have more knowledge regarding biological control of citrus pests than is true for any other crop.

It is also well recognized that citrus pest problems (as with such problems on other crops) may differ between habitats within a state, between states, and between countries, so that no single or simple solution is expected. This situation means that basic ecosystem studies in each major region are required. This is especially true where the crop, as with citrus, has a large complex of pest species. The biology of most key citrus pests is fairly well known but it is obvious that we still know too little in many instances.

In line with the aforestated point that we already have much basic information that is not being used in commercial control of citrus pests, the Citrus Subproject set out to do three things primarily: (1) to explore for and introduce new natural enemies for use against insect and mite pests of citrus, (2) to evaluate programs of pest control for the complex of citrus pests that would interfere least with biological control of citrus insect and mite pests while reducing costs and pesticides loads in the environment, and (3) to develop a more sophisticated understanding of the crop itself, the pests, the natural enemies, and the interactions among these various factors and the processes in citrus culture and marketing, including the economic costs and benefits, both internal to the industry and external to it.

Because there existed substantial information supporting the concept that many otherwise key insect pests of citrus could be controlled by natural enemies already introduced or naturally resident in United States citrus if the pesticide programs used in citrus could be modified (or even deleted in certain areas or seasons), biological control was taken as the basic cornerstone of the Citrus Subproject IPM effort.

Many major citrus pest problems have been solved by importation of exotic natural enemies. Complete biological control of all insect and mite pests on citrus has even been attained in some instances in individual groves or areas in various countries. This represents the ultimate in ecological pest management—permanent regulation of potential pests at a subeconomic level without the necessity of continuing control costs. In most citrus areas, however, many problems remain to be solved before general extension or application of satisfactory pest management can be achieved and, currently, unilateral chemical pest control continues to be the vastly dominant method used, to the detriment of long term interests of the growers, the consumers, and the environment.

Examples of formerly very serious citrus pests in the United States that have been reduced and are being permanently maintained at subeconomic levels by purposely* imported natural enemies are as follows:

1. The cottony cushion scale, *Icerya purchasi* Maskell: completely controlled in California, Florida, Texas, and Arizona by *Rodolia cardinalis* (Mulsant) from Australia.
2. The Florida red scale, *Chrysomphalus aonidum* (Linnaeus): completely controlled in Florida and Texas by *Aphytis holoxanthus* DeBach from Hong Kong.
3. The purple scale, *Lepidosaphes beckii* (Newman): completely controlled in Texas and Florida, and substantially controlled in California by *Aphytis lepidosaphes* Compere from S. China.
4. The black scale, *Saissetia oleae* (Olivier): completely controlled in certain California areas, substantially in others, by *Metaphycus helvolus* (Compere) from South Africa.
5. The citrophilus mealybug, *Pseudococcus fragilis* Brain: completely controlled in California by *Coccophagus gurneyi* Compere and *Tetracnemus pretiosus* Timberlake from Australia.
6. The California red scale, *Aonidiella aurantii* (Maskell): completely controlled in certain California areas, substantially in some, only partially in others, by *Aphytis melinus* DeBach, *A. lingnanensis* Compere, *Comperiella bifasciata* How. (red scale strain), and *Prospaltella perniciosi* Tower from India, Pakistan, and S. China.
7. The yellow scale, *Aonidiella citrina* (Coquillett): completely controlled in California (when not upset by pesticides) by *C. bifasciata* (yellow scale strain) and *A. melinus* from Japan and India-Pakistan, respectively.

These successes truly were the key to the limited amount of pest management that existed prior to initiation of the project. If all of these pests had continued at their formerly serious status, so much chemical treatment would be required that it is doubtful if any significant degree of real ecological pest management would be possible. Hopefully, with the future biological control of certain of our current major or key pests, such as the snow scale, *Unaspis citri* (Comstock), in Florida, true ecological pest management is made easier.

Many ecological studies on citrus in California, Florida, Texas, and abroad have shown that biological control may be negated by the adverse influence of environmental factors on natural enemies. Such factors principally include pesticides of many types (both direct application and drift), climatic extremes, airborne dust which acts like a toxin, certain cultural practices, and ants which interfere with natural enemies. There is little question but that

*Note that many others are maintained under natural biological control, either by indigenous enemies, or by ones that become accidentally imported.

certain pesticides constitute the most adverse factor of all. Their responsibility for upsets of nontarget organisms and above-normal resurgences of target pests has been well documented in citrus. The initial use of pesticides tends to accentuate and perpetuate their continued use—thus presenting the aspect of an "ecological pesticides addiction."

The problem of how to back away from the use of chemicals and to develop good ecologically based pest management is a major one. We know from field tests on citrus that various means of manipulation of established natural enemies as well as their conservation has proven practical or shows real promise in this regard. Mass production and periodic colonization is one approach. Field trials make it clear, however, that very careful monitoring of the ecology of the entire faunal complex is essential for reliable pest management decisions and yet this had not been done prior to establishment of this project. Further, this points up the necessity of having trained bio-ecologists as the final decision makers. Such decisions, however, must be based on a complex of inputs, adequately structured and molded, and embracing economic and systems analysis of the entire pest control and crop production system (above).

We have lacked the scientific information or tools (such as natural enemies) to solve certain key problems; we lacked sound economic data and adequate analysis of the results of alternative approaches; also, the research disciplines have been operating more or less independently, and this precluded systems analysis. What was lacking and vitally needed was a unified multidisciplinary well organized team effort to integrate and properly apply the various ingredients. To obtain this understanding we needed substantially more funds as well as some on-the-job retraining of personnel. Furthermore, we were and are hampered by "social" constraints to the achievement of pest management, especially in citrus—that is, there are artificial industry-set standards (often "cosmetic") for rating insect damage, and the chemical control method is very strongly entrenched both economically and traditionally; and finally there are some external relations to other crops or practices which adversely influence pest management on citrus. The latter category includes drift of pesticides from other crops (very serious in Texas), airborne dust from outside sources, and occasionally immigration of pests from other crops or habitats into citrus.

OBJECTIVES AND OVERALL PLAN

Our broad goals in citrus pest management, as in the general proposal, are to optimize the cost-benefit ratio and at the same time minimize adverse environmental effects. To accomplish this, the more precise goals are to (1) maximize biological control and other nonchemical control methods and thus reduce chemical costs to the grower and external costs to society in general in terms of environmental pollution, and cheaper produce; (2) use chemicals only when necessary and then in a highly selective manner so that they are least disruptive to the entire beneficial faunal complex and so that the amount used is the minimum required.

Precise knowledge of what constitutes a pest or pest damage is critical in determining the allowable population level for citrus pests. We needed to determine by appropriate economic and field plot studies what constitutes the treatment threshold level and to raise these levels when they have been arbitrarily set too low. Much of this has been done arbitrarily. The California red scale may be judged a pest by the industry if only a single scale is present on some of the fruit per tree. Entomologists often make intuitive decisions as to what the economic injury threshold is. The single or specific threshold density at which a pest population requires treatment may differ at diverse locations and with different citrus varieties in which a natural enemy is not established versus those in which the natural enemy is or can be established.

The principal methods of citrus insect and mite pest limitation that appeared practical for immediate experimentation, manipulation, or employment in the citrus program are clear-cut. They are (1) use of natural enemies (parasites, predators, and pathogens) by (a) importation and colonization of new ones from abroad, (b) conservation of established enemies by mitigation of adverse influences or provision of lacking requisites, and (c) augmentation of established enemies by periodic colonization or selective breeding for improved fitness; (2) use of selective chemicals and selective use of chemicals to achieve the necessary control (which of necessity should be clearly established) of target pests in such a way as to have minimal effects on non-target organisms and the environment; (3) use or modification of cultural techniques either to directly achieve pest limitation or to indirectly do so by favoring or conserving natural enemies; and (4) use of pheromone traps (now being tested against California red scale males).

Although plant resistance is an elegant tool for pest limitation, it does not appear to have much possibility for short-term application in citrus. Selection for new lines in citrus usually requires many years and then market acceptance of new varieties seems almost impossible to achieve.

The broad research lines that have been most emphasized in the Citrus Subproject are as follows:

1. Evaluation of the extent of biological control occurring in long-term untreated plots in order to determine which are real pests actually lacking effective established natural enemies as distinguished from "apparent" pests which have been induced to pest status by pesticidal upset or otherwise.

2. Determination or development of selective pesticides for use against real pests in order to minimize the side-effects on the natural enemies, on environmental pollution and on overall natural balance in the ecosystem.
3. Development of improved biological control, or of alternative biological (nonchemical) methods, for regulation of pest populations at subeconomic densities. This involves importation and colonization of new natural enemies from abroad, experiments on conservation of natural enemies by habitat manipulation, tests on the augmentation of natural enemies by mass production and periodic colonization or by selective breeding, as well as tests on cultural or habitat manipulations that may directly or indirectly reduce pest populations and tests involving use of pheromones.
4. Determination of possible interactions between various production and protection practices (i.e., cultivation, fertilization, use of minor elements, deficiency sprays, pruning, treatment for plant pathogens, etc.) and optimal pest management.
5. Study of the economics of pest control, pest damage, consumer acceptance, injury thresholds, and the basis for and uniformity of packing house, association, and industry market screening standards.
6. Methods of extension and application of scientific pest management programs.

In Florida, the snow scale, the citrus rust mite, *Phyllocoptruta oleivora* (Ashmead), and other mites have received major emphasis. The snow scale in Florida lacked any significant natural enemies so new ones were sought in its original home areas in the Orient. (See below.) Selective acaracide use as well as effects of natural enemies have been studied. Special attention has been given to developing models for citrus growth and phenology and of rust mite dynamics (Chapter 11).

In Texas, the citrus rust mite, the Texas citrus mite, *Eutetranychus banksi* (McGregor), the chaff scale, *Parlatoria pergandii* Comstock, California red scale, and several pests of secondary pest upset nature, constituted the major problems. The brown soft scale, *Coccus hesperidum* Linnaeus, normally completely controlled by parasites, has been severely upset by drift of chemicals from cotton. Subsequently, attempts to control this scale with carbaryl applications led to increases in Texas citrus mites, chaff scale, and California red scale which tended to lead to further increased use of chemicals. Tests on selective use of pesticides have also been stressed. A major approach both in Texas and Florida has involved at least one large ecosystem study block such as those described for the California program in the next paragraph.

In California, pest problems differ markedly in different areas, primarily because of the great diversity in climate. In addition to several necessary subsidiary efforts, the major effort has involved a series of ecosystem study plots which have been studied in at least three different climatic zones—the Central Valley, the Southern Coast, and the Southern Interior. Major or key pests differ in these districts. In the Central Valley, the California red scale, the citricola scale, *Coccus pseudomagnoliarum* (Kuwana), the citrus thrips, *Scirtothrips citri* (Moult.), and the citrus red mite, *Panonychus citri* (McGregor), have received emphasis; in the Southern Coast the citrus red mite, the bud mite, *Eriophyes sheldoni* Ewing, on lemons and citrus thrips have been most important; in the Southern Interior area, the California red scale, the citrus red mite, and the citrus thrips are the major pests. Chemical treatment for each of the major pests mentioned causes upsets among minor pests and their natural enemies; hence, greater understanding of the ecology of the major pests and their natural enemies has been the key to the approach to achievement of better, ecologically oriented pest management.

RESULTS

Studies in commercial citrus groves in California, Florida, and Texas are herein reported and partially evaluated following three consecutive harvest periods ending with the 1975 harvest (early 1976). Because of differences in soil, climate, varieties, cultural, harvesting, and marketing factors, and in the key pests, and their natural enemies, very few direct comparisons can be made between the results obtained in the different states. The results in each state must be evaluated within the context of its own parameters. Furthermore, even within California there are four climatically distinct citrus growing areas, each with its own complex of pest problems. Largely for this reason, progress of the work in these states will be reported separately.

California Results

Ecosystem Study Plots

Ecosystem study plots were established in commercial navel orange groves in four localities in California. Major plots consisting of six subplots, each 5 acres in size, were established near Woodcrest, in the vicinity of Riverside (Southern Interior district) and near Orosi (Central Valley district). These major plots provided for the testing of six different pest treatment-management regimes. Other plots consisting of three subplots, each 5 acres in size, were established near Fillmore (intermediate Southern Coastal district) and near Famoso (lower Central Valley district) and each included three of the treatment-management regimes used for the major plots. Additional satellite, or observation, plots furnished information for other citrus varieties and for other areas. Unless otherwise noted, this report deals only with the Woodcrest study where the research found an integrated pest management program applicable by the citrus industry to orange and grapefruit orchards from the coastal to the interior zones of southern California.

Both the size of the ecosystem subplots and a virtually square shape were chosen to provide an adequate buffer band to protect the central area, or core acre, of the subplot against drift from the perimeter of pesticides from adjoining differently treated subplots. Of particular importance in the biological control subplots, the core acre was totally shielded from the influence of treatments outside its own subplot and data from this core acre and its surrounding buffer trees within the same subplot were kept separate. The square design used was also more compatible with the cultural practices used by the grower. The treatment-management regimes used were superimposed on the growers' usual irrigation, fertilization, and weed control practices.

The data collected include (1) counts of populations of pests and of natural enemies, (2) the amount of yield of subplots on both a per tree and a commercial pack basis (for the subplot), and (3) external quality of the oranges as graded to provide information on the marketability and market use of the crop. The relationships among these various data form the bases for the conclusions presented here. The results through 1975 in two locations in southern California, near Fillmore, and particularly in the vicinity of Riverside (Southern Interior district) demonstrate that one of the programs employed can serve as a prototype, or basic model, for a practical integrated pest management program for both oranges and grapefruit in the southern California area, excluding the desert district. At present, in the area indicated, the combined total acreage of oranges and grapefruit is 76,000. For this acreage, on the basis of the current average cost of control of insect and mite pests, use of the integrated pest management program could result in an estimated savings to growers of approximately $4 million annually.

California recommendations for control of insect and mite pests of citrus (Div. Agr. Sci., Univ. Calif., 1976) lists 30 species separately and groups of citrus aphids and mealybugs. Of these, 10 species are not or are only rarely present on navel oranges in the Riverside area and natural enemies control one armored scale, three unarmored scales, and four species of mealybugs. Of the remaining 16 species and citrus aphids, the following three, California red scale, *Aonidiella aurantii* (Mask.), citrus red mite, *Panonychus citri* (McG.), and citrus thrips, *Scirtothrips citri* (Moult.), are generally considered to be key pests. California red scale is the principal one because without suppression it can kill the citrus tree in a relatively short span of time.

DeBach, Rosen, and Kennett (1971) note that *Aphytis melinus* DeBach was introduced against California red scale in southern California in 1956 to 1957 and show its geographic distribution as generally established in interior southern California districts by 1965. *A. melinus* was established as a parasite of this scale in the citrus orchards at Woodcrest at the time the integrated pest management experiment was begun at that location.

Treatment Programs at Woodcrest. The general schedules of the annual programs of foliar treatments used in the ecosystem study subplots at Woodcrest are given in Table 10.1. A treatment at pre-bloom was not used routinely but was made only if required for control of citrus red mite. The petal-fall treatment is scheduled routinely to control citrus thrips. The mid-June treatment of Program I is for control of California red scale. The September treatment in Programs V and VI is primarily for control of citrus red mite. All of the treatment programs include applications of the nutritional minor elements of $ZnSO_4$ and $MnSO_4$ in the spring and in the fall. These materials are water soluble and are applied either by mist spray (MS) or low volume (LV) techniques. These applications do not seriously upset populations of natural enemies, especially *A. melinus*, the key parasite of California red scale.

In Table 10.1, Program I is the reference program of recommended organochemical pesticides that accords with general pest control practices of the area. The parathion treatment in mid-June is in accord with recommendations as listed (Div. Agr. Sci., Univ. Calif., 1976); in the Woodcrest orchard this treatment would apply 7 lbs of parathion (active ingredient— AI) per acre. For the treatment the air-blast machine was selected for spraying; in other orchards having taller trees or denser foliage an oscillating boom sprayer would be needed. For our purposes an exception to general practice was made and a variation was actually used; an air-blast sprayer operated at a ground speed of 1.0 mph was used to apply a spray of 3.75 lbs. of parathion AI in 1000 gal of water per acre. The compensating factor is the slower ground speed, which produces a more effective distribution of spray into the interior of the tree. In this treatment it is essential to keep exactly to the specified ground speed and consequently to proper nozzling; while a longer time is taken to complete the spraying (1.0 vs. 1.4 mph) there is a compensation in the use of less insecticide, for example, parathion. Program II was planned as a means of investigating the results of integrated pest management using an organochemical acaricide for citrus red mite control and with no treatment for California red scale to be made until it became necessary as ascertained by monitoring, in which case the treatment to be with parathion. Programs III and IV were exclusively biological control; foliar insecticide and acaricide treatments were not used. Programs V and VI used a foundation treatment of oil spray in September primarily for citrus red mite control (above); the difference between them is the treatment used for citrus thrips control, that is, with dimethoate in Program V as opposed to Ryania in Program VI.

In addition to the reference given in footnote *d* of Table 10.1, information

Table 10.1. Foundation Treatments of the Treatment-Management Regimes Used for Insect and Mite Control in the Ecosystem Study Plots at Woodcrest—1973–1975.

Period	Program I Conventional Recommended Organochemical Insecticides & Acaricides	Program II I.P.M. with Organochemical Acaricides & Botanical Insecticide	Programs III & IV Biological Control	Program V I.P.M. with Spray Oil & Organochemical Insecticide	Program VI I.P.M. with Spray Oil & Botanical Insecticide
Pre-bloom April	Acaricide® [a] MS[c] Minor Elements: (ZnSO$_4$ & MnSO$_4$)[b] at 5 lbs/acre MS	Acaricide® [a] MS Minor Elements: (ZnSO$_4$ & MnSO$_4$)[b] at 5 lbs/acre MS	Minor Elements: (ZnSO$_4$ & MnSO$_4$)[b] at 5 lbs/acre MS	Minor Elements: (ZnSO$_4$ & MnSO$_4$)[b] at 5 lbs/acre	Minor Elements: (ZnSO$_4$ & MnSO$_4$)[b] at 5 lbs/acre MS
Petal-fall May	Dimethoate 1.34 lbs/AI/acre MS	Ryania proprietary prep 15 lbs + 6 lbs sugar/acre MS		Dimethoate 1.34 lbs AI/acre MS	Ryania proprietary prep 15 lbs + 6 lbs sugar/acre MS
Mid-June post petal-fall	Parathion 3.75 lbs AI in 100 gal water Air-blast machine spray application TDC[c]				
September	Minor Elements: (ZnSO$_4$ & MnSO$_4$)[b] at 5 lbs/acre MS Acaricide MS	Minor Elements: (ZnSO$_4$ & MnSO$_4$)[b] at 5 lbs/acre MS Acaricide MS	Minor Elements: (ZnSO$_4$ & MnSO$_4$)[b] at 5 lbs/acre MS	N-R 415 spray oil[d] 10 gal/acre + 22 ml 2,4-d-e to spray oil Minor Elements: (ZnSO$_4$ & MnSO$_4$)[b] at 5 lbs/acre 100 gal/acre LV[e]	N-R 415 spray oil[d] 10 gal/acre + 22 ml 2,4-D[e] to spray oil Minor Elements: (ZnSO$_4$ & MnSO$_4$)[b] at 5 lbs/acre 100 gal/acre LV

[a] Oxythioquinox (Morestan®) at 1.875 lb AI/acre; Propargite (Omite®) at 4.5 lb AI/acre; or, 2,3 hexakis (β,β-dimethyl-phenethyl) distannoxane (Vendex®) at 2.0 lb AI/acre.
[b] Proprietary mixture of ZnSO$_4$ & MnSO$_4$ containing 18% Zn and 13½% Mn.
[c] MS = mist spray; TDC = thorough distribution coverage-median gallonage; LV = low volume types of spray coverage (see text).
[d] N-R 415 spray oil (see text).
[c,d] Div. Agr. Sci., Univ. Calif. 1976. 1976–1978 Treatment Guide for California Citrus Crops, pp. 2–95, Berkeley.
[e] 2,4-D, 44% isopropyl esters, Citrus Formulation.

on the narrow-range 415 (N-R 415) type of spray oil used is presented by Riehl (1969). The treatment listed for September in Programs V and VI was made by the low volume method with application of 100 gal of spray mixture per acre. This method uses an air stream pattern of high velocity as the primary medium for distributing the spray mixture to the surface of the tree. The machines used in the ecosystem subplots at Woodcrest from 1973 through 1975 were the low silhouette model. Models of these machines are now available commercially which have spray discharge heads positioned at two heights on an " air tower " (Carman, 1977). The modification adding on the upper spray discharge heads gives a great improvement in the spray distribution coverage in the upper and top areas of citrus trees. This is an important dimension in the control of California red scale and citrus red mite, key pests which may require treatment in an integrated control program. The droplets of the deposit of low volume oil spray are very small, having a media volume diameter (MVD) of less than 150 microns, and are discrete. These properties give a reduced adverse effect on natural enemies, such as the predaceous mites which feed on citrus red mites and, particularly, on *A. melinus*, the California red scale parasite—that is, less adverse effect than that occasioned by complete surface wetting to the point of runoff, as occurs with dilute oil sprays applied with sufficient volume and pressure to accomplish complete film wetting of the foliage surfaces.

California Red Scale.   As mentioned previously California red scale is considered the most serious pest of citrus in California and, therefore, controlling it is the central focus of an integrated pest management program for citrus in this area. The results from the annual treatment-management programs, shown in Table 10.1, on the densities of California red scales in the ecosystem subplots in the fall months are listed in Table 10.2 for the years 1973 (Yr 1) through 1975. The months selected for the data reported in Table 10.2 are at the end of the warm season in California when the highest densities of this scale are generally found in orchards. For our purpose the data for these months seem to offer a concise way of comparing the net results of the different regimes relative to control of California red scale. The data of Table 10.2 are the mean number of unfertilized (" gray " adult) and fertilized adult female California red scales per twig unit of 3 inches in length (samples of 24 twig sections about 1.5 years old) chosen in the peripheral foliage area at a height of 3 to 6 ft above the ground, otherwise at random, from 5 sample trees selected in a pattern representing the area of the core acre of the treatment program subplots (experimental orchard at Woodcrest in the interior district of southern California).

The results of Table 10.2 show good control of California red scale under Program I and relatively low numbers of scales under Programs II, V, and

Table 10.2.   Mean Numbers of Adult Female California Red Scales per Twig Unit, Each 3 in. Long (Peripheral Foliage Area) on Sample Trees in the Core Area of Ecosystem Study Subplots at Woodcrest—August to November, 1973–1975

Time	Program I Conv. Rec. Org-chem. Ins. & Acar.	Program II I.P.M. Org-chem. Acar. Bot. Ins.	Programs III Biol. Contr.	IV Biol. Contr.	Program V I.P.M. Oil & Org.-chem. Ins.	Program VI I.P.M. Oil & Bot. Ins.
1973						
August	0	0	0.01	0.025	0.075	0
September	*a*					
October	0	0	0.02	0.15	0	0
November	0	0.05	0.27	0.40	0.09	0
1974						
August	0	0.02	0.16	0.61	0.20	0
September	0	0.03	0.15	0.50	0	0
October	0	0.22	1.42	1.65	0	0.03
November	0	0.25	1.50	1.89	0.02	0.025
1975						
August	0	0.05	0.13	0.34	0.05	0.02
September	0	0.04	0.21	0.42	0.09	0.05
October	0	0.07	0.46	0.87	0.08	0.05
November	*a*					

*a*Counts were not made because treatments were applied to some subplots.

VI. The principal difference in Table 10.2 relating to numbers of California red scales is between Programs III and IV (the biological control programs) and the other four programs in which certain chemicals were used but some of which preserved a higher degree of biological control than others. The difference is illustrated in Figure 10.1 with comparisons of the numbers of scales per 10 twigs for the selected programs I, III, IV, and VI in the first year, 1973, versus the third year, 1975. It is seen in Table 10.2 that at the end of the second year (1974), the numbers of California red scales had increased on the order of 5-fold under the exclusively biological control programs (III and IV) and at the same rate of increase to a noticeable density under Program II; yet, under these same programs continued, the densities in the third year (1975) had subsided considerably. The reasons responsible have not yet been isolated, associated with available data, and interpreted. In any event, with regard to this key pest, Program VI offers possibilities of use in commercial integrated programs. While the data of Table 10.2 indicate that the

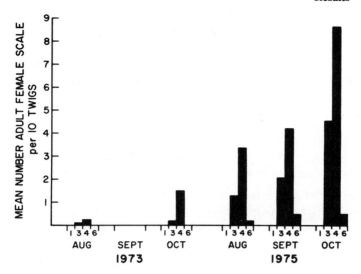

Figure 10.1. Mean number of adult female California red scale per 10 twigs, each 7.6 cm long, of peripheral foliage of sample trees in the core area of ecosystem study plots I, III, IV, and VI at Woodcrest.

numbers of scales under Programs II and V are nearly as low as those under Program VI, data obtained subsequent to that presented in this report show that under Program V the numbers of scales increased to distinctly more than those under Program VI, and that under Program II the density of the scale increased to a level requiring treatment with parathion, as considered in the plan for that regime.

The amount of deposit of N-R 415 spray oil by the low volume treatment given in Table 10.1 for programs V and VI is not enough in itself to produce effective control of California red scale. However, Ebeling (1936) reported that small amounts of spray oil cause high mortality of immature stages of California red scale and the efficiency of spray oil deposit as related to scale mortality was much greater on the immature stages, even through the "gray" adult, than on the fertile adult female scale. Therefore, the low volume N-R 415 spray oil treatment caused mortality and thereby removal of a portion of the immature stages from the population of scales in September. This appears to exert sufficient suppression of the populations, so that the percentage of parasitization of the remaining live scales achieved by *A. melinus* subsequent to the treatment is sufficient to maintain effective regulation of the scale population at low levels as shown in Program VI of Table 10.2 and Figure 10.1.

In October 1975 the condition of the trees in subplots III and IV in regard to the amount of " dead wood " of twigs and small branches in the peripheral foliage area and inward to some extent, resulting from damage by California red scale, led to a decision to use a spray treatment to reduce the number of scales. An N-R 415 spray oil treatment by low volume technique was chosen for application on November 4, 1975.

Data collected on parasitization of California red scale by *A. melinus* in subplots III and IV show 35.5 percent parasitization in III in October 1973, 18.0 percent in IV in November 1973, 11.5 percent in III and 11.0 percent in IV in September 1974, 6.0 percent in III and 4.3 percent in IV in November 1974, and 2.0 percent in III and 16.1 percent in IV in October 1975.

Citrus Red Mite.   Populations of citrus red mite were measured by counting the adult female mites on the terminal four leaves of 8 twigs (32 leaves) per tree in accord with the technique described by Jeppson et al. (1969); the

Table 10.3.   Mean Number of Adult Female Citrus Red Mite Per Leaf of Sample Trees in the Core Area of Ecosystem Study Plots at Woodcrest— Selected Months, 1973–1975

Time	Treatment Programs					
	I	II	III	IV	V	VI
1973						
August	0	0	0	0	0	0
September	0.18	0.05	0.83	0.04	0.15	0.02
October	0.83	0.31	1.62	0.16	0	0.01
November	1.34	1.08	3.47	0.41	0.04	0.03
1974						
August	0.79	0.13	0.49	0.86	5.22	1.88
September	2.21	1.40	1.44	1.57	0.02	0.81
October	1.95	4.81	3.05	2.55	0	0.02
November	1.67	0	3.87	1.70	0.04	0.07
1975						
April	0	0.02	3.63	2.83	1.16	1.21
May	0.01	0.75	4.77	2.08	2.99	3.15
June	0.11	0.91	0.94	1.25	3.60	1.42
July		0.11	0.38	0.20	6.11	0.06
August	0.26	0.10	2.10	0.52	6.33	0.03
September	0.66	0.76	1.21	0.30	4.09	0.15
October	5.66	2.92	4.84	2.53	0.03	0
November	10.71	7.91	8.01	4.44	0.08	0
December	0.01	0	0.23	0	0.07	0.18

counts were made on eight sample trees selected in a pattern representing the area of the core acre of the ecosystem subplot. The mean number of adult female citrus red mites per leaf under the treatment-management programs of Table 10.1 in the ecosystem subplots in the fall months of 1973 and 1974 and the spring and fall months of 1975 are listed in Table 10.3 and illustrated in Figure 10.2.

In the plan for the conduct of Program I it was decided to treat for citrus red mite when the count reached a level of two mites per leaf. The first application was made April 14, 1973, with oxythioquinox (Morestan®) used at 1.875 lbs AI per acre on mature trees with no fruit present. The applications following were propargite (Omite®) at 4.5 lbs AI per acre on November 26,

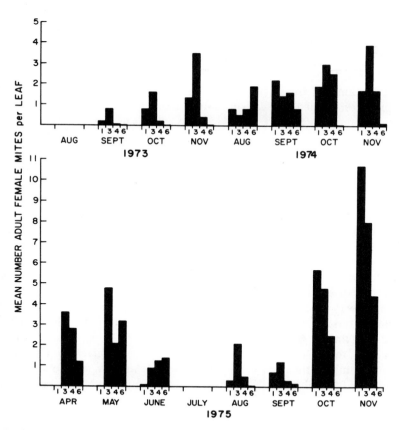

Figure 10.2. Mean number of adult female citrus red mite per leaf of sample trees in the core area of ecosystem study plots I, III, IV, and VI at Woodcrest.

1973, dicofol (Kelthane®) at 4.0 lbs AI per acre on July 29, 1974, propargite at 4.5 lbs AI per acre on November 15, 1974, dicofol at 4.0 lbs AI per acre on July 29, 1975, and propargite at 4.5 lbs AI per acre on November 4, 1975. Although development of resistance by citrus red mite to dicofol at the Woodcrest location is probable, it was used in July, 1974 and 1975, because the only other organochemical acaricide available then, propargite, cannot be used during periods when the daily maximum temperature is above 90°F. In the treatment times given it can be seen that the applications were made at intervals of about 8 months; this is the reason an acaricide application is listed twice each year in Programs I and II in Table 10.1. The data of Table 10.3 indicate that in the fall of 1975 the treatment for Program I should have been made in October but an unavoidable delay occurred until November, when the increase in numbers of mites was plainly high.

In Program II the numbers of mites remained below two per leaf during 1973. Applications were required and propargite was applied at 4.5 lbs AI per acre on May 2, 1974, and October 16, 1974. A level of two mites per leaf did not occur again until October 1975, and a treatment was made on November 4, 1975, with propargite at 4.5 lbs AI per acre.

In Program V a treatment with oxythioquinox at 1.875 lbs AI per acre was made on April 14, 1973, on mature trees with no fruit present. After that, in Programs V and VI, N-R 415 spray oil was used to control citrus red mite with application each year in mid-September, as listed in Table 10.1.

Results in Table 10.3 show that control of citrus red mite in Program VI was as satisfactory, in general, as that in Program I. In Program VI in May, 1975, a judgment situation occurred when the number of adult mites per leaf exceeded two. The decision was then made to delay treating; while the cause was not identified, the numbers of mites did not increase as would be expected. In Program V the mite counts also rose above two per leaf in May, 1975, and remained above two per leaf throughout the summer; this may have been the result of a difference in production of new flush growth on the trees which furnishes the most important supply of food. However, summer temperatures can exert a depressing effect on citrus red mite populations, so treatment was postponed and, since a serious continuing increase in the population did not develop, treatment was not made until the scheduled time in mid-September.

For 1975, in Table 10.3, counts of citrus red mite are given for December. The low numbers then present in Programs III and IV subplots were the result of the low volume N-R 415 spray oil treatment on November 4, 1975, which was made to reduce California red scale, rather than citrus red mite.

Feeding by the citrus red mite on leaf tissue can be responsible for leaf drop which may progress to general defoliation. This is most evident on the current or last flush of growth and in the tops of trees. Feeding on the green bark also causes weakening of fruit stems, causing fruit drop. These effects and their relation to economic injury have been observed many times through the years and there is ample evidence that the citrus red mite can be a serious economic pest in southern California. However, the extent of damage in given orchards is variable and influenced by various factors, independently and in combinations—such factors as high temperature, low humidity, predatory action, disease, low soil moisture, and so on. The data of Table 10.3 show the differences obtained in control of citrus red mite between the different treatment-management programs. However, it must be noted that within the conditions of the experiment in the orchard at Woodcrest from 1973 through 1975, the differences in numbers of mites do not correlate well with the differences in fruit production given in Table 10.4.

Rind Damage Grading and Fruit Yield.   Yield data for the various programs, packing house commercial grading, and grading of oranges placed in the by-products category to identify the types of rind damage responsible for diversion to by-products are given in Table 10.4. The 1972 crop was picked prior to beginning the pest management experiment; the 1972 crop is included to provide background information on the yield in the same subplots. The 1973 crop was lost following two periods of abrupt change from moderate climatic conditions to high temperatures with low relative humidity and wind on May 27 to 29 and June 27 to 29, 1973. This caused the young fruits to drop.

In Table 10.4, under the category of Packing and Grading, the crop is divided into two main groups, percent packed and percent to by-products. In the listing of percentages of various types of rind damage as by-products fruit, the rind damage caused by citrus thrips and California red scale are the most interesting from the standpoint of degree of control of these pests. However, types of damage from mechanical causes such as wind, punctures, and so on, and from intrinsic conditions (such as coarse texture, puffiness, and color blemish) are generally prevalent, and these sorts of rind damage collectively exceeded that caused by insects alone. An overall estimate of this type of damage may be made from the 1974 crop data wherein the numbers diverted because of citrus thrips and California red scale were relatively small.

The data of Table 10.4, presenting percentages of oranges with thrips scarring, are used as the record for degrees of control of citrus thrips under the different treatment-management programs I to VI. Citrus thrips feeding on newly set fruits causes scarring of the rind around the stem end and, also, progressively down the sides. The most susceptible period is when the flower petals wither and fall, which is a period of a week or longer, thereby exposing some of the young fruits before petal-fall is complete. To protect honey bees,

Table 10.4.   Yield, Commercial Packing Grading, and Percentages of Oranges in By-Products Fruit with Citrus Thrips Scarring or California Red Scale, of Ecosystem Study Plots at Woodcrest

	Treatment Program					
	I	II	III	IV	V	VI
	Conv. Rec. Org.-chem.	I.P.M. Org.-chem. Acar. & Bot. Ins.	Biol. Contr.	Biol Contr.	I.P.M. Oil & Org.-chem. Ins.	I.P.M. Oil & Bot. Ins.
Total Cartons per acre						
1972	130.1	215.8	285.7	195.8	113.8	111.7
1974	156.2	50.1	55.7	88.4	194.5	229.4
1975	62.3	400.3	290.7	131.3	99.8	286.7
Packing and Grading						
1974						
Percent packed	46.4	51.9	60.0	72.2	64.4	72.2
% "premium" grade	40.8	43.6	53.1	64.2	59.1	64.9
% "standard" grade	5.6	8.3	6.9	8.0	5.3	7.3
Percent to by-products	53.6	48.1	40.0	27.8	35.6	27.8
% thrips scars only	2.4	0.2	1.1	0.3	0.3	1.0
% comb. sum gen. rind dam^a % thrips scars	15.2	7.0	13.7	6.2	5.9	2.9
% thrips scars & red scale						
% with red scale only						
up to 3 per orange						0.1
more than 4 per orange	0.3					0.1
% comb. sum gen. rind dam^a & red scale						
1975						
Percent packed	61.4	63.8	62.9	58.7	65.3	67.5
% "premium" grade	28.3	40.6	40.3	38.0	45.3	44.6
% "standard" grade	33.1	23.2	22.6	20.7	20.0	22.9
Percent to by-products	36.8	36.2	37.0	41.3	34.7	32.5
% thrips scars only	3.7	8.2	3.1	3.4	3.7	5.2

Table 10.4. *(Continued)*

	Treatment Program					
	I	II	III	IV	V	VI
	Conv. Rec. Org.-chem.	I.P.M. Org.-chem. Acar. & Bot. Ins.	Biol. Contr.	Biol. Contr.	I.P.M. Oil & Org.-chem. Ins.	I.P.M. Oil & Bot. Ins.
% comb. sum, gen. rind dam[a] & thrips scars	13.9	7.0	8.9	5.9	8.3	9.6
% thrips scars & red scale		0.7	1.6	1.3	1.0	0.2
% with red scale only up to 3 per orange	0.3	2.4	1.3	2.7	1.4	0.7
more than 4 per orange		5.5	6.1	12.2	3.7	0.4
% comb. sum gen. rind dam.[a] & red scale	0.2	5.1	19.2	19.2	0.4	0.2
Sum of all % thrips scars as % of total crop	6.4	5.8	5.0	4.3	4.5	4.9
Sum of all % red scale as % of total crop	0.2	5.0	10.4	14.6	1.8	0.4

[a]Sum of combinations of other general types of rind damage with either thrips scarring or Calif. red scale.

application of certain insecticides, including dimethoate, is commonly restricted from the time blossoms begin to open until 90 percent or more of the petals have fallen. At the Woodcrest ranch, experience has shown that it is more practical to check at petal-fall for the presence of citrus thrips on the fruits, and to withhold treatment until the thrips are positively present, rather than to routinely treat at petal-fall.

The data of Table 10.4 indicate that scarring from citrus thrips was more severe in 1975 than in 1974. This appears to be just a variation between years. However, under Program I, with a treatment of dimethoate for citrus thrips, there were more, rather than less, thrip-scarred oranges compared to Programs III and IV, left untreated, or to Programs II and VI, treated with Ryania. This result should not be viewed *a priori* as a deficiency in either the performance of dimethoate or the timing of the treatment contrasted to the use of Ryania, since generally equivalent percentages of scarring to those under Program I were found in similar data obtained in the experiment near Orosi (Central Valley district) where treatment with dimethoate was applied just at 90 percent petal-fall. Rather, the results for thrips scarring indicate that at Woodcrest in 1974 and 1975 citrus thrips was not a significant pest. Under these conditions no advantage accrued from using dimethoate. More important, comparisons for the 1975 crop show that the percentages of oranges culled for California red scale presence were higher under Program V than VI. This suggests a deleterious effect of dimethoate on *A. melinus*. There is other support for this suggestion from laboratory experiments and field data on parasitization following treatments with dimethoate.

The data for California red scale in Table 10.4 show population increases from 1974 to 1975. The ranking of the different treatment-management programs from least to most scales in 1975 is I, VI, V, II, III, IV. The differences in the data for percentage of oranges with scales present (Table 10.4) are larger than the differences in the data given in Table 10.2. The data in Table 10.2 present counts of red scales collected only from the lower parts of the trees, whereas the data in Table 10.4 represent samples from the whole trees, including oranges from the top parts of the trees which are more favorable to the scale. While the control would be judged generally satisfactory in Program I in 1975, the percentages of oranges with scales would usually be sufficient reason to watch the infestation, as would the percentages of oranges with scales under Program VI. Likewise, the percentages under Program V would be more evident and those under Program II would be cause for even closer scrutiny. The percentages of oranges with California red scales present under Programs III and IV for the 1975 crop would customarily be regarded as indicative of a serious infestation requiring treatment at the earliest opportunity in order to avoid loss of the next crop and, as well, more lasting damage to the trees.

In Table 10.4 the percentages shown for thrips scarring and scales present are obtained from samples of by-products fruit. For the 1975 crop about 36 percent of the oranges were sorted into the by-products grade, but it is to be noted that this grade contains all of the oranges with unsatisfactory qualities of the rind, not just that due to citrus thrips and scales. Therefore, for any grading unit the percentage determined for a sample of by-products fruit will be larger than if it were based on all the oranges of the crop from the subplot. While the percentages shown in Table 10.4 are useful directly for comparisons of the degree of control of the pests under the treatment-management programs, an interesting and practical perspective is provided by converting the percentages found for by-products fruit to their size in terms of the total crop. These are given for thrips scarring and scales present in the two lines at the bottom of the table. Estimated percentages of such categories of rind damage based on the total crop are the customary practical information given to growers, consultants, field men, and so on.

Under Program I the parathion treatment that was actually used in mid-June was a minimal treatment and it provided satisfactory control of California red scale. The air-blast machine was a low silhouette model and the coverage deficiency in the top center area of the tree, as described by Carman (1977), may have contributed to the oranges having scales, the small percentage listed in Table 10.4. Adjustments to improve control of California red scale can be made easily by using different spray equipment and by raising the dosage of parathion since the dosage used here was at the low end of the recommended range.

As mentioned earlier, the low volume machine used in Program VI was a low silhouette model. The models having air towers (Carman, 1977) now available give greatly improved coverage in the top center area of the tree. This enhances control of California red scale. Also, the amount of N-R 415 spray oil listed in Table 10.1 can be raised, as needed, within the provisions of the recommendation for its use in California, for control of citrus red mite (footnote *d*, Table 10.1).

While the data in Table 10.2, Figure 10.1, and Table 10.4 show markedly higher populations of California red scale under Programs III and IV, and also under Program II, the yields in total cartons per acre for 1975 were not reduced in these subplots in comparison with yields under Programs I and VI. This evidence varies from expected based on the recognized fact that California red scale can kill foliage and branches of citrus trees and cause leaf and fruit drop. Higher populations of California red scale (Table 10.2, number of live adult females per twig) developed in the subplots of Programs III and IV in the fall of 1974 and continued through 1975. As of now, until the data are studied further, it is interesting that the numbers of scales observed did not cause decreases in yield, although the reasons are not apparent. The principal visual manifestation of California red scale in subplots II, III, and IV is its infestation of the fruits, but this did not represent an economic loss. The percentages of fresh fruit cartons from the different treatment-management programs were reasonably close. The numbers of packed cartons obtained (the base unit for returns) were in direct proportion to total cartons obtained for the different treatment programs.

The number of total cartons per acre, by years, presents differences within some of the programs from year to year; such variation, called "alternate bearing" is commonly manifested by oranges in southern California. Such alternation (Table 10.4) is not correlated with specific years and so is not a response *per se* to the pattern of climate. The yields of the experimental subplots at Woodcrest were less than the average for the Riverside area. For reasons given previously the subplots are large separate units and the experimental design did not provide the customary type of replication for comparisons of yield. These comparisons are complex and will be given further study. However, as seen in Table 10.4, the yield for Program VI in terms of total cartons, percent packed, and grading data is as good as that for Program I, the designated reference program.

Further study is indicated on the apparent lack of direct correlation between California red scale infestation and reduction in yield in Programs III and IV at Woodcrest. Such data will become available from the 1976 crop season. In contrast to these indications at Woodcrest, we know from results of the experiment near Orosi (Central Valley district) in 1974 that infestations of this scale killed many branches of the trees and reduced the crop on those trees in direct proportion to the amounts of foliar parts of the trees that were killed. It has not yet proved possible to effectively establish *Aphytis melinus* in commercial orchards in the Central Valley and parasitization in those subplots following heavy mass releases was virtually nil, so its parasitization did not complicate the relationship of the effect of the scale on reduction of the crop.

Annual Costs of Treatment-Management Programs. The overall annual costs of insecticides, acaricides, and nutritional minor elements, and their application at general commercial rates differed substantially at Woodcrest among the several treatment-management programs (Table 10.1). The cost of Program I, if fully used, would have been $203 per acre but the modified program actually used cost $158 per acre. On the other hand, if a treatment were required for citrus red mite in the summer months and if the boom sprayer were used instead of the air-blast machine for the treatment for California red scale, the cost could be as much as $239 per acre. The cost of Program VI was only $54 per acre. The cost of Program II, as used, was $82 per acre, but a treatment for California red scale in the third year would have raised the average cost per year to $106 per acre and if in addition a treatment for citrus red mite were necessary in the summer months, the cost would be increased to $176 per acre. While the costs of Programs V and VI were essentially the same, the pest management results for California red scale discussed earlier indicate that Program V has less utility. The biological control Programs III and IV used only the basic nutritional spray of minor elements common to all programs, at a cost of $22 per acre.

Satellite Plot on Mites on Lemon Trees in a Coast District. An experiment was initiated in 1973 in a lemon orchard near Oceanside (Coastal District) to determine the long term effects of several acaricidal spray programs on the citrus red mite and its predators. The results in replicated subplots of one acre in relation to untreated subplots indicated that an application in the summer of 8 to 10 gal of N-R 440 spray oil in 100 gal by the low volume method provided effective control of citrus red mite. Properties of N-R 440 spray oil are given in Div. Agr. Sci., Univ. Calif. (1976). The selective action of the treatment results in greater suppression of citrus red mite than of its major predator, the phytoseiid mite, *Amblyseius hibisci* (Chant). The low volume spray oil treatment thus provided better long term control and greater stability of populations than the acaricides dicofol or propargite, which gave a high initial kill of citrus red mite but also of *A. hibisci*.

California red scale was observed on some trees, but remained under effective biological control in both treated and untreated subplots. Citrus bud mite, *Eriophyes sheldoni* Ewing, was present but did not become a problem of economic importance.

It appears probable that the program of spray oil by low volume application is applicable to 4700 acres of lemons in Orange and San Diego Counties (southern Coastal District). Further field investigations will be needed to ascertain the utility of the program in lemon acreage in other coastal districts in southern California.

Production records taken in 1974 showed no significant differences in fruit size or fruit yield between the treated and untreated subplots, even though high populations of citrus red mite were present in the untreated subplots. Thus, it appears that treatments for citrus red mites in lemon orchards in coastal San Diego County may not be necessary if adequate numbers of *A. hibisci* are present.

Releases of introduced species of phytoseiid mites have been made in two untreated subplots since 1973. *Amblyseius stipulatus* Athias-Henroit is now established and has spread virtually throughout one of the release subplots, displacing *A. hibisci* on most trees. Populations are being monitored to compare the abundance and effectiveness of *A. stipulatus* with that of *A. hibisci*.

Importation of Natural Enemies

Foreign exploration for natural enemies of California red scale has emphasized attempts to discover parasites better adapted to the more severe climatic areas in which citrus is grown in California, such as the interior and desert districts of southern California and the Central Valley. Although the primary aim of foreign exploration is to secure new species, a strong secondary objective is the importation of already established parasites from different foreign areas in the hope of obtaining biotypes better adapted to the more severe climatic areas. Thus, *Aphytis melinus, A. lingnanensis* and *Comperiella bifasciata* continue to be imported from diverse foreign areas. Hopefully, these importations are adding to the gene pool of each of the species already established in California and might result in natural selection of a complex of strains collectively more tolerant to the whole spectrum of variations in climate.

A number of other species of natural enemies were also imported into the quarantine facility at Riverside. Between January 1973 and June 1977, 261 shipments containing 32 species of natural enemies of seven pests of citrus were received from 26 countries. In addition, natural enemies were imported of two insects that have invaded California and are a threat to citrus; these include 25 species of parasites and predators of the woolly whitefly, *Aleurothrixus floccosus* (Maskell), and 15 natural enemies of the Comstock mealybug, *Pseudococcus comstocki* (Kuwana).

The numbers of natural enemies released and of species represented in the releases made against infestations in field situations are given in Table 10.5 by years for 1973 to 1976 for proven and potential pests of citrus in California.

Foreign exploration was undertaken in May, 1977, in Saudi Arabia, Iran, Pakistan, and India for parasites of California red scale. Saudi Arabia was the main target because of its severe climate and because study of museum specimens indicated that perhaps a species of *Aphytis* was present there on *Aonidiella orientalis* Newstead, the oriental yellow scale. Material received from Saudi Arabia yielded *Aphytis* to start a culture on California red scale. This culture has been expanded for production of parasites for release in

Table 10.5. Number of Individuals and of Species of Natural Enemies Reared in Insectaries and Released Against Established or Potential Pests of Citrus in California by Years—1973–1976

Established Pest	Number of Individuals and of Species ( ) of Natural Enemies			
	1973	1974	1975	1976
California red scale	2,200,000 (8)	3,115,000 (5)	1,810,000 (7)	189,000 (4)
Citrus red mite	100,000 (13)	53,000 (9)	56,000 (19)	75,000 (15)
Citricola scale	0	6,700	3,900 (2)	0
Potential Pest				
Comstock mealybug	235,000 (8)	2,570,000 (5)	7,130,000 (11)	62,000 (8)
Woolly whitefly	8,000 (8)	86,000 (6)	37,000 (5)	3,000 (6)
Citrus whitefly	0	0	40 (1)	77 (1)

field situations. According to DeBach (1977) this *Aphytis* may be a new species closely related to *A. melinus* and *A. lingnanensis*. Since it comes from a very hot desert area it offers a good chance of being adapted to the more severe climatic areas of citrus in California where existing species of *Aphytis* and other parasites established in California are currently ineffective.

California Red Scale Sex Pheromone Trap Study

Sex pheromone traps containing live virgin California red scale females were proven to be an efficient and economical tool for the detection and survey of insidious populations of California red scale in the field where suppression measures were practiced or eradication was attempted. However, the trap had not been substantially utilized in citrus orchards with well established populations of the scale. Therefore, this trap was used to sample California red scale infestations by trapping males in subplots of the ecosystem study plots near Orosi and near Woodcrest. Trap cards were replaced weekly or biweekly and virgin females were exchanged biweekly (personal communication, D. S. Moreno).

Data from the various subplots indicated that male activity was synchronous for the various programs unless disrupted by spray treatments. Generally there were 3 to 3.5 peak flights per year representing the number of generations occurring during the year. Their occurrence shifted depending upon meterological conditions for the year but generally peak flights occurred in late June, late July or early August, and mid-September, with a separate smaller or partial flight in October. Also, there was generally a small peak flight in May representing the overwintering population. It was also found that by changing the color of the traps from white to yellow, the parasitoid populations of *Aphytis* sp. could be sampled and their seasonal activity determined. Weather permitting, as many as seven peaks of activity of *Aphytis* sp. were recorded near Woodcrest; the predominant peaks occurred in May, June, September, October, and November. In the ecosystem study plot near Orosi *Aphytis* sp. were not found in the traps because *Aphytis* is not established in the area.

Recent evidence from the IPM ecosystem study plots and ancillary studies in other plots indicates that the numbers of California red scale males caught in the traps per peak flight are positively correlated with subsequent numbers of infested fruits in the orchard; by fitting regression lines for these data, the numbers of males trapped can be used to predict levels of infestation for selected intervals of time (personal communication, D. S. Moreno and C. E. Kennett).

The California red scale sex pheromone can be prepared synthetically now. Since the synthetic product is much easier to use and is available for field work, the use of sex pheromone traps will surely increase for the purposes of detection, survey and monitoring. In addition, the effect of trapping males by synthetic sex pheromone as a means of directly controlling the scale is an interesting possibility that needs to be investigated.

Toxicities to Natural Enemies of Acaricides and Insecticides for Citrus

The most widely distributed and generally the most abundant phytoseiid mite that feeds on the citrus red mite in California is *Amblyseius hibisci* (Chant) (McMurtry, 1969). This species and certain other mite predators are absent in many orchards where pesticides have been regularly applied. In order to utilize mite predators in the development of pest management programs on citrus, it appears essential to acquire information on the toxicities to these predators of acaricides and insecticides used in citrus. Therefore, the relative toxicities of eight acaricides, twelve insecticides, and a fungicide to adult predaceous phytoseiid mites, in relation to their toxicities to plant feeding mites were evaluated in laboratory tests by Jeppson et al. (1975). The toxicity factors for *Typhlodromus occidentalis* Nesbitt, *Amblyseius hibisci* (Chant), *Amblyseius stipulatus* Athias-Henriot, *Iphiseius degenerans* (Berlese), and *Phytoseiulus persimilis* Athias-Henriot were compared to those for the citrus red mite, *Panonychus citri* (McGregor), and the Pacific spider mite, *Tetranychus pacificus* McG. *T. occidentalis* was more resistant to many of the pesticides than the plant feeding mites. The other four species of predaceous phytoseiids were generally more susceptible to the pesticides than the citrus red mite or the Pacific spider mite. Azinphosmethyl and Vendex® had an order of toxicity to the four predaceous species similar to that of the plant feeding mites, and tricyclohexylhydroxytin (Plictran®) was less toxic to all five predaceous mites than to either the citrus red mite or the Pacific spider mite.

A. hibisci eggs were exposed to concentrations of citrus pesticides by utilizing the Potter Tower sprayer. A wide range of susceptibility of both eggs and later immature stages was shown to exist to the various pesticides evaluated. The egg stage was relatively tolerant to a number of the pesticides. Propargite and Vendex® residues were the least toxic to the immature mites.

Similar laboratory evaluations are in progress of the relative toxicities of pesticides used on citrus to several parasites of California red scale.

Florida Results

The Florida work in the first years was concentrated on an effort to obtain biological control of citrus snow scale, *Unaspis citri* (Comstock), an oriental insect which by 1970 was posing a real threat to the whole Florida citrus industry. Later, citrus ecosystem study plots were established, or previously existing ones reoriented and incorporated into the study. In these plots, treatment-management programs, particularly for control of scale insects, citrus rust mite, *Phyllocoptruta oleivora* (Ashmead), and two plant diseases, melanose, *Diaporthe citri* (Fawcett), and greasy spot, *Mycosphaerella citri* (Whiteside), have been designed and tested. Other pests that received attention were the citrus whitefly, *Dialeurodes citri* (Ashmead), the cloudywinged whitefly, *D. citrifolii* (Morgan), the citrus mealybug *Planococcus citri* (Risso), the Texas citrus mite *Eutetranychus banksi* (McGregor), the citrus red mite *Panonychus citri* (McGregor), and a few less important species. In addition, a major effort was begun still later to develop a model of citrus tree growth and phenology and for the population dynamics and management of citrus rust mite populations by Jon C. Allen (reported in Chapter 11).

Biological Control

Biological control has long been recognized as important to the Florida citrus pest control program. Hubbard (1885) reported on the importance of "ladybirds feeding on aphids and coccids" and the "several internal parasitic Hymenoptera that prevent outbreaks of armoured and naked (soft) scale insects." Watson and Berger (1937) pointed out the importance of biological control or, as they termed it, "natural checks" in controlling many of Florida's citrus pests. They emphasized the importance of entomogenous fungi, particularly the use of *Aschersonia* sp. for the control of citrus whitefly. Although Griffiths and Thompson (1953) placed greater emphasis on chemicals for control of the then major citrus pests (purple scale, citrus rust mite, and Florida red scale) they nevertheless recognized the importance of biological control in maintaining low populations of other potentially serious citrus pests in Florida.

Prior to 1960, purple scale and Florida red scale were considered the first and third most important pests of Florida citrus. However, with the subsequent enigmatic introduction of *Aphytis lepidosaphes* Compere and the colonization of *Aphytis holoxanthus* DeBach these scale insects have now been reduced to the role of minor pests (Muma, 1969). However, as is often the case when one insect species is reduced, another may arrive from abroad or spread and increase. This was the situation facing the Florida citrus industry in 1970 with the increase and spread of the citrus snow scale (Brooks, 1973).

Citrus snow scale was introduced into Florida some time around the turn of the century. By 1916 it was being intercepted and quarantined on citrus nursery stock (Newell, 1916). Yothers (1949) and Thompson (1950) reported it to be a serious problem from time to time, but mainly in the Geneva and Oviedo sections of Seminole County, Florida. Brooks and Thompson (1963) reported that heavy infestations were weakening and killing entire trees. By 1970, this insect had developed into a major pest of Florida citrus, infesting some 45 percent of the groves and with approximately 25 percent of all groves requiring chemical control (Brooks & Whitney, 1973). Brooks (1964; Brooks & Whitney, 1973) demonstrated that two annual high volume applications of phosphatic insecticides were necessary to control this insect.

In 1972, the parasite *Aphytis lingnanensis* Compere (HK-1) was introduced from Hong Kong as part of the IBP Integrated Pest Management Program supported by NSF and EPA (Chapter 1). The parasite was cultured in the laboratory on oleander scale, *Aspidiotus nerii* Bouche (Vallot), and released against citrus snow scale in the late fall of 1972. The initial releases were immediately successful in suppressing the scales in seriously infested citrus groves of Lake and Polk Counties of Florida.

However, a problem was soon encountered that slowed the biological control campaign. It was found that citrus snow scale was not present in every grove; rather, the infestations were scattered and, in most cases, occurred several miles apart. In some cases, infestations were present only in pocketlike areas separated by whole counties. This required a more extensive rearing and release program than was originally anticipated because the parasites did not have an " open highway " of continuous host infestation through which they could readily spread on their own.

A. lingnanensis (HK-1) is a highly effective natural enemy of citrus snow scale. The data presented in Table 10.6 illustrate the rather rapid decline in host populations following release of the " Hong Kong Wasp," as it is referred to by Florida citrus growers. The rate of parasitism on citrus snow scale females is presented in Table 10.7. The data in these two tables show the rapid buildup of parasites and the subsequent decline in the host populations. Thus far this parasite has been released at 71 locations in 14 counties

Table 10.6. Populations of Citrus Snow Scale, *Unaspis citri* (Comst.), Following Release of the Parasite *Aphytis lingnanensis* Compere (HK-1)

	Number of Citrus Snow Scales per Square Inch of Bark Surface				
	Before parasites were released	After parasites were released			
Location		8 weeks	16 weeks	30 weeks	42 weeks
Ruskin, Fla.	723.3	5.3	0.75	0.0	0.0
Arcadia, Fla.	460.0	45.6	64.9	2.1	0.4

Table 10.7.   Populations of *Aphytis lingnanensis* Compere (HK-1) After Release Against the Citrus Snow Scale

Location	Number—or Percent—of Parasites per 100 Scales After Release			
	6 weeks	8 weeks	11 weeks	17 weeks
Ruskin, Fla.	34	14	56	10
Arcadia, Fla.	84	4	10	3

where citrus is grown commercially. Populations of this scale have been significantly reduced at 55 of these locations.

Bioassay studies with *A. lingnanensis* (HK-1) and the chemicals used in Florida's typical citrus pest management program have shown that most materials commonly used have some initial adverse effect on the parasite but little residual activity after 14 days of weathering in the field in Florida, where summer rains are frequent, in contrast to California. The materials listed in Table 10.8 include the nutritional materials such as zinc and manganese and the fungicides, as well as the acaricides and insecticides. Materials that continued to be toxic to *A. lingnanensis* for 14 days or longer were azinphosmethyl, carbophenothion, diazinon, dioxathion, ethion, malathion, methidathion, and wettable sulfur. Insecticides that have been tested in small plot field tests and show promise of being incorporated into "Florida's Spray and Dust Schedule" were also screened against *A. lingnanensis* (HK-1) and the results are presented in Table 10.9. These data show chlordimeform and chlorpyrifos to be toxic for 14 days or longer.

From these studies, it would appear that *A. lingnanensis* could survive, though reduced in activity, as the *Aphytis* parasites of purple scale and Florida red scale have survived, under use of most of the chemicals commonly applied in Florida's Extension sponsored pest management program. *A. lingnanensis* is estimated to be saving the Florida citrus grower $8 to $10 million per year in the cost of additional treatments that would otherwise be required (Meade, 1976). Biological control of the three major scale insect pests of Florida citrus is currently estimated to be saving in excess of $25 million annually in additional spray costs that are thereby not required.

Ecosystem Study Plot

About 25 years ago, Griffiths (1951) and Griffiths and Thompson (1953) compared three pest control strategies in a Valencia orange and seedy grapefruit grove to determine the feasibility of using reduced spray programs for citrus planned for canned products. Justification for their study was based

Table 10.8. Toxicity of Materials Recommended in the "Florida Citrus Spray and Dust Schedule" to *Aphytis lingnanensis* Compere (HK-1), the Introduced Parasite of Citrus Snow Scale

Material, Formulation	Rate (oz AI) per 100 gal.	Use[a]	No. times tested	Days with sig. mort.[b]
azinphosmethyl, 2SC	4.0	I	2	21
benomyl, 50W	1.6	F	3	0
carbophenothion, 8E	4.0	A,I	3	14
chlorobenzilate, 4E	2.0	A	3	0
copper hydroxide (Kocide®), 54%	12.3	F	1	0
copper oxide, 75%	12.0	F	1	1
copper sulfate, 53%	12.3	F	1	1
demeton, 2SC	4.0	I	2	1
dialifor, 4E	6.0	A	2	7
diazinon, 4E	8.0	I	2	14
dicofol, 4E	4.0	A	3	0
dioxathion, 8E	4.0	A	2	21
ethion, 4E	4.0	A,I	2	14
ferbam, 95 WP	18.0	F	2	0
formetanate hydrochloride, SP-92%	1.0	A	2	1
lead arsenate, 96%	6.0	P	2	0
malathion, 5E	20.0	I	2	14
manganese sulfate, 27.3%	4.0	N	2	0
methidathion, 2E	4.0	I	3	21
oxamyl, 2E	2.0	A	2	7
oxydemeton-methyl, 2SC	4.0	I	2	7
petroleum spray oil, 95%	1%[c]	A,F,I	2	1
parathion, 8E	4.0	I	2	7
phosphamidon, 8E	16.0	I	2	7
Plictran®, 50W	2.0	A	2	0
propargite, 6.75E	4.2	A	2	1
sodium borate, 66%	3.5	N	1	0
sodium molybdate, 46%	0.5	N	1	0
sulfur, wettable, 95%	80.0	A	3	14
Vendex®, 50W	2.0	A	1	0
zinc sulfate, basic, 52%	4.0	N	2	0

[a]A = acaricide; F = fungicide; I = insecticide; N = nutrition; P = physiologic.
[b]Days with mortality of *A. lingnanensis* significantly higher than the untreated check. Data subjected to analysis of variance and Duncan's multiple range test at the 5% level.
[c]Spray oil is applied as an emulsion diluted to a specified concentration (0.5 to 1%).

Table 10.9.   Toxicity of Promising Candidate Insecticides for Florida's Citrus Pest Management Program Against *Aphytis lingnanensis* Compere (HK-1), the Introduced Parasite of Citrus Snow Scale

Material, formulation	Rate (oz AI) per 100 gal.	No. times tested	Days with signif. mortality[a]
chlordimeform, 4E	4.0	2	14
chlorpyrifos, 4E	4.0	2	21
difluron, 25W	0.5	3	0
Chemagro 9306, 6E	4.0	2	7
FMC-35001, 4E	4.0	1	7
ICI PP-199, 25 Col.	0.8	2	7
ICI PP-199, 25 Col.	1.6	2	7
H.L.R. Ro-10-3108 50 %E	1.3	3	0
Shell 43775, 2.4E	1.6	2	0

[a]Days with mortality of *A. lingnanensis* significantly higher than the untreated check. Data subjected to analysis of variance and Duncan's multiple range test at the 5 percent level.

on the assumption that external quality is of little consequence where the fruit is produced for canning. Approximately 90 percent of the Florida orange crop is normally canned. Therefore, spray programs primarily designed for the production of fresh fruit could be altered at considerable savings to the grower (and now of course with reduced environmental consequences). They assumed that some external fruit blemish caused by citrus rust mite was acceptable. Little attention was given to melanose control and other disease problems. The pest control strategies they used to reduce production expenses consisted of sulfur dust only or no treatment at all. After four years, the above researchers found that although external (cosmetic) quality of the fruit varied and was considered inferior in the unsprayed plots, yield and internal quality were as good as under the more traditional spray programs. Under the sulfur dust program, yield and fruit quality, both external and internal, were similar to results under the traditional or the "no treatment" plots; however, purple and Florida red scale populations were slightly higher, and citrus red mite populations were significantly higher. Greasy spot disease was severe in the unsprayed control and the sulfur plots, but yield was not affected.

Today, sulfur remains economical and effective in control of citrus rust mite. However, it is known to destroy a wide range of natural enemies that attack and may be effective in control of other citrus pests (Griffiths and

Fisher, 1950). Therefore, the use of sulfur in a citrus pest control program in Florida is considered unwise, particularly following the securing of successful biological control of both Florida red scale and purple scale by introduction of parasites and now of citrus snow scale.

In view of the increasing production costs facing growers, innovations in harvesting, and restrictions on the use of some pesticides, the need exists for a new strategy employing biological control as much as possible, and using reduced amounts of pesticides, perhaps even more acutely today than in the 1950s. Emphasis is being placed on a strategy which integrates the tactics of biological, cultural, and chemical control rather than one utilizing the cheapest available pesticide(s). An overview of a 4-year integrated control study conducted on Valencia oranges is here presented. In this study, the integrated control strategy was compared with a conventional program and a " no spray " (unsprayed) program. The primary system variables measured to assess the results were yield and fruit quality.

From 1972 to 1976, three different pest control programs were maintained in 30-acre sections of three commercial Valencia orange plantings in central Florida. Each 30-acre section was divided into adjacent 10-acre plots with each plot to receive a specific program for insect, mite, and disease control (melanose and " greasy spot "). All plots received other routine horticultural practices such as nutritional elements, fertilization, irrigation, and cultivation. Groves were not " hedged " during the study. Pesticides were applied both by ground and air at the direction of the cooperators or research personnel.

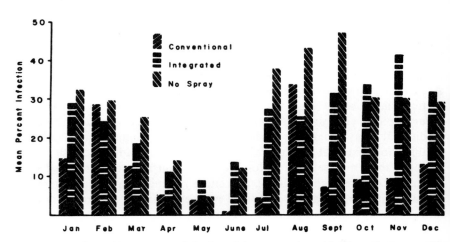

Figure 10.3. Seasonal mean percent infection of citrus rust mites with *Hirsutella thompsonii* in different management programs for 3 " Valencia " orange groves from 1972–1976.

Population densities of citrus rust mite and scale insects were monitored throughout the study; sampling frequency was varied with the relative past abundance of the pests. Whitefly, mealybug, spider mites, and other minor pests were monitored when their densities increased to easily detectable levels. The fungus *Hirsutella thompsonii* Fisher which attacks citrus rust mite was also monitored throughout the study. The incidence of greasy spot infection on the leaves and leaf drop during the winter were also monitored each year. Sample locations used for monitoring the different pests were replicated three times within each 10-acre plot.

Each year, 20 trees from the center of each 10-acre plot in each grove were harvested to obtain an estimate of yield and fruit quality. From the harvested fruit, a random sample was taken from each plot for grading in the packing-house. "Packout" (U.S. No. 1 grade) was determined on fruits from each plot. Eliminations were sorted according to the grade lowering factors responsible for the loss. Soluble solids and acids, important in citrus quality, were also determined for each sample.

Pest Control Strategies. The three strategies designated conventional, integrated, and "no spray" (untreated) differed in the following ways (Table 10.10):

1. In the *conventional program*, where fruit was usually grown for the fresh market, a postbloom copper spray was applied for melanose control. Various recommended fungicides were applied for greasy spot disease control during the summer. Various recommended insecticides or acaricides were applied for insect and mite control. Treatments for citrus rust mite were applied on a calendar basis or, occasionally, according to recommendations based on 10 to 15 percent of the leaves being infested.

Table 10.10. Comparison of Pest Control Strategies Used in Three Valencia Orange Groves—1972–1976

Production problem	Conventional	Integrated	No spray
Melanose	Copper to postbloom	None	None
Greasy spot	All recommended fungicides Mid-June to mid-July	Oil Mid-June to mid-July	None
Citrus rust mite	All recommended acaricides 10 to 15% leaf infestation	Selective acaricides 1% fruit w/75 mites/cm^2	None
Scale insects	All recommended insecticides	None	None

Foliar treatments for minor nutritional deficiencies were applied regularly in the postbloom spray.

2. In the *integrated program*, no treatment for melanose was applied. The first treatment was applied for citrus rust mite when approximately 1 percent of the fruits harbored 75 mites/cm^2 of surface (approximate mite density required to cause visible injury). If minor nutritional deficiencies were visible on leaves, a foliar spray without copper was included with the acaricide. The summer treatment was applied between mid-June and mid-July to achieve greasy spot and citrus rust mite control. After the first year, regardless of its population density, treatments were not made for citrus rust mite control in July and August in the integrated plot. With one exception, fall mite control was applied only when 1 percent of the fruits harbored 75 mites/cm^2 of fruit surface. Fungicides such as copper, which causes increases in citrus rust mite populations in the field (McCoy et al., 1976*b*), and benomyl (Benlate®), which is toxic to entomopathogenic fungi in the laboratory, were not used for greasy spot control. Rather, 1 percent spray oil [meeting FC 435–66 (Simanton & Trammel, 1966) specifications], which is innocuous to parasites, predators, and entomopathogens, was used exclusively in the integrated control program. Acaricides known to be least disruptive to entomophagous parasites (spray oil and Acaraben®) were generally used to control citrus rust mite. Chlorobenzilate (Acaraben®), spray oil, formetanate hydrochloride (Carzol®) and dicofol (Kelthane®) (only where citrus snow scale was absent) were applied at different times in the integrated program.

3. In the *"no spray"* program, no pesticides were applied at any time during the study. Foliar treatments for minor nutritional deficiencies were applied when visible symptoms appeared on the leaves.

Results From the Three Strategies.   During the 4-year study, greasy spot disease and citrus rust mite were the only pests that required control in all groves (all three programs). Young tree decline (YTD) or blight was severe in two groves; however, no differences attributable to the specific pest control programs were seen. Spider mites, citrus snow scale, and other potential pests were present in all groves (all three programs), but they appeared to remain below economic injury levels and never required treatments.

Results indicate that greasy spot disease had a significant effect on defoliation and reduced yield in the untreated plots (Table 10.11). Loss of yield occurred the year following a high incidence of disease in all "no spray" plots. In the conventional plot of Grove I, failure to control greasy spot in 1974 probably explained the mean loss in pounds of solids per acre, comparing 1975 to 1976 (Table 10.11). In the integrated control plot of Grove II, the application of 1 percent spray oil by air for two years gave poor control

Table 10.11.   Comparison of Mean Percent Leaf Infection by Greasy Spot and Estimated Yield in Pounds of Solids Per Acre of Valencia Orange in the Following Year for Three Groves, Each Under Different Management Programs

		1972–74		1973–75		1974–76	
Location	Management program	Percent infection[a]	Mean lb solids per acre	Percent infection	Mean lb solids per acre	Percent infection	Mean lb solids per acre
Grove I	Conventional	25.6 a[b]	3761.6 a	NS	4248.1 a	80.0 a	2340.6 a
	Integrated	29.6 a	2797.5 b	NS	4024.9 a	13.1 b	3659.0 b
	No spray	72.0 b	2044.9 c	NS	3429.1 b	94.7 a	1998.0 a
Grove II	Conventional	13.6 a	3188.6 a	13.8 a	6082.1 a	30.6 a	3972.7 a
	Integrated	29.2 b	3603.3 a	34.8 b	5289.9 b	75.8 b	2990.8 b
	No spray	37.0 b	2665.9 b	52.2 c	4980.9 b	97.6 b	3084.0 b
Grove III	Conventional	21.2 a	2044.9 a	13.4 a	3555.9 a	15.4 a	2469.9 a
	Integrated	24.6 a	3494.1 b	22.8 ab	3736.0 a	50.4 b	2805.3 a
	No spray	37.9 a	1951.5 a	35.2 b	3126.9 b	95.3 c	2416.1 a

[a]Percent leaf infection by greasy spot disease based on a random sample of 500 or more leaves per treatment taken in January.
[b]Values with the same letter are not significantly different at the 5 percent level using Duncan's multiple range test.

of greasy spot. In the following two years even where more thorough coverage with spray oil was obtained via ground application, incidence of greasy spot infection remained high and significant loss of yield occurred (Table 10.11). Apparently, spray oil alone was inadequate for control once the infection pressure reached a high level. Similar results have been noted by Whiteside (1973).

Citrus rust mite injury to the fruit varied considerably from year to year and between pest control programs. The greatest loss in U.S. No. 1 grade "packout" from citrus rust mite injury occurred in the fall of 1974 when weather conditions were extremely dry from October through December. Generally, the mean percent injury during early and late season was highest in the untreated programs (Tables 10.12 and 10.13). By comparison, virtually no difference in citrus rust mite injury was found between the conventional and integrated control programs, even though the number of acaricide applications for mite control, particularly in the fall, was fewer under the integrated program (Tables 10.12 and 10.13). In the conventional pest control program,

Table 10.12.   Number of Acaricide Applications and Mean Percent Fruit With Excessive Early Season Rust Mite Injury in Different Management Programs—1972–1976

Manage-ment program	1972–73		1973–74		1974–75		1975–76	
	No. trmt.	% inj.[a]	No. trmt.	% inj.	No. trmt.	% inj.	No. trmt.	% inj.
Conventional	2.3	5.1 a	2.7	2.6 a	2.3	1.4 a	2.0	3.6 a
Integrated	2.0	8.6 a	1.3	5.5 a	1.3	3.0 a	2.0	3.4 a
No spray	0.0	20.4 b	0.0	4.7 a	0.0	8.0 b	0.0	8.9 b

[a]Numbers followed by same letters are not significantly different at the 5 percent level by Duncan's multiple range test.

an average of 3.4 sprays per year were applied for the total pest control program compared to 2.2 sprays in the integrated control plots. This reduction in number of treatments resulted in savings of $30 per acre in labor and material costs.

The reduction in number of sprays applied for mite control appeared to be influenced by the activity of the fungus *H. thompsonii*, which attacks citrus rust mites in the summer and fall. The mean number of rust mites having the fungus disease was higher in the integrated control program. The greatest difference in percent infection existed in the summer and fall. As expected, the incidence of rust mite disease by *H. thompsonii* was highest in the untreated ("spray") plots.

Table 10.13.   Number of Acaricide Applications and Mean Percent Fruit With Excessive Late Season Rust Mite Injury in Different Management Programs—1972–1976

Manage-ment program	1972–73		1973–74		1974–75		1975–76	
	No. trmt.	% inj.[a]	No. trmt.	% inj.	No. trmt.	% inj.	No. trmt.	% inj.
Conventional	1.7	2.2 a	1.3	2.6 a	0.7	14.0 a	1.3	0.6 a
Integrated	1.0	5.6 a	0.0	2.6 a	1.0	39.8 b	0.0	1.8 a
No spray	0.0	10.1 b	0.0	4.0 a	0.0	19.0 a	0.0	1.8 a

[a]Numbers followed by same letters are not significantly different at the 5 percent level by Duncan's multiple range test.

Table 10.14.   Mean Percent Packout From Different Management Programs—1972–1976

Management program	Mean percent packout[a]				
	1972–73	1973–74	1974–75	1975–76	Mean
Conventional	81.0 a	47.7 a	52.2 a	73.9 a	63.7 a
Integrated	73.7 ab	45.7 a	23.4 b	71.5 a	53.6 a
No spray	62.4 b	45.7 a	35.9 ab	65.3 a	52.3 a

[a]Represents an average from three Valencia groves. Numbers followed by same letters are not significantly different at the 5 percent level by Duncan's multiple range test.

As reported by Albrigo and McCoy (1974), leaf injury by citrus rust mite was generally localized to a specific area of the tree. On occasion, yellowish de-greened patches, brown spotting, and bronzing were observed on leaves. Injury was generally highest in the untreated plots. However, in the fall of 1974, when the trees were under severe moisture stress, leaf injury in the integrated control plots was severe enough to cause scattered defoliation (McCoy, 1976), although it had no apparent effect on subsequent yield.

Percentage of U.S. No. 1 quality fruits packed varied considerably from year to year and was generally lowest in the untreated programs (Table 10.14). The lowest pack of quality fruit was in 1974 and 1975, when late season injury by citrus rust mite was severe. However, wind damage appeared to have a greater overall effect on losses (Table 10.15). Melanose injury was generally low even in the plots not receiving a postbloom fungicide for this pest. Loss

Table 10.15.   Percent Fruit with Excessive Wind Scar From Different Management Programs—1972–1976

Management program	Mean percent wind scar[a]			
	1972–73	1973–74	1974–75	1975–76
Conventional	9.5 a	28.4 a	22.6 a	15.0 a
Integrated	8.4 a	27.5 a	19.2 a	15.1 a
No spray	6.4 a	25.8 a	29.1 a	18.0 a

[a]One-third of surface in aggregate or more affected. Represents an average from three Valencia groves. Numbers followed by the same letter are not significantly different at the 5 percent level using Duncan's multiple range test.

in top quality grade from excessive melanose injury averaged from 0.9 to 4.1 percent for the four years in the plots under the untreated programs.

There was no significant difference in mean pounds of solids per box per year between the different management programs (Table 10.16). No difference in solids was detected where one to three spray oil treatments were applied annually. Preharvest "off-flavors" were detected in juice from selected fruits having excessive surface bronzing and peel shrinkage from Grove II where fall injury to fruits exceeded 30 to 40 percent in 1974 (McCoy et al., 1976a). No flavor problem was detected, however, in a random sample collected at harvest.

Mean pounds of solids per acre varied annually according to the typical alternate bearing characteristics of Valencia oranges (Table 10.17). After the first year, yields in mean pounds of solids per acre in the untreated program were considerably lower than in the other programs (Table 10.17). As previously mentioned, yield was apparently influenced by defoliation from greasy spot.

Greasy spot disease was the most important pest causing excessive defoliation, loss of tree vigor, and subsequent lower yield in all Valencia groves where treatments for it were not made or were ineffective (inferior). These findings substantiate the results of Whiteside (1975) regarding the importance of greasy spot disease on oranges. In the integrated pest control program, the use of spray oil as a selective fungicide and the allowance of higher citrus rust mite densities in the summer significantly increased the natural control of rust mites during the other seasons of the year by the parasitic fungus, *H. thompsonii*, and external fruit quality at harvest was not greatly affected. Since other potential insect and mite pests remained innocuous where the selective pesticides were applied under the integrated program, this program

Table 10.16.   Mean Pound Solids Per Box Per Year From the Different Management Programs—1972–1976

Management program	Mean pound solids per box[a]			
	1972–73	1973–74	1974–75	1975–76
Conventional	6.5 a	6.3 a	6.6 a	6.1 a
Integrated	6.7 a	6.7 a	6.5 a	6.2 a
No spray	6.9 a	6.4 a	6.4 a	5.7 a

[a]Represents an average from three Valencia groves. Numbers followed by the same letters are not significantly different at the 5 percent level using Duncan's multiple range test.

Table 10.17.   Estimated Mean Pound Solids Per Acre Per Year From the Different Management Programs of Three Groves—1972–1976

Management program	Mean pound solids per acre[a]				
	1972–73	1973–74	1974–75	1975–76	All years
Conventional	5291.8	3096.9	4628.7	2927.7	3986.2 a
Integrated	5024.2	3022.2	4350.3	3151.7	3887.1 a
No spray	5156.3	2496.8	3845.6	2499.4	3499.5 b

[a]Numbers followed by the same letter are not significantly different at the 5 percent level using Duncan's multiple range test.

would appear acceptable for cannery fruit production, the regular market outlet for approximately 90 percent of the crop, as noted previously. However, the use of spray oil alone for greasy spot control can present serious problems in groves having a high greasy spot potential or where the spray oil application results in poor foliar coverage. In these situations, a summer treatment for greasy spot disease may require a less selective fungicide than spray oil, such as benomyl, until the infection intensity has been reduced.

Texas Results

The citrus rust mite has been the number one pest of citrus in Texas almost every year since the beginning of the industry in 1920, and its control is a top priority expense to Texas citrus growers. The feeding of the mite on the rind of the fruit causes russeting that greatly impairs appearance. Contrary to the case in Florida, about 50 percent of grapefruits and 40 percent of the oranges produced in Texas are marketed through fresh fruit channels for which the price received ranges from two to five times that for fruit destined for processing. Some of the acaricides available for use against this mite upset the good biological control which exists for scale insects and disrupt parasite activity during the postbloom season, resulting subsequently in sharp increases in numbers of diaspid scales. However, studies in Texas showed that chlorobenzilate and dicofol can be used to control citrus rust mite without upsetting the biological control of pest scales. Also, Villalon and Dean (1974) reported the occurrence of the fungus *H. thompsonii* on the citrus rust mite in Texas in 1972. This fungus was described in Florida by Fisher (1950). McCoy and Kanavel (1969) describe its isolation from the citrus rust mite and cultivation on various synthetic media. This hyphomycetous fungus is the first biological control agent shown to provide significant control of the citrus rust mite. In the work in Texas, procedures were devised to maintain cultures of the fungus in the laboratory and to produce material for use in spray applications to citrus trees. Application of hyphae to grapefruit trees caused very definite reductions of the populations of citrus rust mite.

Introduction of the parasite *Aphytis lepidosaphes* Compere by Dean (1975) provided complete control of the purple scale, formerly a major pest of citrus in Texas. Two other *Aphytis*, *A. lingnanensis* Compere and *A. holoxanthus* DeBach, introduced earlier, became established and are effective natural enemies, respectively, of California red scale and Florida red scale. The chaff scale, *Parlatoria pergandi* Comstock, in Texas, has two established imported parasites, *Aphytis hispanicus* (Mercet) and *Prospaltella fasciata* Malenotti, but they have not kept this scale at subeconomic levels. Chaff scale presently rates as the most important scale pest of Texas citrus.

Reinking (1964) reported that brown soft scale, *Coccus hesperidum* Linnaeus, changed from a minor to a damaging pest in Texas between 1957 and 1960. He suggested that a contributing factor was the large amount of methyl parathion used in cotton fields which were frequently adjacent to the citrus orchards. The brown soft scale is a troublesome pest both because of its direct feeding damage to the tree and from the effects of its coating the leaves with "honeydew" which supports the sooty mold fungus, *Meliola camelliae* (Catt.), which develops on this sugary material excreted by the scale. Hart et al. (1966a) showed that methyl parathion caused explosive buildup of brown soft scale, and Hart et al. (1969) discussed drift of methyl parathion from aerial applications to cotton as a contributing factor to the buildup of brown soft scale in citrus orchards. These authors also reported the numbers, abundance, and effectiveness of parasites and predators of brown soft scale in Texas; they found that *Microterys flavus* (How.) was the most abundant parasite of the scale in the summer months. Recent results in Texas show that early detection of brown soft scale followed by releases of *M. flavus* will hold the scale in check.

While the natural enemies mentioned above are well established in Texas, disruptions of biological control occur frequently in citrus orchards because the adjacent fields on more than one border of the orchard are commonly planted with other crops, frequently cotton, which are treated with disrupting pesticides. Studies show that increases of pest populations on this citrus from upsets of the natural enemies develop because of drift of pesticides from these adjacent crops, from indirect effects of pesticides applied in the citrus orchards themselves, and from effects of adverse weather conditions. Nonetheless, many Texas citrus growers are well aware of the importance of these beneficial arthropods in the economic production of citrus and are willing to modify their pest control programs to better utilize them.

ACKNOWLEDGMENTS

The authors wish to express sincere appreciation to the Church of Jesus Christ of Latter Day Saints Mormon Woodcrest Citrus Project, Riverside, California for generous cooperation in the use of their orchards and for assistance in various operations in the conduct of the experiment, and to the Corona-College Heights Orange and Lemon Association, Riverside, California for cooperation in handling the fruit in packing house operations and for packing reports.

The California studies were planned and conducted under the guidance of the California Citrus Project's Executive Committee and the authors express their gratitude for coperation to its members: A. S. Mostafa, P. DeBach, W. H. Ewart, L. R. Jeppson, R. F. Luck, and J. A. McMurtry (University of California, Riverside), C. E. Kennett (University of California, Berkeley), D. S. Moreno and J. G. Shaw (Agricultural Research Service, Boyden Laboratory, Riverside) and D. K. Reed (Agricultural Research Service, Fruit and Vegetable Insects Research Station, Vincennes, Indiana).

The authors also wish to thank the following cooperators and cooperating agencies in Florida: Mr. A. G. Selhime, USDA-ARS; Dr. J. O. Whiteside, Dr. William Grierson, and Dr. W. F. Wardowski (University of Florida, IFAS, AREC at Lake Alfred); Dr. Jerry Ting (Florida Department of Citrus); The Coca-Cola Fruits Division; Adams Packing Corporation; Waverly Growers Cooperative and Golden Gem Growers Cooperative.

LITERATURE CITED

Albrigo, L. G., and C. W. McCoy. 1974. Characteristic injury by citrus rust mite to orange leaves and fruit. *Proc. Fla. State Hortic. Soc. 87*:48–55.

Brooks, R. F. 1964. Control of citrus snow scale *Unaspis citri* in Florida. *Proc. Fla. State Hortic. Soc. 77*:66–70.

Brooks, R. F., and W. L. Thompson. 1963. Investigations of new scalicides for Florida. *Fla. Entomol. 46*(4):279–284.

Brooks, R. F., and J. D. Whitney. 1973. Citrus snow scale control in Florida. *I. Congreso Mundial de Citricultura 2*:427–431. International Society of Citriculture.

Carman, G. E. 1977. Evaluation of citrus sprayer units with air towers. *Citrograph 62*(5): 134–139.

Dean, H. A. 1975. Complete biological control of *Lepidosaphes beckii* on Texas citrus with *Aphytis lepidosaphes*. *Environ. Entomol. 4*:110–114.

DeBach, P. 1958. The role of weather and entomophagus species in the natural control of insect populations. *J. Econ. Entomol. 51*:474–484.

DeBach, P. 1977. New red scale parasite imported from Saudi Arabia. *Citrograph 62*: 321–322.

DeBach, P., E. J. Dietrick, C. A. Fleschner, and T. W. Fisher. 1950. Periodic colonization of *Aphytis* for control of the California red scale, Preliminary tests, 1949. *J. Econ. Entomol. 43*:783–802.

DeBach, P., D. Rosen, and C. E. Kennett. 1971. Biological control of coccids by introduced natural enemies, in C. B. Huffaker (ed.), *Biological Control*. Plenum, New York, pp. 165–194.

Div. Agric. Sci., Univ. Calif. 1976. 1976–1978 Treatment guide for California citrus crops. Leaflet 2903, Berkeley.

Ebeling, W. 1936. Effect of oil spray on California red scale at various stages of development. *Hilgardia 10*(4):95–125.

Fisher, F. E. 1950. Two new species of *Hirsutella* Patouillard. *Mycologia 42*:290–297.

Griffiths, J. T. 1951. Possibilities for better citrus insect control through the study of the ecological effects of spray programs. *J. Econ. Entomol. 44*:464–468.

Griffiths, J. T., and W. L. Thompson. 1953. Reduced spray programs for citrus for canning plants in Florida. *J. Econ. Entomol. 46*:930–936.

Griffiths, J. T., Jr., and F. E. Fisher. 1950. Residues on citrus in Florida: Changes in purple scale and rust mite populations following the use of various spray materials. *J. Econ. Entomol. 43*:298–305.

Hart, W. G., S. Ingle, M. Garza, and M. Mata. 1966a. The response of brown soft scale and its parasites to repeated insecticide pressure. *J. Rio Grande Val. Hortic. Soc. 20*:64–68.

Hart, W. G., J. W. Balock, and S. Ingle. 1966b. The brown soft scale, *Coccus hesperidum* L. (Homoptera: Coccidae) in citrus groves in Rio Grande Valley. *J. Rio Grande Val. Hortic. Soc. 20*:69–73.

Hart, W. G., S. Ingle, and M. Garza. 1969. Current status of brown soft scale in citrus groves of the Lower Rio Grande Valley. *Ann. Entomol. Soc. Amer. 62*:855–858.

Hubbard, H. G. 1885. Insects affecting the orange. U.S. Dept. Agric., U.S. Government Printing Office, Washington, D.C.

Jeppson, L. R., W. E. Westlake, and F. A. Gunther. 1969. Toxicity control, and residue studies with DO-14 [2-(*p-tert*-Butylphenoxy) cyclohexyl 2-Propynyl Sulfite] as an acaricide against the citrus red mite. *J. Econ. Entomol. 62*:531–536.

Jeppson, L. R., J. A. McMurtry, D. W. Mead, M. J. Jesser, and H. G. Johnson. 1975. Toxicity of citrus pesticides to some predaceous phytoseiid mites. *J. Econ. Entomol. 68*:707–710.

McCoy, C. W. 1976. Leaf injury and defoliation caused by the citrus rust mite, *Phyllocoptruta oleivora* (Ash.). *Fla. Entomol. 59*:403–410.

McCoy, C. W., and R. F. Kanavel. 1969. Isolation of *Hirsutella thompsoni* from the citrus rust mite, *Phyllocoptruta oleivora*, and its cultivation on various synthetic media. *J. Invetebr. Pathol. 17*:270–276.

McCoy, C. W., R. F. Brooks, J. C. Allen, and A. G. Selhime. 1976a. Management of arthropod pests and plant diseases in citrus agro-ecosystems. *Proc. Tall Timbers Conf. Ecol. Anim. Control by Habitat. Manage. 6*:1–17.

McCoy, C. W., P. L. Davis, and K. A. Munroe. 1976b. Effect of late season fruit injury by the citrus rust mite, *Phyllocoptruta oleivora* (Prostigmata: Eriophyoidea), on the internal quality of Valencia orange. *Fla. Entomol. 59*:335–341.

McMurtry, J. A. 1969. Biological control of citrus red mite in California. *Proc. First Int. Citrus Symp. 2*:855–862.

Meade, F. W. 1976. Economic insects in Florida. 1975. Fla. Dept. of Agric. Consumer Services, Div. of Plant Industry.

Muma, M. H. 1969. Biological control of various insects and mites on Florida citrus. *Proc. First Int. Citrus Symp. 2*:863–870.

Newell, W. 1916. The quarterly bulletin of state plant board of Florida, pp. 51–53.

Riehl, L. A. 1969. Advances relevant to narrow-range spray oils for citrus pest control. *Proc. First Int. Citrus Symp. 2*:897–907.

Reinking, R. B. 1964. Brown soft scale in Texas. *Proc. Fla. State Hortic. Soc. 77*:70–71.

Simanton, W. A., and K. Trammel. 1966. Recommended specifications for citrus spray oil in Florida. *Proc. Fla. State Hortic. Soc. 79*:26–30.

Thompson, W. L. 1950. Scales, whiteflies and mealybugs—why do we control them and how. *Citrus Mag. 13*(3):31–33.

Villalon, B., and H. A. Dean. 1974. *Hirsutella thompsonii* a fungal parasite of the citrus rust mite, *Phyllocoptruta oleivora*, in the Rio Grande Valley of Texas. *Entomophaga 19*:431–436.

Watson, J. R., and E. W. Berger. 1937. Citrus insects and their control. *Fla. Agric. Ext. Serv. Bull. 88*:1–135.

Whiteside, J. O. 1973. Evaluation of fungicides for citrus greasy spot control. *Plant Dis. Rep. 57*:691–694.

Whiteside, J. O. 1975. Planning of spray programs for the control of fungal diseases in citrus groves. *Proc. Fla. State Hortic. Soc. 88*:49–55.

Yothers, W. W. 1949. Citrus odds and ends. *Citrus Industry 30*:14.

11

A SYSTEMS APPROACH TO RESEARCH AND DECISION MAKING IN THE CITRUS ECOSYSTEM

R. F. Luck

Division of Biological Control
University of California, Riverside, California

J. C. Allen

Agricultural Research and Education Center
University of Florida, Lake Alfred, Florida

D. Baasch

Division of Biological Control
University of California, Riverside, California

INTRODUCTION

A systems approach to research and decision making in the citrus ecosystem is in a less advanced state than that of some other crop ecosystems (e.g., alfalfa and cotton) due to the more recent application of systems analysis to the crop. The development of this framework has been initiated in two states, Florida and California. Although the methods of citrus culture, the pests involved, the climate, the citrus varieties grown and the economics of the citrus industry differ in various ways in the states that grow citrus commercially, the principles learned from these efforts are applicable to citrus management in general. As we gain more understanding of the citrus plant, for example, the citrus plant model will be applicable not only in Florida and California but also in Arizona and Texas, or wherever else citrus is grown. Since it is the impact of the various pests, climatic factors, and cultural practices upon the crop that we wish to assess, we must have a workable plant model through which we can match these various impacts. Thus, we first turned our attention to the plant model.

CITRUS PLANT MODEL

The citrus tree model is under initial development at Lake Alfred, Florida. Its basic mathematical framework is composed of a system of nonlinear, *rate equations*:

$$\dot{x}_1 = -\mu_1(\bar{x}, t)x_1 + \lambda_1(\bar{x}, t)x_5 \tag{1}$$

$$\dot{x}_2 = -\mu_2(\bar{x}, t)x_2 + \lambda_2(\bar{x}, t)x_5 \tag{2}$$

$$\dot{x}_3 = -\mu_3(\bar{x}, t)x_3 + \lambda_3(\bar{x}, t)x_5 \tag{3}$$

$$\dot{x}_4 = -\mu_4(\bar{x}, t)x_4 + \lambda_4(\bar{x}, t)x_5 \tag{4}$$

$$\dot{x}_5 = P(\bar{x}, t)x_1 - R(t) \sum_{i=2}^{4} - x_5 \sum_{i=1}^{4} \lambda_i(\bar{x}, t) \tag{5}$$

$$\dot{x}_6 = W(t) - \varepsilon(x, t)x_6 \tag{6}$$

where

x_1 = leaf mass (kg dry wt)

x_2 = stem mass (kg dry wt)

x_3 = fruit mass (kg dry wt)

x_4 = root mass (kg dry wt)

x_5 = usable carbohydrate mass (kg dry wt)

x_6 = soil moisture (hectare–cm) to a depth of 152 cm

$\dot{x}_i$ = the rate of change in the ith component and $\bar{x}$ indicates the vector of all system variables.

$W(t)$ = rainfall input

$\varepsilon(\bar{x}, t)$ = evapotranspiration output

Figure 11.1 represents the flow diagram for this system. The *matrix form* representing the equations of the model (above) is as follows:

$$= \begin{bmatrix} -\mu_1(\bar{x}, t) & 0 & 0 & 0 & \lambda_1(\bar{x},t) & 0 \\ 0 & -\mu_2(\bar{x}, t) & 0 & 0 & \lambda_2(\bar{x}, t) & 0 \\ 0 & 0 & -\mu_3(\bar{x}, t) & 0 & \lambda_3(\bar{x}, t) & 0 \\ 0 & 0 & 0 & -\mu_4(\bar{x}, t) & \lambda_4(x, t) & 0 \\ P(x, t) & -R(t) & -R(t) & -R(t) & -\sum_{i=1}^{4} \lambda_i(\bar{x}, t) & 0 \\ 0 & 0 & 0 & 0 & 0 & -\varepsilon(\bar{x}, t) \end{bmatrix} \begin{bmatrix} x_1 \\ x_2 \\ x_3 \\ x_4 \\ x_5 \\ x_6 \end{bmatrix} + \begin{bmatrix} 0 \\ 0 \\ 0 \\ 0 \\ 0 \\ w(t) \end{bmatrix}$$

wherein growth of components $x_1 \cdots x_4$ is by allocation of carbohydrate (x_5) through allocation functions $\lambda_1 \cdots \lambda_4$; mortality of the components $x_1 \cdots x_4$ is affected by mortality functions $\mu_1 \cdots \mu_4$; the μ's are, in general, time and system dependent; these mortality functions depend, in part, on the pest population densities; ET is the evapotranspiration rate in hectare–cm/

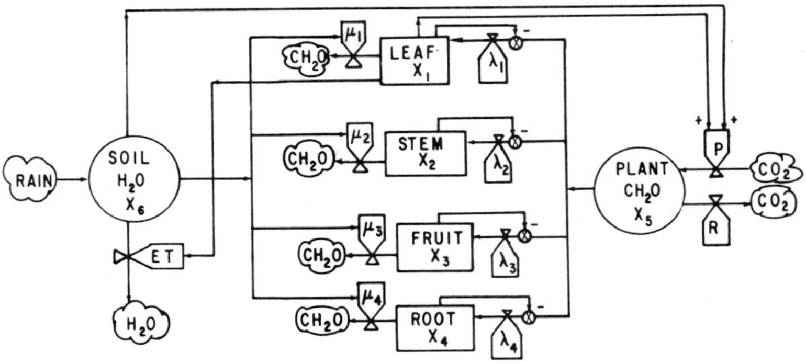

PLANT MODEL FLOW DIAGRAM

Figure 11.1 Flow diagram for the citrus tree model with soil moisture effects.

day; P is the net photosynthetic rate in kg glucose/(kg of leaf); and R is the total respiration rate for all components except leaves, in kg glucose/kg/day.

Before simulating the system described above and in Figure 11.1, the model's stability characteristics were considered. By representing the model as a matrix equation (above), we gain some insight into its stability (Siljak, 1972, 1974, 1975; Patton, 1975). For example, even though the components of the matrix vary both with time and the state of the system, the behavior of growth, mortality, photosynthesis, respiration, and so on, is largely bounded and oscillatory. Systems of this sort will tend to be stable if they have large negative diagonals relative to off-diagonal elements (Siljak, 1974, 1975). This is precisely the sort of matrix used to describe the citrus plant. In other words, the tree model should have a rather strong tendency to stabilize its mass of leaves, stems, roots, and fruits.

Model Simulation

The form of the tree simulation program is shown in Figure 11.2. The main program CITRUSIM, calls SUBROUTINE RKGS (IBM 1970) to solve the system of rate equations in SUBROUTINE FCT. SUBROUTINE FCT implicitly calls a list of function subroutines to calculate the coefficients in the rate equations. These time and system dependent coefficients correspond to the coefficients in the mathematical matrix model discussed above. The function subroutines for these coefficients are separate and independent and

STRUCTURE OF THE CITRUS TREE SIMULATION PROGRAM

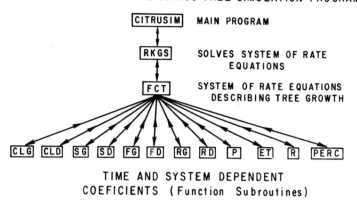

TIME AND SYSTEM DEPENDENT
COEFICIENTS (Function Subroutines)

Figure 11.2. Diagram of the citrus tree simulation program (FORTRAN).

can, therefore, be easily changed as our knowledge of citrus plant physiology increases. These functions are:

1. FUNCTION CLG (Citrus Leaf Growth Rate). The leaf growth rate is currently defined as

$$\left(1 - \frac{x_1}{k_1}\right)\left[A \sin\left(\text{Day} \frac{2\pi}{365} - \frac{\pi}{2}\right) + \overline{\text{CLG}}\right] \tag{7}$$

where x_1 is the current leaf mass; k_1, the optimal leaf mass for trees of this age; A_1, the amplitude of the annual oscillation in leaf growth rate; $\overline{\text{CLG}}$, the average value for the annual growth rate; and DAY, the Julian date. Function (7) oscillates about an average $\overline{\text{CLG}}$ with amplitude A and has a maximum in June and a minimum in December. Growth increases when leaf mass (x_1) drops below the optimum (k_1). The optimal value of leaf mass (k_1) was computed from the data presented by Turrell et al. (1969) $[k_1 = 1.426(\text{AGE})**0.8095]$.

Although the function used here is simply a guess, it permits us to develop the modeling structure. The function (component) can be replaced when better information becomes available. Studies are underway utilizing data from 130 citrus groves taken over a period of 16 years (Simanton, 1970) to gain some insight into those factors which are associated with leaf growth. Polynomials (up to the ninth order) have been fitted to these data, using both Julian and degree days as independent variables (Fig. 11.3A,B) for each of the five citrus districts. Julian days appear to be a better predictor of

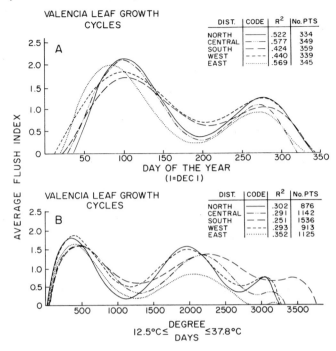

Figure 11.3. Polynomial (up to 9th order) curves relating leaf growth to day of the year (A) and degree-days between 13.5 and 37.8°C (B). [Data of W. A. Simanton.]

growth as judged by the larger R^2 values. Such a relationship implicitly incorporates temperature, photoperiod, and other variables which show a seasonal pattern in variation, suggesting that growth is not a simple function of temperature. This is further suggested by Moss's (1969) data which indicate that use of *chilling* units together with a following warming trend appears to be a valuable predictor of growth. Simanton's (1970) data are also being used to investigate the possibility.

2. FUNCTION CLD (Citrus Leaf Death Rate). Leaf mortality rate is currently defined as

$$A \cos \left(\text{DAY} \, \frac{2\pi}{365} \right) + \overline{\text{CLD}} \qquad (8)$$

where A and DAY are defined as before, and $\overline{\text{CLD}}$ is average leaf mortality. As with leaf growth, leaf mortality oscillates about an average value $\overline{\text{CLD}}$, with amplitude A, and has a maximum on 1 January and a minimum on about

30 June. Studies are currently underway to measure leaf mortality as a function of age, temperature, moisture stress, and greasy spot disease. Whiteside's (1970) data indicate that leaf drop is maximal (measured as recently fallen leaves) in March and April, somewhat later than described by function (8); however, leaf tagging studies (Allen, unpublished data) show that maximal drop sometimes occurs during November and December. Rate of leaf drop is obviously influenced by several factors, three of the most important being greasy spot disease (in Florida), age distribution of leaves, and moisture stress.

3. FUNCTION SG AND FUNCTION SD (Stem Growth and Death Rate). The rates for stem growth and death are defined by functions analogous to these for leaves, since growth at new flush growth periods involves both stems and leaves.

Stem growth is defined as

$$\left[1 - \frac{x_2}{k_2} \right] \left[A \sin \left(\text{DAY} \left(\frac{2\pi}{365} - \frac{\pi}{2} \right) + \overline{\text{SG}} \right) \right] \tag{9}$$

where x_2 is the current stem mass; k_2, the optimal stem mass for trees of a specified age; A, the amplitude of the annual oscillation in stem growth; SG, the average annual stem growth rate; and DAY is defined as before. Data from Turrell et al. (1969) was used to calculate optimal stem mass [$k_2 = 0.0513(\text{AGE})^{**}2.632$].

Stem mortality, defined as

$$A \cos \left(\text{DAY} \frac{2\pi}{365} \right) + \overline{\text{SD}} \tag{10}$$

is typically taken as an order of magnitude lower than leaf mortality, since information on these processes is lacking. ($\overline{\text{SD}}$ is average stem mortality; A and DAY are defined as before).

4. FUNCTION FG (Fruit Growth Rate). Fruit growth rate is defined as

$$A \cos \left(\text{DAY} \frac{2\pi}{365} - 3.347 \right) + \overline{\text{FG}} \tag{11}$$

Function (11) oscillates about the average growth rate $\overline{\text{FG}}$ with amplitude A and the growth peak in mid-July (DAY = 194). Previously, fruit growth in terms of surface area (cm^2) in Florida was defined as

$$153/[1 + \exp(4.8 - 0.02264 \text{ DAY})] \tag{12}$$

by Allen (1976) where DAY is defined as before. Equation (12) gives a maximum growth rate at DAY = 212 (31 July). In contrast, Bain's (1958) data indicates that on a dry weight basis, maximum growth occurs about three months after bloom, that is, mid-June in Florida. Additional data are being sought to determine the timing of this peak more precisely.

5. FUNCTION FD (Fruit Drop Rate). This function is now tentatively set at zero, since information is unavailable concerning the processes leading to fruit drop. Studies are underway, however, to measure the effects of moisture stress and pest damage on the drop rate. Rust mite damage increases the drop rate, especially in the presence of moisture stress (Allen, 1978).

6. FUNCTION RG and FUNCTION RD (Root growth and root death rates). Root growth and death rates for citrus are poorly understood; therefore, the functions for these rates are, temporarily, assumed to be identical to those for leaf growth and leaf death. What information is available suggests that roots regenerate between May and October (Halma & Compton, 1936) and that their growth is stimulated by warm temperatures and rainfall (Crossman, 1940). Somewhat conflicting information exists as to the relationship between root and leaf growth. Crider (1927) observed alternate periods of root and leaf growth while Reed and MacDougal (1937) found that these periods of growth, while different, overlapped greatly. However, the data suggest a mid-June peak in root growth.

7. FUNCTION P (Net photosynthesis rate for leaves). The function relating photosynthesis and light intensity is taken from the data of Kriedemann (1968) and is described by:

$$P = 12(1 - \exp(-2I)) \qquad (13)$$

where I is light intensity in ergs cm^{-2} sec^{-1} $\times$ 10^{-5} and P is in mg CO_2/ dm^2/hr^1. Average light intensity for the day at the canopy top is calculated by averaging the sine of the solar elevation angle (Baker et al., 1973; Stapleton et al., 1973; Shultze, 1976) during the photoperiod and then applying the atmospheric attenuation (Shultze, 1976). Photosynthesis is calculated at each of seven levels in the canopy as a function of average light intensity at each level. Light intensity is assumed to follow "Beer's Law" of exponential decay as more leaf area index (leaf area/ground area) is penetrated, that is,

$$I = R_0 \exp(-0.4CLAI_i) \qquad (14)$$

where I is light intensity, R_0 is the radiation at the canopy top, and $CLAI_i$ is the cumulative leaf area index to level i. Photosynthesis in each of the seven levels is then summed to give the total for the whole tree in kg glucose/ (kg of leaf)/day.

The effects of temperature on photosynthesis are introduced by multiplying the summed value for photosynthesis by a bell-shaped function for temperature effect fitted to Kriedemann's (1968) data. This function has a maximum value at 23°C and the temperature value used is the average temperature which occurred *during the photoperiod.*

The effects of moisture stress are included in a similar manner to that of temperature. The effect of moisture stress on photosynthesis is mimicked by including a term

$$1 - \exp[-5(s + 0.1)]] \tag{15}$$

where s is the status of soil moisture (X_6) measured to a depth of 152 cm as a fraction between the wilting point (WP) and field capacity (FC), that is,

$$s = 1 - [(FC - x_6)/(FC - WP)] \tag{16}$$

8. FUNCTION R (Respiration rate). In spite of its importance, whole tree citrus respiration rate is the least known of the model parameters. Respiration rate is currently calculated as:

$$A \cos \left(DAY \frac{2\pi}{365} - 3.303 \right) + \bar{R} \tag{17}$$

where $\bar{R}$ is the average value for respiration and A and DAY are defined as before. Function (17) oscillates about an average $\bar{R}$ with an amplitude, A, and has a peak in mid-June. In order to prevent the simulated tree from collapsing due to carbohydrate depletion, it has been necessary to assume an $\bar{R} \simeq 0.0003$ and an $A \simeq 0.00025$ (kg glucose/(kg of tree)/day). These values may seem to be low but they include the large mass of dead xylem in the trunk and branches. For example, Bain (1958) reported a value for R of 30 mg CO_2/kg (fresh wt)/hr for mature fruits. This would convert to ≈ 0.003 kg glucose/kg (dry wt)/day, which is an order of magnitude above the assumed $\bar{R}$. It seems reasonable, however, that this value should be reduced by an order of magnitude if trunks and branches are included.

Function (17) will be modified by partitioning it into maintenance and growth fractions (Hesketh et al., 1971; Baker et al., 1972) and by constructing weighted values for the different tree components.

9. FUNCTION ET (Evapotranspiration rate). A large body of literature exists on predicting evapotranspiration based on solar radiation, vapor pressure deficit, and wind (Pennan, 1948; Linacre, 1967; Messem, 1975). This method has been applied to citrus with limited success (Hashemi & Gerber, 1967). A better fit to the data has been obtained by Koo (1953, 1969) using only mean daily temperature. Therefore we used "Koo's curve" (Fig. 11.4) to predict evapotranspiration in the tree model by multiplying the value obtained from the curve by x_6/FC, the fraction of the field capacity which exists in the soil.

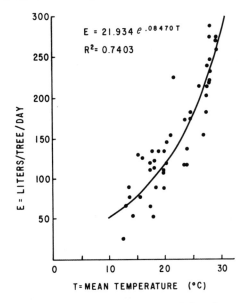

Figure 11.4. Relationship between mean daily temperature (°C) and evapotranspiration. [Data of R. C. J. Koo.]

10. FUNCTION PERC (Percolation rate). Percolation accounts for approximately 30 percent of the moisture loss on Astatula fine sand soil in Florida (Koo, 1953); thus, a function was derived from Koo's (1953) data in which

$$PERC = 0.00316x_6 \qquad (18)$$

Percolation is in units of ha-cm/tree/day. Below field capacity (4.01 ha-cm), percolation is assumed to be zero.

Simulation Results

Depletion of the carbohydrate pool (x_5) in the model, due either to low photosynthetic rate or to high respiration rate, results in a collapse of the tree (decline of all tree components). Furthermore, strong cyclic behavior in the occurrence of carbohydrate levels within this pool is typical of the model. This sort of behavior is characteristic of many citrus varieties. For example, Murcott tangerine trees exhibit such strong alternate bearing cycles that in heavy bearing years they appear visually "dead" due to the absence of leaves (Stewart et al., 1968; Smith, 1976). The carbohydrate (starch) in all tree components oscillates inversely with these fruit bearing cycles (Smith, 1976). A similar pattern in carbohydrate reserves also occurs in Valencia oranges (Jones et al., 1970) although the trees do not exhibit visible collapse symptoms like those of Murcott tangerines.

An output for a simulation run of an 18-year-old tree is given in Figure 11.5. An average daily Florida temperature [5.56 sin (Day $(2\pi/365) - 1.915$) + 22.2] and rainfall cycle {0.254 sin [Day $(2\pi/365) - 1.894$] + 0.3503} was used as input. These values, in C° and cm (their amplitudes and averages), were obtained by curve fitting to temperature and rainfall data recorded at Lakeland, Florida, between 1934 and 1974. On this basis, the model produces a yield of 28 kg (dry wt) of fruit after one year, or 186.7 kg (fresh wt), assuming fruit to be 15 percent dry wt (Chapman, 1968).

To assess their impact on yield, the values of several variables were varied in a simulation run. For example, doubling the leaf mortality rate caused an 11.9 percent *decrease* in yield, increasing temperature by 3°C *reduced* yield by 13.2 percent; while *decreasing* it by 3°C *increased* yield by 5.4 percent, and finally increasing rainfall by 50 cm increased yield by 2.8 percent, while decreasing it by 50 cm *decreased* yield by 29.6 percent. These results should be viewed cautiously, however, since the model remains crude and has yet to be

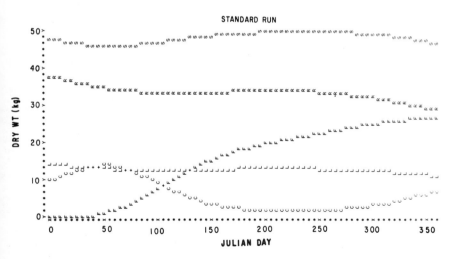

Figure 11.5. Example of "CITRUSIM" output. L = leaf, S = stem, F = fruit, R = root, C = carbohydrate storage (dry wt kg).

validated. Still, these results do permit preliminary comparisons of simulation results with those from the field, and as such are very important for further development of the model and eventually a much better understanding of the whole process.

CITRUS RUST MITE DAMAGE MODEL

The citrus rust mite *Phyllocoptruta oleivora* (Ash.) has been an important pest of Florida citrus for some 100 years. The mite damages the fruits by puncturing the rind and feeding on the epidermal cell contents (McCoy & Albrigo, 1975). Fruit surface discoloration (russetting) results and is associated with lignin formation and probably oxidation of substances contained within the cytoplasm of the damaged epidermal cells. A high puncture frequency results in death of these cells (McCoy & Albrigo, 1975). It is this damage, when extensive (as when a high mite population is present), which visually reduces the commercial grade of fruits, reduces fruit size (Yothers & Mason, 1930) and increases water loss (Ismail, 1970). This latter factor increases fruit drop during periods of severe water stress (Fig. 11.6) (Allen, 1978).

Models were derived which relate rust mite densities as a function of time to percentage of damaged area on a fruit (Allen, 1976). The first model assumes that the rate of damage is proportional to mite density; the second

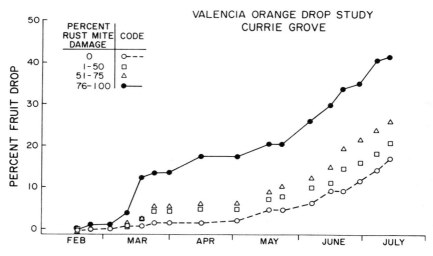

Figure 11.6. Relationship between percent rust mite damage (categories) and percent fruit drop.

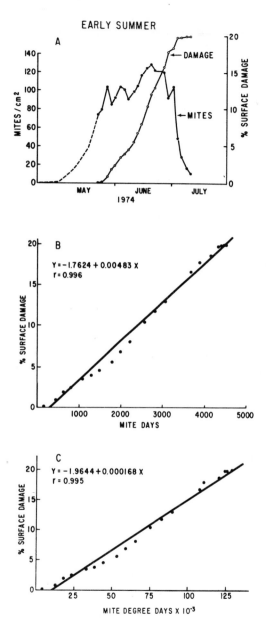

Figure 11.7. Relationship between (A) rust mite population and percent surface damage on Valencia orange for early summer, (B) mite-days and percent surface damage, (C) mite degree-days and percent surface damage.

assumes that the "proportionality constant" varies with time; and the third incorporates the amount of surface area available on a growing fruit. These models were then compared with field data for agreement.

Figure 11.7 indicates that accumulated mite-days of infestation (Fig. 11.7b) and mite-day degrees of the infestation (Fig. 11.7c) are equally good predictors of percent surface damage (as measured by the regression coefficient, r). Both these relationships appear linear. The fact that accumulated mitedays appear to be linearly related to percent damage implies that fruit damage rate is proportional to mite densities. Moreover, the slopes used to describe the 1974 data were also observed to increase with time. This implies that the "proportionality constant" varied with respect to time during different periods of fruit growth. Thus, the amount of damage that can be expected from a given density of mites varied with Julian date (Fig. 11.8).

The percent of surface damage on Valencia orange fruits can be predicted with the following equation:

$$P(t) = 0.0115 \int_0^t m(\tau) \, d\tau/(1 + \exp(6.92 - 0.03592\tau)) \qquad (19)$$

where $P(t)$ is the percent surface area damaged at time (τ), $m(\tau)$ is the observed (or simulated) mite density,

$$0.0115/[1 + \exp(6.92 - 0.03592\tau)]$$

is the relationship described in Figure 11.8, and τ is the Julian date (Allen, 1976). This equation (19) is the first step in developing a pest management system for citrus rust mite. The next step, currently underway, is determining an "economic threshold" for this species.

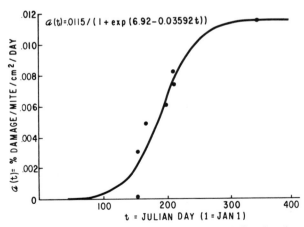

Figure 11.8. Damage rate, at (t), from eqn. (3) as a Julian time function.

CALIFORNIA RED SCALE/*APHYTIS* MODELS

Background

The objective of the California modeling effort has been to investigate the influence of macroclimatic factors on the population dynamics of citrus arthropod pests and their natural enemies, particularly for the key pest, California red scale, *Aonidiella aurantii* (Mask.).

Climate plays a major role in determining the severity and composition of California's citrus pests. Citrus is grown in four more or less distinct climatic regions within the state (Chapter 10): (1) the coastal area, marked by mild temperatures and less severe humidities; (2) the interior valleys of southern California, marked by greater differences in daily and seasonal temperatures and humidities; (3) the Great Central Valley (the San Joaquin and Sacramento Valleys), also marked by greater differences in daily and seasonal temperatures and humidities; and (4) the desert area, marked by extremes in temperature and by rather consistent, low humidities. The interior southern California valleys differ from the Great Central Valley in the type of temperature and humidity patterns they experience. During the fall and winter the interior valleys are subjected to periodic northeast winds which cause both compressional heating and very low humidity. Conversely, during the winter, the Great Central Valley is usually subjected to extensive periods of temperature inversion which produces fog and low temperatures throughout the diel period.

These regional differences in climatic patterns are known to affect the distribution of parasitic natural enemies and, by inference, their effectiveness in regulating host populations (Ebeling, 1959; DeBach et al., 1955; DeBach & Sisojevic, 1960; DeBach & Sundby, 1963). The best documented example of this is seen in the interaction between several species of aphelinid wasps of the genus *Aphytis*, and their host, the California red scale, *Aonidiella aurantii* (DeBach & Sisojevic, 1960; DeBach & Sundby, 1963). This scale is present in all citrus regions of the State, but it is now a major pest only in the desert and Great Central Valley areas, although it was formerly severe in all areas. In the other two regions (coastal and interior valleys of southern California) its populations are generally limited to subeconomic densities by introduced parasitoids, the most important being *Aphytis lingnanensis* Compere and *Aphytis melinus* DeBach. *Aphytis chrysomphali* was for some years the most prevalent species. Two other parasitic species, *Comperiella bifasciata* Howard (red scale strain) and *Prospaltella perniciosi* Tower, may also be significant. All of these parasites are aphelinids in the Order Hymenoptera. Most of them were purposely introduced over a period of about 80 years in an attempt to obtain economic control of this key citrus pest.

Competition among *Aphytis* species, following their sequential introductions, resulted in a sequence of one species displacing another so that their current geographical distributions can be summarized as follows (DeBach & Sundby, 1963; DeBach et al., 1971): *A. lingnanensis* occurs only along the coast, *A. chrysomphali* is encountered infrequently in coastal samples, while *A. melinus* occurs in the interior valleys. *A. melinus* also occurs in the Great Central Valley on ornamental and dooryard plantings, but it has failed to establish itself in the valley's commercial groves even though it has been extensively released. DeBach and Sisojevic (1960) and DeBach and Sundby (1963) studied the effects of temperature on various attributes of the *Aphytis* species in laboratory and field experiments. Their results, although incomplete, correlate well with these parasitoids' geographical distributions as related to prevailing temperature patterns, but the mechanism(s) of this competition, evidenced as competitive displacement, remains unidentified.

Our initial objective has been to investigate the influence of macroclimatic factors on the population dynamics of California red scale and its parasitoid complex. California red scale was chosen for central focus for several reasons. First, it is the key pest in the Great Central Valley where 55 percent of the State's orange crop is produced. Second, it is potentially a serious pest in the coastal and interior valley regions of southern California, should pest management practices disrupt the efficiency with which the parasitoids now regulate this scale's populations at subeconomic densities. Third, population data and field observations were already available, with respect to scale and parasitoid ecology, age structure and densities, and host-parasitoid interactions, in far greater detail than for any other citrus pest. Fourth, California red scale is a pest whose laboratory culture techniques have been well worked out. And fifth, it exhibits a number of characteristics common to many citrus pests: multiple generations, overlapping stages, and age specific mortalities, all of which interact with climate and evidence themselves in the dynamics of the pest. It is the dynamics of such pest populations that we must better understand if we are to develop sound integrated pest management systems for citrus. Thus California red scale was the logical choice for the initial modeling effort.

The Requirements for Modeling

In designing a population model to mimic the dynamics of California red scale and *Aphytis* the following characteristics were considered essential: (1) a method of mimicking population age structure, since scale mortality from parasitism as well as from certain other unknown causes is "age" (maturity or instar) specific, (2) a method of mimicking scale phenology as observed in the field, (3) a method of analyzing variable climatic conditions characteristic of the regions in which citrus is grown, and (4) a method of realistically coupling a parasite population with that of its host (both species having overlapping generations and continuous reproduction and development).

Part of these requirements are met by the choice of temperature as the initial climatic factor to be investigated. Since rate of development of insects is a function of temperature, it provides the dominant "timing mechanism" for the population interactions (Shelford, 1929; Hughes, 1963; Gutierrez et al., 1974). The diel and annual patterns also present major differences between the various citrus regions within the State. Furthermore, it is an important parasitoid mortality factor and thus may limit the distribution or effectiveness of such parasites (DeBach, 1965; DeBach & Sisojevic, 1960). It is also a significant mortality factor for certain scale instars as well (Abdel-rahman, 1974).

A model designed to incorporate age structure, mimic phenology, and include variable temperatures requires the following structure and information: (1) stage (age) specific developmental rates for the scale and for the immature stage of each parasite species; (2) age specific fecundity rates for the scale converted to a temperature based physiological time scale (similar to the time–temperature derived developmental rates); (3) temperature dependent attack rates for the parasite; (4) an insect developmental model for host and parasites which (a) reflects variation in individual development rates, yet (b) maintains a within-instar age structure to allow for age specific parasitization (only younger scales within the second and third instars are susceptible to attack); and (5) the input of real temperature data for conversion to a relevant "timing scale" for both the scale insect and parasite populations, and for assessing temperature effects on population survival of scale and parasite.

California Red Scale Developmental Submodel

The California red scale developmental submodel has initially employed a heat accumulation algorithm that is a synthesis and modification of methods described or used by Baskerville and Emin (1969), Eubank et al. (1973), and Gutierrez et al. (1974). Laboratory derived data on mean developmental time, in days, for each instar at a series of constant temperatures were obtained from Willard (1972). These data were fitted with second order least-square regressions to obtain a series of instar specific equations of the following form:

$$\text{Devel}(i) = T(b - c \cdot T) \tag{21}$$

where Devel(i) is the amount of time, in days, required by an average individual of stage i (instar, male or female scale) to complete its development at a specified temperature, T in deg C. The inverse of these equations are the stage specific development rates (Fig. 11.9)

$$\text{Growth}(i) = 0.041667/[a - T(b - c \cdot T)] \qquad (22)$$

where Growth(i) is the proportion of the total growth of stage i that occurs during an hour at a specified temperature, temp. (1/24 hrs = 0.041667).

Willard's (1972) Australian data for the developmental rates of this insect are similar to those obtained by Jones (1935) in Southern Rhodesia, Munger and Cresson (1948) in California, and Bodenheimer (1951) in Persia.

But California red scale is exposed to variable temperatures in the field; hence, the average physiological age accrued by a population during a day (24 hours) is a function of the temperature pattern for that day and its appropriate stage specific developmental rate (eqn. 20). The daily temperature pattern is approximated in two parts by fitting a sine curve to the day's minimum and maximum temperatures and to that day's maximum and next day's minimum temperatures. Thus, the curve for a time period is a continuous curve of sinusoidal shape between adjacent minimum–maximum and maximum–minimum temperatures. Using these curves, an approximate temperature for the midpoint of each hourly period is easily calculated.

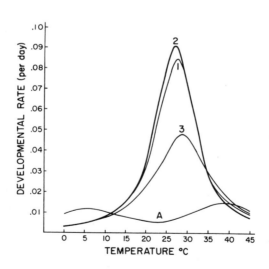

Figure 11.9. Relationship between the rate of development and temperature for immature scale stages 1–3 and the rate of ageing and temperature for adult female California red scale (A).

Thus the midpoint temperature is calculated from the time, t, by

$$T = T_m + a(-\cos(t)) \tag{23}$$

where T is the temperature at the midpoint of a specified one hour interval; T_m, is the mean temperature for a given twelve hour period, i.e.,

$$T_m = (T_{min} + T_{max})/2.0 \tag{24}$$

a is the amplitude of the temperature pattern for a given twelve hour period, that is,

$$a = (T_{max} - T_{min})/2.0 \tag{25}$$

and time is the midpoint of the one hour interval $(1, 2, \ldots, 12)$ expressed in radians:

$$t = \frac{\pi}{12}(\text{interval}) - \frac{\pi}{24} \tag{26}$$

The growth per day can then be calculated for each instar by summing the hourly growth obtained by evaluating the inverse of the stage specific developmental curves. The temperature $T_{(k)}$ at the midpoint of each hour is used as input, that is,

$$T\text{Growth}(i) = \sum_{k=1}^{24} [0.04167/(a - T_{(k)}[b - c \cdot T_{(k)}])] \tag{27}$$

Thus, the average daily growth of a given stage in the life cycle is the sum of 24 hourly rates.

With adequate food and tolerable humidity, an insect's growth rate is largely a function of temperature and the given individual's genetic constitution. Thus, genetic variability is also a factor affecting an insect's growth rate, and this variability can explain some of the differences observed among a population for any given temperature regime. This variability is easily observed in laboratory data relating temperature and developmental time: all insects of a cohort do not emerge or molt simultaneously. An empirical distribution of developmental times can be determined and then used in a developmental model to reflect this inherent variability. Thus, for example, 50 percent of the population has molted to the next stage when 100 percent development is achieved. Because Willard's (1972) data are expressed in the form of means and variances we have initially chosen an algorithm which approximates a cumulative normal distribution to move individuals within any cohort from one stage (instar) to the next. Transfers from one stage to the next begin two standard deviations from the mean. The algorithms used to approximate this are taken from Abramowitz and Stegun (1964). To obtain the number of scales from a cohort that must be transferred to the next stage, the total number of scales that began a cohort is subtracted from what actually remains of that cohort (due to transfers from earlier growth), thereby giving a negative value. This value is then added to the number of scales that should have transferred at that physiological age ($cp \cdot$ TOT). The result gives the number of scales that should be moved to the next stage i, that is,

$$\text{transfers}(i, j) = \text{ACT}(i, j) - \text{TOT}(i, j) + cp \cdot \text{TOT}(i, j) \qquad (28)$$

where j is the cohort. The total transfers, then, summed for all cohorts, j, of stage i, are

$$\text{TOT transfers}(i) = \sum_{j=1}^{n} \text{transfers}(i, j) \qquad (29)$$

Adult female life expectancy is handled in a manner similar to that for immature development: half the adult females are dead when the average female has lived its expected life span. Life expectancy is assumed to be a function of temperature. Data on the average life expectancies for California red scale at several different constant temperatures are taken from Willard (1972). The inverse of these values are regressed against temperature, giving an equation for the average rate of aging per day as a function of temperature (Fig. 11.9). This is then converted to the average aging rate per hour:

$$\Delta_{age} = 0.041667/(9.37 \times 10^{-3} + 1.0 \times 10^{-3} T + 1.42$$
$$\times 10^{-4} T^2 + 5.48 \times 10^{-6} T^3 - 6.25 \times 10^{-8} T^4) \qquad (30)$$

Crawler production is considered to be a function of temperature and physiologic age (relative life expectancy); therefore, at each temperature the age specific value (an average) is transformed to expected births per individual based on physiologic age. These values are stored in tabular form (Table 11.1). The model uses this table to obtain average births per individual by means of two-dimensional linear interpolation between table values (which represent specific combinations of temperature and physiologic age). The effect of variable temperature on birth rate (especially the time lag) is unknown for California red scale; hence, we have arbitrarily chosen a three day temperature average as the temperature input. Fecundity is evaluated for each adult age group once every iteration (i.e., each diurnal temperature cycle). The value obtained is then multiplied by the number of adult female scales in each age group and by the percent life expectancy used during the iteration [represented in 1 percent life expectancy units (E)]. The products are then summed over all age groups

$$\text{transfers } (i, \text{egg}) = \sum_{j=1}^{n} [\text{FEC}(j)][\text{ACT}(i, j)][\text{E}] \qquad (31)$$

Table 11.1.   Life Expectancy and Temperature-Specific Fecundity Values for California Red Scale (From Willard, 1972; See discussion for table's derivation)

Age (percent aver. life expt.)	Temperature, °C							
	5	10	15	20	25	30	35	40
0.0	0.0	0.2	1.4	2.2	3.1	4.5	6.0	0.0
0.3	0.0	0.2	0.7	1.8	2.3	3.6	4.5	0.0
0.6	0.0	0.1	0.3	1.4	1.7	2.7	3.5	0.0
0.9	0.0	0.1	0.2	1.1	0.9	1.0	0.9	0.0
1.2	0.0	0.05	0.2	0.6	0.2	0.3	0.1	0.0
1.5	0.0	0.05	0.1	0.3	0.0	0.0	0.0	0.0
1.7	0.0	0.05	0.1	0.0	0.0	0.0	0.0	0.0
2.1	0.0	0.0	0.0	0.0	0.0	0.0	0.0	0.0

where transfers (i, egg) are the number of crawlers produced (California red scale's eggs develop internally). ACT(adult, j) is the actual number of adult female scale in cohort j. E is the amount of adult female life expectancy (in 1 percent life expectancy units) used during the 24 hour interaction.

Mortality, except that associated with parasitism, is applied as a stage specific constant to the transfers moving from one stage to the next. Thus we assume that scale mortality, other than that due to parasites, occurs when the scales molt. We have not found scale predation to be a significant source of mortality in the field. Information on stage specific scale mortality is obtained from field censusing.

The functions we have included in the model represent information obtained from the literature, from discussions with experienced researchers, or from our best guesses on what seems biologically reasonable. Thus the model represents a synthesis of current information and understanding of the population dynamics of California red scale. It identifies hypotheses that need testing or areas of ignorance that need further investigation. The model has been designed in such a way that the components (submodels) can be easily changed without much modification to the structure of the rest of the model. The model is programmed in a language (PASCAL) which easily facilitates these changes (see Wirth, 1971, or Jensen & Wirth, 1974).

In summary, then, the California red scale model is a complex computer algorithm which attempts to mimic the scale population's phenology, age structure, and population growth processes.

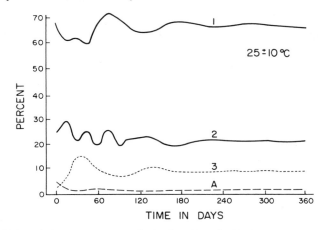

Figure 11.10. Population age structure obtained under a fluctuating temperature of constant amplitude. Numbers 1–3 are immature scale stages 1–3 and A refers to mature adult female scales.

Simulation Experiments With the Model

Preliminary simulation experiments using different constant temperatures yielded California red scale populations with the expected differences in the stable age distribution for each temperature. Similar results were obtained when cyclic temperatures of constant amplitude ($\pm 10^{\circ}$C) were used (Fig. 11.10). Simulation experiments using varying daily temperatures (daily maximum and minimum values) from the United States Weather Bureau Station at Riverside, California for the years 1970 to 1972, indicated an average of 3.1 generations per year (settler to a mature reproducing female). This is similar to the number of generations empirically determined by Dickson and Lindgren (1974) for the scale at this location (3.4 in 1939–40; 3.1 in 1940–41). Furthermore under varying temperatures the simulated scale population showed marked changes in the proportion of ages present at different times of the year (Fig. 11.11). These changes in age distribution are due to the varying daily temperature pattern operating on different instar specific, nonlinear, developmental rates. It appears, then, that regional temperature patterns affect the scale's age structure, independent of any scale mortality they might cause.

Parameterization and Validation Experiments
on the Scale Model

Experiments to validate the heat accumulation algorithm used to mimic California red scale growth have been initiated. Thermocouples, placed in

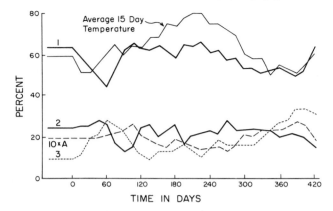

Figure 11.11. Red scale age structure using daily maximum and minimum temperatures recorded during 1972 at Riverside, California.

the immediate vicinity of a number of scale individuals located on navel orange leaves, stems and fruits (Lindcove Field Station near Exeter, California), record the temperatures to which the scales have been exposed. These temperatures are fed to a data acquisition system (Esterline Angus®, Model D2020) and recorded on magnetic tape (Kennedy® Model 1600) every 20 minutes. Scale development is followed with daily photographs of individual scales (Polaroid Land Camera® Model CU-5 fitted with a 3 × lens and strobe flash unit mounted as a ring around the lens). The developmental times for scales in each stage will be compared to that predicted by the algorithm using the recorded temperature near the individual scale. A number of different growth algorithms can be tried (see Stinner et al., 1974).

The daily temperature pattern within the canopy of two navel orange trees at different times of the year (20 locations/tree) are to be compared to that characterized by a sine or skewed sine curve.

The polaroid pictures are also being used to obtain detailed partial life table data for known scale cohorts. Individual scales will be followed throughout their lives with sequential photographs and the stage at which mortality occurs noted. An example of a partial life table from the crawler stage is presented in Table 11.2. Note that survivorship (1_x), especially during the early scale stages, is higher on fruits than on either leaves or branches. The reduced survivorship later in the scale's life cycle on fruits is due to parasitism by *Comperiella bifasciata* (red scale strain). This parasite also attacks the scale on branches and leaves (Table 11.2). The degree of parasitism, however, varies with the season as well as with the host substrates. Scale mortalities other than those associated with this parasite are

Table 11.2. Partial Life Table for a California Red Scale Cohort on Leaves, Stem and Fruit Which Settled During the Interval 7/28 to 8/19, 1976 (Data from navel orange trees, Lindcove Field Station)

Stage X	Fruit $n = 117^a$ $n = 108^b$			Stem $n = 138^a$ $n = 138^b$			Leaf $n = 162^a$ $n = 158^b$			
	l_x	d_x	$100q_x$	l_x	d_x	$100q_x$	l_x	d_x	$100q_x$	dF_x
First Stage	1000			1000			1000			
First Moult	923.1	76.9	7.7	652.2	347.8	34.8	703.7	296.3	29.6	unk
Second Stage	879.9	43.1	4.7	521.7	130.4	20.0	561.7	142.0	20.2	unk
	(791.1)	88.9	10.1	(384.1)	137.7	26.4	(434.0)	127.7	22.7	unk
	2 × (631.1)	160.0	20.2	2 × (188.4)	195.7	50.9	2 × (114.9)	319.1	73.5	♂♂
Second Moult	1262.1	73.2	5.8	376.8	115.9	30.8	229.8	38.3	16.7	unk
Third Stage	1189.0	56.6	4.8	260.9	29.0	11.1	191.5	76.6	40.0	unk
		652.5	57.6		58.0	25.0		43.1	37.5	C. bifasciata
♀♀	479.8			173.9			71.8			

^{a} Number of scale at the start of the study.
^{b} Number of scale accounted for at the end of the study.

388

used as background mortality for the modeled scale population. Recruits, or crawler production and loss, are not being validated, since we have been unable to devise a reliable method of estimating these variables under field conditions.

The Parasite Model

Developmental algorithms for the immature parasite stage (egg to adult stages) were obtained by fitting a second order least-squares regression to the mean developmental times determined for each parasite species at several constant temperatures. These algorithms were then used in developmental and growth submodels similar to those used for their host, the California red scale (eqn. 21–30). Adult female longevity is handled in a similar manner to that for California red scale (eqn. 34), but transfers are dead female parasites. Parasite fecundity is included in the attack submodel.

Scale mortality caused by *Aphytis* is of two types, that caused by host feeding of adult female parasites, and that caused by oviposition and subsequent development of the immature parasite. *Compleriella bifasciata* only causes scale mortality of the latter type. Host feeding by *Aphytis* is currently handled as a constant per parasite individual until more information can be obtained on the relation between host feeding and parasite age, parasite density, host density, and host age structure.

The *area of discovery*, or the attack rate, is obtained by using the Hassell-Varley submodel (Hassell and Varley, 1969)

$$a = QP^{-m} \tag{32}$$

where a is the " area of discovery "; Q, the " quest constant " or the " area of discovery " when the parasitoid density is unity; and m is the mutual interference constant.

This submodel has been chosen for several reasons. First, it has been used to explore the properties of a host–parasite interaction as part of a host–parasite model (Hassell & May, 1973; Hassell et al., 1976; Hassell, 1976). Second, it has been used to define two important behavioral components of a parasite's search efficiency, Q, and how Q decreased with increasing parasite–host ratios, m. The searching efficiency defines the equilibrium density of the interaction. The mutual interference constant defines the degree and type of stability exhibited by the host parasite (model) interaction (Hassell & May, 1973; Varley et al., 1974). Finally, the submodel has been used to assess the consequences of a parasite–host interaction involving several co-inhabiting species of the winter moth, *Operophtera brumata* (L.), in England and to predict the outcome of an interaction in eastern Canada when the

English parasites were introduced into Canada for control of the winter moth.

In this model we assume that the parasites search at random, that the host population is regularly distributed (or that the parasite population does not concentrate its searching on host aggregations), and that the parasite has an unlimited egg supply. The effect of host density is included in the model by including equation (31) in the Holling disc equation (Holling, 1959) as suggested by Rogers (1972; see Hassell & May, 1973, their eqn. 15). Thus

$$H_{par(t+1)} = H_{(t)}\{1 - \exp[-QP_{(t)}^{1-m}/(1 + QbH_{(t)})]\} \qquad (33)$$

where H_{par} are the hosts parasitized; H, the susceptible hosts available; P is the adult female parasite population; Q, the "quest constant"; m, the "mutual interfence constant"; and b, the handling time. The number of susceptible hosts is determined by summing the susceptible stages

$$H_{(t)} = \sum_{i=m}^{l} \sum_{j=n}^{k} N_{scale(t)}(i, j) \qquad (34)$$

where $N_{scale}(i, j)$ is the number of scales in the ith instar and jth cohort. Only cohorts n through k of instars m through l are susceptible, however.

Model Parameterization

Estimates of the quest and mutual interference constants, Q and m, respectively, (eqn. 36) were obtained by fitting least-square linear regressions to area of discovery data for several parasite densities at a constant host density (see Hassell, 1976). The area of discovery is calculated from the following equation:

$$a = \frac{1}{P_t} (\ln N_s - \ln N_t) \qquad (35)$$

where N_s is the scale density that escaped parasitization; N_t, the initial scale density; and P, the parasite density. Q and m are then obtained by fitting the following model to the data:

$$\ln a = \ln Q + m \ln P \qquad (36)$$

The handling time, b, (eqn. 36) is determined by observing a female parasite during her ovipositional behavior and recording the time it takes her to "handle" a host, that is, the time of initial encounter until the time she withdraws her ovipositor from the scale.

Investigating the interaction between *Aphytis lingnanensis* and *A. melinus* has led to several interesting results. The relation between ln a and ln parasite density is clearly nonlinear (Fig. 11.12), as has been recognized by several workers (e.g., Royama, 1971; Rogers & Hassell, 1974; Beddington, 1975; Hassell, 1976). However, Hassell (1976) reported that *A. melinus* lacked a mutual interference component, suggesting this species can be characterized by the Nicholson–Bailey equation (Nicholson & Bailey, 1935). Under our experimental design it has *m* values, which range between -0.3 and -0.8 depending on the temperature (Table 11.3). We see a similar relationship between temperature and the Quest constant. Thus the values obtained to parameterize equation (36) are largely dependent on experimental conditions

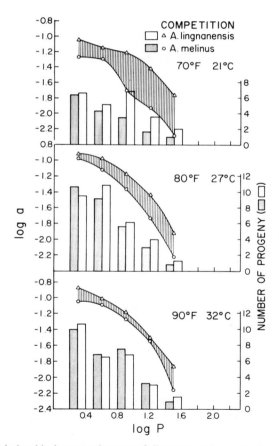

Figure 11.12. Relationship between the area of discovery and parasite density at 3 constant temperatures for *Aphytis lingnanensis* and *A. melinus* and the production of parasite progeny per parasite female at various parasite densities.

Table 11.3.   Values for the Quest and Mutual Interference Constants Obtained at Three Constant Temperatures for *Aphytis lingnanensis* and *A. melinus*

	21°C	70°F	27°C	80°F	32°C	90°F
	Q	m	Q	m	Q	m
A. melinus	−2.46	−0.37	−1.35	−0.835	−2.42	−0.297
A. lingnanensis	−2.17	−0.40	−1.33	−0.678	−1.23	−0.714

which fail to duplicate those in the field. Other than as a method for initially exploring the parasite-host interaction and as a method around which research on the interaction in the field can be organized, the coupling of the parasite–host interaction is unrealistically mimicked by equation (36). We are attempting to reformulate the relationship based on more detailed behavioral observations.

Our observations of *Aphytis* ovipositional behavior have indicated a number of interesting host–parasite and interspecific parasite–parasite interactions which explain the pattern of displacement in *Aphytis* sp. following their sequential introduction into Southern California. Furthermore, these observations explain why *A. melinus* dominates the warmer interior areas. Tentatively the gregarious habit of *A. melinus* (around 50 percent of the scales it parasitizes harbor two parasite eggs as opposed to 15 percent for *A. lingnanensis*) and the increasing female progeny production per female by *A. melinus* (compared to *A. lingnanensis*) with increasing temperature gives it a competitive edge at warmer temperatures (≥90°F). The low vagility of *A. lingnanensis* relative to *A. melinus* prevents it from reinvading the areas from which it was excluded (in periods of high temperature) during the seasonal periods when the temperatures again favor it in competition (≥80°F) (Kfir & Luck, unpublished data, and Fig. 11.12).

CONCLUSION

The above discussed material suggests the scope of the systems analysis effort being applied to the citrus program. It is clear that these efforts are still in their infancy, but nevertheless, substantial progress has been made in both plant and pest modeling. The approach has synthesized the biological information already available and focused attention on those areas where such information is inadequate. The efforts to obtain biological information on the plant/pest/natural enemy systems must be continued in order to complete the effort and to move to the next stage, that is, the development of optimization models for identifying those decisions which would lead to rational pest management in citrus.

Currently, however, citrus pest management decisions are confounded by arbitrary market standards which fluctuate with crop size: this allows more heavily infested fruits to reach the market during years of reduced crop production. The standards of fruit quality currently used are also largely cosmetic in nature. This implies two different "economic thresholds" for California red scale: the first is the infestation levels on fruits, which lead to grade reduction because of appearance; and the second is related to the reduction in yield of the crop in quantity and quality and to branch and twig mortality, which may affect yields the next year(s). It is the arbitrariness of the former which makes optimization modeling very difficult.

ACKNOWLEDGMENTS

We (Robert F. Luck and David Baasch) thank Mr. J. Barbieri and B. Ewing for many helpful discussions and to Mr. J. Barbieri for critical review of the manuscript.

LITERATURE CITED

Abdelrahman, I. 1974. The effect of extreme temperatures on California red scale, *Aonidiella aurantii* (Mask.) (Hemiptera: Diaspididae) and its natural enemies. *Aust. J. Zool.* 22:203–212.

Abramowitz, M. and I. A. Slegun. 1961. Handbook of mathematical functions. U.S. Dept. Commerce Natl. Bureau Standards. Appl. Math. Series 55 U.S. Government Printing Office, Washington, D.C.

Allen, J. C. 1976. A model for predicting citrus rust mite damage on Valencia orange fruit. *Environ. Entomol.* 5:1083–1088.

Allen, J. C. 1978. The effect of citrus rust mite damage on citrus fruit drop. *J. Econ. Entomol.* 71:746–750.

Bain, J. M. 1958. Morphological, anatomical and physiological changes in the developing fruit of the Valencia orange, *Citrus sinensis* (L.) Osbeck. *Aust. J. Bot.* 6:1–24.

Baker, C. H., R. D. Hurrocks, and D. W. Decker. 1973. A computer simulation of solar radiation. *Trans. Missouri Acad. Sci.* 7:181–189.

Baker, D. N., J. D. Hesketh, and W. G. Duncan. 1972. Simulation of growth in cotton: I. Gross photosynthesis, respiration and growth. *Crop Sci.* 12:431–439.

Baskerville, G. L., and P. Emin. 1969. Rapid estimation of heat accumulation from maximum and minimum temperatures. *Ecology* 50:514–517.

Beddington, J. R. 1975. Mutual interference between parasites or predators and its effect on searching efficiency. *J. Anim. Ecol. 48* : 331–340.

Bodenheimer, F. S. 1951. *Citrus Entomology*. Dr. W. Junk. The Hague.

Chapman, H. D. 1968. The mineral nutrition of citrus, in W. Reuther, L. D. Batchelor, and H. J. Weber (eds.), *The Citrus Industry*. Univ. of Calif., Div. of Agric. Sci. *2* : 177–289.

Crider, F. J. 1927. Root studies of citrus trees with practical applications. *Citrus Leaves* *7*(4) : 1–3, 27–30.

Crossman, K. F. 1940. Citrus roots: Their anatomy, osmotic pressure and periodicity of growth. *Pal. J. Rot. 3* : 65–103.

DeBach, P. 1965. Weather and the success of parasites in population regulation. *Can. Entomol. 97* : 848–863.

DeBach, P., T. W. Fisher, and J. Landi. 1955. Some effects of meterological factors on all stages of *Aphytis lingnanensis*, a parasite of the California red scale. *Ecology 36* : 743–753.

DeBach, P. and P. Sisojevic. 1960. Some effects of temperature and competition on the distribution and relative abundance of *Aphytis lingnanensis* and *A. chrysomphali* Hymenoptera: Aphelinidae. *Ecology 41* : 153–160.

DeBach, P., and R. A. Sundby. 1963. Competitive displacement between ecological homologues. *Hilgardia 34* : 105–166.

DeBach, P., D. Rosen, and C. E. Kennett. 1974. Biological control of coccids by introduced natural enemies, in C. B. Huffaker (ed.), *Biological Control*. Plenum Press, New York, pp. 165–194.

Dickson, R. C., and D. L. Lindgren. 1947. The California red scale. *Calif. Citrograph 32* : 524, 542–544.

Dickson, R. C., and D. L. Lindgren. 1947. New data on California red scale. *Citrus Leaves 27* : 6–7, 34.

Ebeling, W. 1959. *Subtropical Fruit Pests*. Univ. of Calif., Div. Agr. Sci.

Eubank, W. P., J. W. Atmar, and J. J. Ellington. 1973. The significance and thermodynamics of fluctuating versus static thermal environments of *Heliothis zea* egg development rates. *Environ. Entomol. 2* : 491–496.

Gutierrez, A. P., D. E. Havenstein, H. A. Nix, and P. A. Moore. 1974. The ecology of *Aphis craccivora* Koch and subterranean clover stunt virus in southeast Australia. II. A model of cowpea aphid population in temperate pastures. *J. Appl. Ecol. 11* : 1–20.

Halma, F. F., and C. Compton. 1936. Growth of citrus trees. *Proc. Amer. Soc. Hortic. Sci. 34* : 80–83.

Hashemi, F., and J. F. Gerber. 1967. Estimating evapotranspiration from a citrus orchard with weather data. *Proc. Am. Hortic. Sci. 91* : 173–179.

Hassell, M. P. 1976. Arthropod predator–prey systems, in R. M. May (ed.), *Theoretical Ecology: Principles and Applications*. W. B. Saunders, Philadelphia, pp. 71–93.

Hassell, M. P., J. H. Lawton, and J. R. Beddington. 1976. The components of arthropod predation. I. The prey death rate. *J. Anim. Ecol. 45* : 135–164.

Hassell, M. P., and R. M. May. 1973. Stability in insect host–parasite models. *J. Anim. Ecol.* 42:693–726.

Hassell, M. P., and G. C. Varley. 1969. New inductive population model for insect parasites and its bearing on biological control. *Nature* (Lond.), 223:1133–1136.

Hesketh, J. D., O. N. Baker, and W. G. Duncan. 1971. Simulation of growth and yield in cotton: Respiration and the carbon balance. *Crop Sci.* 11:384–398.

Holling, C. S. 1959. Some characteristics of simple types of predation and parasitism. *Can. Entomol.* 91:385–398.

Hughes, R. D. 1963. Population dynamics of the cabbage aphid, *Brevicoryne brassicae* (L.). *J. Anim. Ecol.* 32:393–424.

IBM. 1970. System 360 Scientific subroutine package versions III. IBM Corp. Tech. Publ. Dept., 112 E. Post Road, White Plains, N.Y.

Ismail, M. A. 1970. Seasonal variation in bonding force and abcission of citrus fruit in response to ethylene, ethephon, and cycloheximide. *Proc. Fla. State Hortic. Soc.* 84:77–81.

Jensen, K., and N. Wirth. 1974. PASCAL user manual and report. Springer-Verlag. New York.

Jones, E. P. 1935. The bionomics and ecology of red scale *Aonidiella aurantii* in Southern Rhodesia. Mazoe Citrus Exp. Sta. Ann. Rept. 1935. British South Africa Publ. Co. Publ. No. 5:13–52.

Jones, W. W., T. W. Embleton, M. I. Steinacker, and C. B. Cree. 1970. Carbohydrates and fruiting of " Valencia " orange trees. *J. Amer. Soc. Hortic. Sci.* 95:380–381.

Koo, R. C. J. 1953. A study of soil moisture in relation to absorption and transpiration by citrus. Ph.D. Diss. Univ. of Fla., Gainesville.

Koo, R. C. J. 1969. Evapotranspiration and soil moisture determination as guides to citrus irrigation. *Proc. 1st Int. Citrus Symp.* 3:1725–1730.

Kriedeman, P. E. 1968. Some photosynthetic characteristics of citrus leaves. *Aust. J. Biol. Sci.* 21:895–905.

Linacre, E. T. 1967. Climate and the evaporation from crops. *J. Irrig. and Drain Div.*, *ASCF Vol. 93 IR 4 Proc.* Paper 5651, pp. 61–79.

Messem, A. B. 1975. A rapid method for the determination of potential transpiration derived from the Penman combination model. *Agr. Meteorol.* 14:369–384.

McCoy, C. W., and L. G. Albrigo. 1975. Feeding injury to the orange caused by the citrus rust mite, *Phyllocoptruta oleivora* (Prostigmata: Eriophyoidea). *Ann. Entomol. Soc. Am.* 68:289–297.

Moss, G. I. 1969. Influence of temperature and photoperiod on flower induction and inflorescence development in sweet orange. *J. Hortic. Sci.* 44:311–320.

Munger, F., and A. W. Cressman. 1948. Effect of constant and fluctuating temperatures on the rate of development of California red scale. *J. Econ. Entomol.* 41(3):424–427.

Nicholson, A. J., and V. A. Bailey. 1935. The balance of animal populations. Part I. *Proc. Zool. Soc. Lond.*, pp. 551–598.

Patton, B. C. 1975. Ecosystem linearization: an evolutionary design problem. *Am. Nat.* 109:529–539.

Penman, H. L. 1948. Natural evaporation from open water, bare soil, and grass. *Proc. R. Soc. London Ser. A. 193*:120–145.

Reed, H. S., and D. T. MacDougal. 1937. Periodicity in the growth of the orange tree. *Growth 1*:371–373.

Rogers, D. J. 1972. Random search and insect population models. *J. Anim. Ecol. 41*: 369–383.

Rogers, D. J., and M. P. Hassell. 1974. General models for insect parasite and predator searching behavior: interference. *J. Anim. Ecol. 43*:239–253.

Royama, T. 1971. A comparative study of models for predation and parasitism. *Res. Popul. Ecol. Suppl. I*, pp. 1–91.

Shelford, V. W. 1929. *Laboratory and Field Ecology*. Williams and Wilkins, Baltimore, Maryland.

Shultze, R. E. 1976. A physically based method of estimating solar radiation from sun-cards. *Agric. Meteorol. 16*:85–101.

Siljak, D. D. 1972. Stability of large-scale structural systems under structural perturbations. *IEEE Trans. on Systems, Man and Cybernetics*. SMC-2:657–663.

Siljak, D. D. 1973. On stability of large-scale systems under structural perturbations. *IEEE Trans. on Systems, Man and Cybernetics*. SMC-3:415–417.

Siljak, D. D. 1974. Connective stability of complex ecosystems. *Nature 249*:280.

Siljak, D. D. 1975. When is a complex ecosystem stable? *Math. Biosci. 25*:25–50.

Simanton, W. A. 1970. Seasonal variation in magnitude of foliage growth in Florida citrus groves. *Proc. Fla. State Hortic. Soc. 83*:49–54.

Smith, P. F. 1976. Collapse of "Murcott" tangerine trees. *J. Am. Soc. Hortic. Sci. 101*: 23–25.

Stapleton, H. N., D. R. Buxton, F. L. Watson, D. J. Nolting, and D. N. Baker. 1973. Cotton: A computer simulation of cotton growth. *Ariz. Agric. Expt. Sta. Tech. Bull.* 206.

Stewart, I., T. A. Wheaton, and R. L. Reese. 1968. "Murcot" collapse due to nutritional deficiencies. *Proc. Fla. State Hortic. Soc. 81*:15–18.

Stinner, R. E., A. P. Gutierrez, and G. D. Butler, Jr. 1974. An algorithm for temperature-dependent growth rate simulation. *Can. Entomol. 106*:519–524.

Turrell, F. M., M. J. Gerber, W. W. Jones, W. C. Cooper, and R. H. Young. 1969. Growth equations for citrus trees. *Hilgardia 39*:429–445.

Varley, G. C., G. R. Gradwell, and M. P. Hassell. 1974. *Insect Population Ecology: An Analytical Approach*. Univ. Calif. Press, Berkeley.

Whiteside, J. O. 1970. Etiology and epidemiology of citrus greasy spot. *Phytopathology 60*:1409–1414.

Willard, J. R. 1972. Studies on rates of development and reproduction of California red scale, *Aonidiella aurantii* (Maskell) (Homoptera: Diaspididae) on citrus. *Aust. J. Zool. 20*:37–47.

Wirth, N. 1971. The programming language PASCAL. *Acta Informatica 1*:35–63.

Yothers, W. W., and A. C. Mason. 1930. The citrus rust mite and its control. *USDA Tech. Bull. 176*.

12

APPROACH
TO RESEARCH AND
FOREST MANAGEMENT FOR
MOUNTAIN PINE BEETLE
CONTROL

R. W. Stark

University of Idaho, Moscow, Idaho

Collaborators:

**A. A. Berryman,   D. G. Burnell,   H. Cabrera,
R. Roelke   and   G. K. White**

Washington State University

**D. L. Adams,   C. Brockway,   N. Crookston,
D. Kulhavy,   R. L. Mahoney,   E. L. Michalson,
J. A. Moore,   A. Partridge   and   J. A. Schenk**

University of Idaho

**G. A. Amman,   W. E. Cole,   W. H. Klein,
V. E. Pace,   L. Rasmussen   and   A. R. Stage**

U.S. Forest Service

INTRODUCTION

The organization of the overall integrated pest management (IPM) project is covered in Chapter 1 of this treatise, and that of the pine bark beetle subproject by several authors (Anderson et al., 1976; Berryman, 1975; Stark, 1973b). The following brief summary will suffice to explain the specific organization, justification, and management features utilized in the mountain pine beetle (MPB) project.

The mountain pine beetle, *Dendroctonus ponderosae* Hopkins, in lodgepole pine was chosen for several reasons. The first is the importance of lodgepole pine.* Lodgepole pine is a western conifer whose distribution ranges from Baja California to the Northwest Territories in Canada and is an important tree species in nine western states. There are more than 13 million acres of commercial lodgepole pine in the United States; it is the fourth largest forest type in the west. It ranks sixth of 17 species in net volume of sawtimber and fifth in growing stock present on commercial forest land. It is the second most ubiquitous species in 70 recognized western forest habitat types. It is equally important as noncommercial forest cover, the acreage of which exceeds that of commercial. The second is the mountain pine beetle. It is generally agreed that the mountain pine beetle is the most important forest insect attacking ponderosa and lodgepole pine in North America. It has been estimated that 58 million trees (77 percent of all trees over 9 inches in diameter) were killed between 1928 and 1936 on 1 to 3 million acres in Montana. Comparable extensive killing of its hosts has occurred frequently up to the present day in

---

*Statistics from "Forest Statistics for the United States by State and Region, 1970," USDA Forest Service, 1972, 96 pp.

various locations in Montana, Idaho, Oregon, Washington, and Canada. The third is the lodgepole pine-mountain pine beetle systems in the Northwest. The extreme range of the host species and of the mountain pine beetle necessitated some constraints on the scope of the study. To determine the underlying relationships which could lead to management of this insect when it reached pest status within the time period of this project, we concentrated on this system in Idaho and Montana. An important consideration was that the insect and the host have been the subject of intensive study for many decades, resulting in a large source of data and information upon which to build an integrated pest management system.

This program has been a cooperative one involving the University of Idaho, Washington State University, and the United States Forest Service, with the grant management administered by the University of California at Berkeley. About 21 professional scientists with diverse specializations and about 12 technical assistants have been involved in the actual studies, but through the compilation of annual reports, periodic discipline-oriented workshops, and the like, interaction with a much larger segment of forestry research has been achieved.

The objectives of this program in general conform to those explained in Chapter 1 for the overall program, but perhaps more than for some other subprojects, the mountain pine beetle program has emphasized the collation, integration, and application of already available knowledge to management problems, rather than the generation of new information. However, it was early recognized that considerable additional research would be necessary. The objectives are three:

1. To develop methods for evaluation of the impact of the mountain pine beetle throughout the range of lodgepole pine, considering *all* values implicit in the resource.
2. To develop insect population and forest stand models and other methodologies which will permit determination of local and regional trends in beetle populations and prognosis of stand development and impacts.
3. To develop management decision models from (1) and (2) which will aid in action decisions, though no action is necessarily implied (these models include choices of tactics and strategies for population suppression or regulation based on the state of the art).

None of the objectives is considered an end unto itself, but each is a part of a continuous closed loop system interfacing with the objectives of forest resource management *in toto* (Fig. 12.1), the pine trees themselves being only one component of the forest.

We describe here the general progress towards meeting the above objectives.

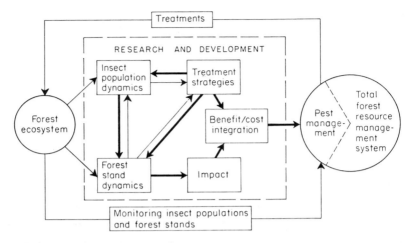

Figure 12.1. Model structure of forest pest management system. [Redrawn from *Integrated summary: Pine bark beetle subproject NSF-EPA Project*, Vol. 1, p. 145, U. C. Berkeley, 1974.]

A realistic review of accomplishments is difficult at any time, particularly before the time period set for achieving certain of the goals has expired. On the other hand, the termination of one project set, particularly this one, will see neither the conclusion of research on the mountain pine beetle nor *the* definitive set of conclusions for management of mountain pine beetle populations.

EVALUATION OF MOUNTAIN PINE BEETLE IMPACT

Impact is in the eye of the beholder. It is only within the past decade or two that we have had to concern ourselves seriously with accounting for the ecological and economic effects of our attempts to regulate insect populations. Until recently, impact on the forest was regarded in only negative terms, that is, the disruptive effects on forest ecosystems in the context of reduction of earning power of that forest ecosystem. It is our belief that in the scope of integrated pest management we are required to divest ourselves of this limited and biased perspective.

Phytophagous insects are common, ubiquitous elements of most terresterial ecosystems (Mattson & Addy, 1975). Normal insect grazing on a forest tree usually does not impair primary production and indeed may accelerate it. Outbreaks occur, probably most commonly, in relatively unproductive forest ecosystems already under stress by other natural factors or because of technological coercion (Schimitschek, 1974). Many ecological effects of such insect grazing can be positive, for example: increased light penetration, reduced competition, improved species composition, increased rate of nutrient leaching, increased amount of nutrient-rich litter, increased water reception, and so on. While within the context of pest management and resource management, emphasis must be placed on evaluating the effects on management objectives, we must keep real ecological effects in mind to avoid conscious or unconscious bias.

Another facet of pest management which must be borne in mind when evaluating impact has been called "axiological impact" (Day, 1976). Day noted, "Axiology is that branch of philosophy dealing with matters of value, as in morals, esthetics, and metaphysics." This concept has some value in pest management since it provides a convenient receptacle for those imponderables which cannot be included under ecology or measured in dollars. The fallout, or consequences, from axiological impact can affect management decisions of public agencies and private corporations. For example, the use of chemicals in pest management has been clearly affected by points of view and rhetoric based on political, moral, esthetic, and emotional grounds. Emotion over esthetics often transcends fact, and terms such as destruction, devastation, and decimation creep into our justification vocabulary. Aided and abetted by scientists, and especially by pesticide advertising, a negative, if not fearful, attitude towards phytophagous insects has been transmitted to the lay public by repeated reference to insects as pests, embellished with alarming predictions of their destructive capabilities.

Having described the mental precautions we have armed ourselves with in approaching the subject of impact of the mountain pine beetle, let me assure the forest managers that our objectives embrace the definition of impact as, "the net effects of a given pest, or pest complex, on the productivity, usefulness, and values of a tree species or forest type with respect to different resource uses and values . . . and management objectives . . ." (Waters, 1977).

We intended initially to evaluate five of the major uses of the forest and we conceptualized the probable effects of mountain pine beetle attack on these uses (Fig. 12.2). These are the five major forest land uses as identified for the United States Forest Service by the Multiple Use Sustained Yield Act passed by Congress in 1960. Shortly after beginning this project, we determined that there was a large effort underway by the United States Forest Service, Region 4, on the first of these, timber productivity, and the decision was made to concentrate our own limited resources on others. In addition to the Forest Service studies, the impact work being done by W. A. Leuschner and his colleagues at Virginia Polytechnic Institute (see Chapter 14) will

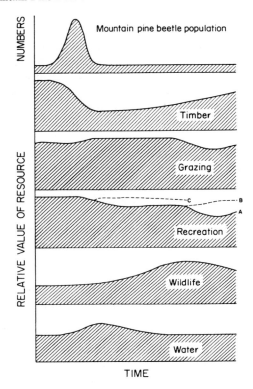

Figure 12.2. Hypothetical impact of mountain pine beetle epidemics on five forest uses over time. [Redrawn from Berryman, 1975.]

provide methodologies applicable to all bark beetles. However, to round out the picture and to present the results of our own impact studies in the context of effects on all forest values, we draw here upon results of others relative to the other values.

Timber Production and Growth

The following is the abstract of a preliminary report by W. H. Klein, United States Forest Service, Region 4, dated April 15, 1976.

Using 35 mm color aerial photography, a double sampling system, and permanently established plots, measurements of stand depletion were made in a 120,222 acre management area on the Targhee National Forest, Idaho in 1972–73. Additional to measurements of standing live and beetle-killed timber, data were also obtained on reproduction forage yield, habitat type, disease (mistletoe), and other factors that will yield insight into the effects of mountain pine beetle damage on other forest resources. A remeasurement of all plots in 1975 showed that although there was some additional tree mortality, there was a net increase in both lodgepole and total stand volume. Plans are to remeasure these plots at periodic intervals as a base (model) for predicting future yield.

This report, previous unpublished studies by Klein, and studies reported by Mattson and Addy (1975) suggest that in some instances the killing of trees may actually have a beneficial effect, not unlike a thinning program. Suppressed trees are released and in residual trees growth rate is generally increased. Thus the loss of timber is offset by increased growth rate, and stand vigor in general may be improved. However, in a large scale outbreak, the real losses of salable timber undoubtedly usually exceed by far these benefits.

Grazing Value

No studies have been initiated by the integrated pest management (IPM) group of mountain pine beetle impact on grazing values of infested forests, but it is anticipated that some indication of this impact may be derived from Klein's work, and additional and more specific studies on this aspect are in progress by United States Forest Service scientists at Bozeman, Montana (W. E. Cole, pers. comm.). We intuitively expect an increase in forage in a beetle-infested area, but we suspect that this would probably be of little significance to cattle grazing. Our observations suggest that cattle prefer to graze in and concentrate in more open areas which are little affected by beetle-caused stand changes. This may not be true, however, of sheep grazing. In those areas where improved grazing may result from beetle outbreaks, we would expect an eventual reduction of the potential benefits when beetle-killed trees begin falling and restricting movement of livestock, damaging fences, adding to the fire hazard, and so on. The clearing of trails, moving of livestock, repair of fence, and the like, constitute a costly expense associated with forest grazing (Berryman, 1975).

Recreation Values

Our original concept of the impact on recreation is diagrammed on Line A in Figure 12.2. We felt there would be an initial negative effect for esthetic reasons, and a later one because of the danger and impediment to human movements resulting from falling and downed trees. Research on campground usage indicated that this is probably true for an informed public, but for uninformed users (and there are many) the impact would be slight, as in line C. Line B indicates a reversal of the impact shown for line A, assuming that the beetle-killed trees are removed before they become a hazard.

Given a public which has been adequately informed of the outbreak, recreational demand models (Michalson, 1975) have shown that significant economic losses can occur from mountain pine beetle outbreaks in public campgrounds. These economic, or dollar, losses portend also human recreation losses beyond those directly implied by these economic measures of loss. Campgrounds in infested and noninfested areas were compared for differences in use, cost per visitor day, and consumer surplus expenditures. Estimated dollar losses in the heavily used recreational area of the Targhee National Forest ranged from about $2 million in infested campgrounds that were studied to an estimated $7 million for all campgrounds, assuming they were all infested. This latter estimate is undoubtedly excessive, since the outbreak was never this severe.

The second approach to measuring impact on recreation was to examine the perceptions and reactions of recreational users to MPB-caused mortality (White, 1976). Photographs taken at three distances and representing two levels of MPB infestation were shown to various types of recreationists, for example, overnighters, fishermen, and campers. It was determined that members of user groups do place varying significance on the various forest characteristics in establishing their preferences for sites, but the broad divergence of preferences suggested that most users were able to select sites within the infested forest that possessed some combination of characters suitable for them, thus minimizing the impact of the mountain pine beetle.

From these two studies it might be extrapolated that if the negative connotations of an outbreak are overdramatized by the press or forest or recreation manager, the economic consequences would probably be negative. On the other hand, if exploited as a natural phenomenon by the forest manager, even an outbreak of "disaster" or "holocaust" dimensions might actually be turned to the advantage (or at least not the disadvantage) of the recreation manager. Guidelines were developed from both studies, particularly the latter, which can assist forest managers in directing recreationists to sites of their preference.

Wildlife Value

No research has been done on the effects of mountain pine beetle outbreaks on wildlife. It is believed that such effects would be largely beneficial (Fig.

12.2). Forage would increase as a result of opening up of the canopy and, within limits, fallen trees would provide abodes, bedding, and protection from predators, hunters, and extremes of weather for many species. However, when large quantities of fallen trees have accumulated and revegetation has increased considerably in size, the stands may then become less accessible or impenetrable for big game (Berryman, 1975) and for sportsmen hunting there.

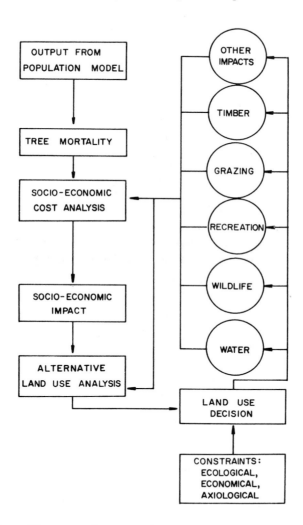

Figure 12.3. A possible structure for an MPB impact analysis and decision model. [Redrawn from Berryman, 1975.]

Water Resources

In 1975 studies were initiated on the effects of mountain pine beetle outbreaks on water runoffs. Three watersheds in southeastern Idaho were being monitored by hydrologists of the USFS-Intermountain Forest & Range Experiment Station. We found that these areas had been infested by the mountain pine beetle during the hydrological observations. Runoff versus precipitation and snow accumulation are now being compared, before and after the infestation. Preliminary analysis indicates some variation, but precise correlation with infestation has not yet been established. More precise determination of the scale of tree loss in the vicinity of monitoring stations was made in 1977; results were inconclusive (see Carter, 1978, p. 33).

Miscellaneous Impacts

There are several other impacts that may result from mountain pine beetle epidemics: increased fire hazard, altered fishery potential, and so on, which need to be identified and defined in terms meaningful to the forest manager. The eventual product of this project group and related studies by others, such as those underway in the Forest Service, should be a model which predicts tree mortality from the population dynamics models and stand structure from stand prognosis models, and interprets these predicted events in terms of economic and other losses or restraints placed on management in the pursuit of forest management objectives. Information flow in such a model may follow that outlined in Figure 12.3. The output from this model would enable the evaluation of total impact and possibly provide guidelines for management decisions on optimal land use options following, or perhaps prior to, bark beetle epidemics.

INSECT POPULATION AND STAND MODELS

The stand prognosis model has been available for some time (Stage, 1973). The stand prognoses are presented in the form of the sequences of yields by time periods (5 years, 10 years) for each forest stand class under each management alternative. These prognoses provide input for long term management decisions, that is, selection of the optimum mix of management activities for each stand class. The stand prognoses developed to date are based upon problem-free stand growth. The input data include standard tree and stand data such as age of stand, growth rate, stocking, and species mixture. The model can describe the future development of the stand under alternative forest management prescriptions. Not yet included are the alternatives necessitated by the presence of a bark beetle outbreak.* However, a hypothetical example of the manner in which the model would operate is illustrated in Figure 12.7 (Stage, 1975).

The major effort in the mountain pine beetle IPM project has been in the difficult area of modeling bark beetle populations. There have been many extensive studies on populations of this insect in both Canada and the United States, from which data have been drawn and various assumptions made. The principal sources of background material have been the long term studies of W. E. Cole and G. Amman (1960–1976) at the USFS Intermountain Forest and Range Experiment Station in Ogden, Utah, and Canadian studies based in Alberta and British Columbia (Reid, 1958, 1963; Reid & Gates, 1970; Reid et al., 1967; Safranyik et al., 1974; Safranyik et al., 1975; Shepherd, 1965).

Raw and summarized data for a number of mountain pine beetle infestations were provided by the above colleagues and comparable data were collected from 35 stands widely scattered in Montana, Oregon, and Idaho from 1973 to 1975. Variables used in the development of the models included tree diameter; height of infestation on tree bole; number of beetle attacks; gross production of beetles (in number of emergence holes per unit); phloem thickness; cm of beetle gallery; cm of beetle gallery surrounded by resin impregnated tissue; gross estimate of extent and distribution of cortical resin canals expressed as proportion of sample, woodpeckering, number of parasite cocoons per unit, proportion of phloem utilized by insects other than the mountain pine beetle (i.e., interspecific competition) and several compound variables; net production [gross production $-$ (1.5 $\times$ number of attacks)], which assumes a 2:1 sex ratio; relative absolute production (gross production $\times$ diameter $\times$ height of infestation); and proportional resinosis (resinosis $\div$ egg gallery length). Analyses were run on both transformed and untransformed data. In most cases, transformation did not significantly improve the coefficient of multiple determination in regression analyses and most analyses were based on untransformed data.

Several approaches have been used to develop simulation models of bark beetle populations. Empirical descriptions of an epidemic were presented by Cole et al. (1976). Berryman (1976a) used a combination of deduction and empiricism to derive a theoretical model of mountain pine beetle dynamics. An explicative model has also been developed utilizing catastrophe theory (Berryman, 1976b). Models are also being developed which will interface with the stand growth and development model of Stage (1973) to be used for long term stand prognosis.

*The prognosis model now includes two pest disturbances, a defoliator, and the MPB.

In general we subscribe to the presence of density-dependent feedback regulators and density-independent control variables acting on mountain pine beetle populations. The basic density-dependent curve (Fig. 12.4) is modified by two sets of control variables: (1) those determining the suitability of the infested tree habitat for beetle reproduction and survival, for example, tree diameter, phloem thickness, resinosis, phloem resin duct density, woodpecker predation, climate, and weather (Amman, 1972, 1975; Safranyik et al., 1975; Cole, 1973, 1975; Berryman, 1976a); and (2) those determining host resistance to attack. For example, stand density and site variables (Amman & Baker, 1972; Schenk et al., unpublished data), root diseases (Partridge, unpublished data), stand age, climate, and weather (Safranyk et al., 1974, 1975). Production of beetles from infested trees is, therefore, determined by the interaction of these variables. The variables, resin canals, phloem thickness, proportional resinosis and woodpeckering, accounted for 40 percent of the variation in the intercept at the Y-axis, the expression of which is considered a function of habitat variations such as climate, food quantity and quality, and other density-independent variables. These forces affect the amplitude of the production curve. For example, phloem thickness has a large positive effect on both gross and net production, while parasitism and competing insects have a low negative effect on mountain pine beetle production (see footnote, p. 411).

Habitat suitability and resistance indicators (*sic*) are being investigated in a variety of ways. Hazard rating of geographical regions for their relative

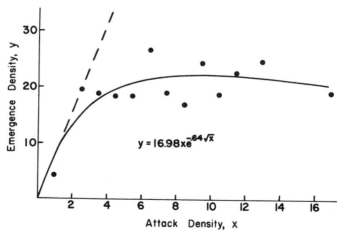

Figure 12.4. Figure 12.4. Relationship between gross production (y) and attack density (x). [Redrawn from Berryman, 1976a.]

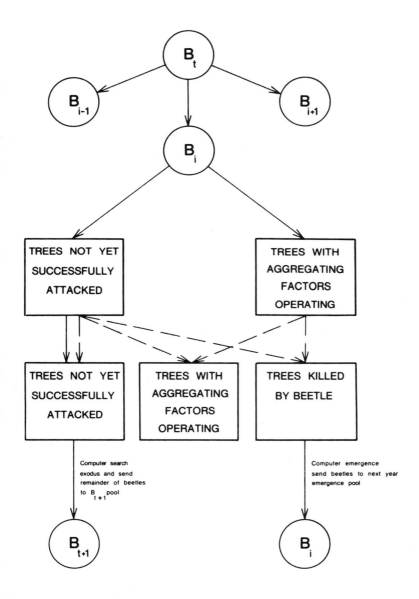

Figure 12.5. Beetle (solid lines) and tree (broken lines) flows for similating dispersal and attack of mountain pine beetles (" B_t " total beetles allocated to i th time interval, " B_i ").

danger of harboring mountain pine beetle infestations in lodgepole pine has been approached using the methods developed in Canada by Safranyik et al. (1974, 1975). The system is climatically based. It assumes a zone is hazardous if it is climatically suitable for the mountain pine beetle. A climatically suitable zone is one which rarely experiences weather conditions detrimental to the mountain pine beetle, or one which is beneficial to lodgepole pine. The process required defining such conditions and then searching the weather records to determine an index of climatic suitability.

Another approach is the development of a hazard rating model based on the Crown Competition Factor (CCF) and the amount of host material

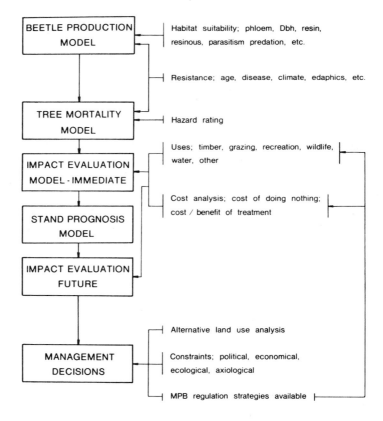

Figure 12.6. Conceptual linkage of models and factors involved in integrated management of beetle affected forests (ca. Nov. 1976).

available, expressed as the percentage or basal area of lodgepole pine in the stand (Schenk et al., 1977). Crown Competition Factor is a measure of average stand competition. A CCF of 100 depicts a stand fully, but not overly, occupied, that is, one where competition is absent (Schenk et al., 1976). There are specific formulas developed for each tree species; data required consists only of number of species per acre by diameter class.

Yet another approach to hazard rating is based on habitat suitability (Roe & Amman, 1970; Schenk et al., 1976). In addition to mensurational data specific to development of the hazard rating model described above, this model uses lodgepole pine site indexes, habitat type and understory plant species common in recurring outbreak areas.

None of the three hazard rating systems described above interface with the beetle production model at present, but they may provide a probability function of attack for the stand prognosis model.

The beetle production model does not account for the interlude between emergence and attack, but rather between attack and emergence. Burnell (1977) has developed a dispersal-aggregation theory culminating in a tree mortality prediction model, the flow chart for which is shown in Figure 12.5. The development of the equations describing quantitatively the inputs and outputs and the rationales and assumptions upon which they are derived are too elaborate to be presented here (interested parties are referred to the above paper and the creator of the model).

The third and most critical component of this modeling process is the model which links the beetle productivity and mortality with the stand prognosis model (Stage, 1973). The development of the linkage model has been our principal task throughout 1976 and will continue to be in 1977.* A diagrammatic representation of this model's likely employment is presented in Fig. 12.6.

MANAGEMENT DECISION AND STRATEGY MODELS

The overriding goal of this project and its counterparts is to learn to "manage" mountain pine beetle populations when it is determined that their activity is counter to forest management objectives. As the need increases to exploit fully the productive potential of our forests, it becomes increasingly important to assess the interactions of such insect pests and forest growth. To accomplish this, we must integrate the science of silviculture with that of

*Positive results are presented in Proceedings of an NSF-sponsored symposium held in Pullman, Washington, April 25–27, 1978.

insect ecology, particularly population dynamics and damage input. However, the forest manager is the ultimate engineer. Describing the future development of forest stands under several management options has been likened to the progress of a railroad car through a switching yard (Stage, 1975). Where the car ends up depends on how the switches are set at each junction. The set may change as goals or values change, or as constraints dictate. The task of the prognosis and impact models is to display the consequences of travelling any particular route—not only in yield but also in terms of ecological and social consequences.

Insects and other potentially destructive forces can act as vandals missetting the switches. The example given by Stage (1975) as cited in Anderson et al. (1976) uses a single beetle regulation strategy, thinning, in an outbreak context (Fig. 12.7). In the example, he postulates two alternatives, thin or not thin, control or not control. Thus, there are four combinations of activities to be evaluated. All that has been described heretofore, the vagaries of

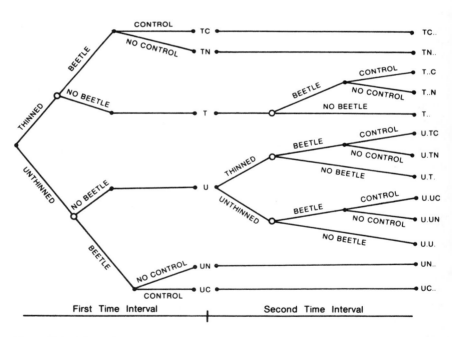

Figure 12.7. Schematic representation of alternative decision paths: "T" is thinned option; "U" is unthinned; "C" is control; "N" is no control. Outcomes that are the same irrespective of strategy have a dot in the place of the irrelevant strategy. [Redrawn from Anderson et al., 1976.]

the insect population explained in probabilities of certain events occurring given the existing stand, under various climatic situations, and so on, must be provided the planner—the engineer. In each option, thin or not thin, the question arises whether the developing stand is susceptible to beetle outbreak. If not, the normal growth cycle is completed; if susceptible, it is still questionable whether an outbreak would occur—so both possibilities must be pursued. Along the latter alternative, we need to switch again, depending on whether treatment is implemented, and this sequence leads to six alternatives. The value of each combination is weighted by its probability of occurrence (Stage, 1975).

The greater the number of time periods, the more complex the process. In each interval with three switches there are eight possible outcomes, so in N time periods there could be 8^N possible outcomes. Even for short term planning this would be a very large computing chore. Also, we have considered only a single strategy; it becomes much more complex when we are faced with a variety of strategies. Fortunately, many outcomes can be eliminated as too rare or too disadvantageous to follow. In this example, instead of 36 outcomes, it is estimated that there are realistically only 13 feasible stand outcomes (descriptions). Fortunately, this apparently complex procedure is implemented quite simply by a computing algorithm (Stage, 1975). The simplistic management model just described is not a product of this project, but it has been modified as a result of it. The project group has not yet directly addressed the strategy portion of its objectives.*

Pest management strategy is directed towards minimizing the impact of the organism involved (in our case, the mountain pine beetle—MPB). Strategy goals might include prevention, suppression, or containment. A fourth, eradication, is often bruited about, but it is difficult to conceive of this strategy as practical in a forest ecosystem. There are a large number of potential tactics which may be utilized in optimizing given strategies. However, each tactic has its side effects and costs, and it is necessary to know these under all conditions of the ecosystem. Our task in the year (and years) ahead is to investigate all available tactics and develop potentially attractive or optimal strategies.

It is our belief, shared by many, that the most promising and realistic tactics for management of the mountain pine beetle are contained within the sciences of silviculture and forest management (Roe & Amman, 1970; Sartwell, 1975; Sartwell & Stevens, 1975; Cole & Cahill, 1976; Amman et al., 1977). Observation has led us to conclude that where ecologically compatible, intensive forest management is practised, bark beetle problems are minimal.

---

*Strategy is addressed in the proceedings of an NSF-sponsored symposium held in Pullman, Washington, April 25–27, 1978.

SUMMARY

The mountain pine beetle subproject of the national NSF-EPA supported project in integrated pest management has been underway for some $4\frac{1}{2}$ years. The greatest expenditure of effort has been in formulating models to describe the impact of the beetle on two values of forest stands, recreation and water production, and the population dynamics of the mountain pine beetle.

From this effort and the extensive additional work done in the past and currently, practical guidelines are now available for managers of recreation facilities in mountain pine beetle (MPB) infested forests. A productivity model has been developed from which the probable direction of the MPB population can be predicted. Given this estimate, an attack-dispersal model has been developed which results in estimates of tree mortality. The remaining time in the project was devoted primarily to quantifying the outputs of these two models into probability estimates which are now entered into an existing stand prognosis model.

Stand hazard ratings based on historical records, climate, habitat, and crown competition factors provide direct inputs to managerial decision models without recourse to the more complex productivity and mortality models described here in specific situations.

LITERATURE CITED*

●Anderson, L. S., A. A. Berryman, D. G. Burnell, W. H. Klein, E. L. Michalson, A. R. Stage, and R. W. Stark. 1976. The development of predictive models in the lodge-pole pine-mountain pine beetle ecosystem, in R. L. Tummala, D. C. Haynes, and B. A. Croft (eds.), *Modeling for Pest Management: Concepts, Techniques and Applications, USA/USSR.* Michigan State University, East Lansing, pp. 149–164.

●Berryman, A. A. 1975. Management of mountain pine beetle populations in lodgepole pine ecosystems: a cooperative, interdisciplinary research and development project, in D. M. Baumgartner (ed.), *Management of Lodgepole Pine Ecosystems.* Vol. I. Symposium proceedings. Washington State University Coop. Ext. Serv., Pullman, pp. 628–650.

●Berryman, A. A. 1976a. Theoretical explanation of mountain pine beetle dynamics in lodgepole pine forests. *Environ. Entomol. 5*:1225–1233.

●Berryman, A. A. 1976b. Dynamics of bark beetle populations: An interpretation using catastrophe theory. Paper presented at XV Int. Congr. Entomol., Washington, D.C. (Available from author.)

*Literature cited includes papers by cooperators and participants in the IPM project. Those published as a direct result of IPM support are marked with a bullet (●). Papers by U.S. Forest Service scientists are listed separately.

●Burnell, D. C. 1977. A dispersal-aggregation model for mountain pine beetle in lodgepole pine stands. *Res. Pop. Ecol. 19*:99–106.

Carter, S. W., Jr. 1978. Potential impacts of mountain pine beetle and their mitigation in lodgepole pine forests, in A. A. Berryman, G. D. Amman, R. W. Stark, and D. L. Kibbee (eds.), *Theory and Practice of Mountain Pine Beetle Management in Lodgepole Pine Forests.* Univ. Idaho, Moscow, pp. 27–36.

Day, B. W. 1976. The axiology of pest control. *Agrichem. Age 19*(5):5–6.

Mattson, W. J., and N. J. Addy. 1975. Phytophagous insects as regulators of forest primary production. *Science 190*:515–522.

●Michalson, E. L. 1975. Economic impact of mountain pine beetle on outdoor recreation. *So. J. Agric. Economics, Dec.*, pp. 43–50.

Reid, R. W. 1958. The behavior of the mountain pine beetle, *Dendroctonus monticolae* Hopk. during mating, egg laying, and gallery construction. *Can. Entomol. 90*:505–509.

Reid, R. W. 1963. Biology of the mountain pine beetle, *Dendroctonus monticolae* Hopk., in the East Kootenay Region of British Columbia. III. Interaction between the beetle and its host with emphasis on brood mortality and survival. *Can. Entomol. 95*:225–238.

Reid, R. W., and H. S. Gates. 1970. Effects of temperature and resin on hatch of eggs of the mountain pine beetle, (*Dendroctonus ponderosae*). *Can. Entomol. 102*:617–622.

Reid, R. W., H. S. Whitney, and J. A. Watson. 1967. Reactions of lodgepole pine to attack by *Dendroctonus ponderosae* Hopkins and blue stain fungi. *Can. J. Bot.45*:1115–1126.

Safranyik, L., D. M. Shrimpton, and H. S. Whitney. 1974. Management of lodgepole pine to reduce losses from the mountain pine beetle. *Can. For. Serv. For. Techn. Rep. 1.*

Safranyik, L., D. M. Shrimpton, and H. S. Whitney. 1975. An interpretation of the interaction between lodgepole pine, the mountain pine beetle and its associated blue stain fungi in western Canada, in D. M. Baumgartner (ed.), *Management of Lodgepole Pine Ecosystems.* Vol. I, Symposium proceedings. Washington State University Coop. Ext. Serv., Pullman, 406–428.

Schenk, J. A., R. L. Mahoney, J. A. Moore, and D. L. Adams. 1976. Understory plants as indicators of grand fir mortality due to the fir engraver. *Entomol. Soc. B.C. 73*:21–24.

Schenk, J. A., J. A. Moore, D. L. Adams, and R. L. Mahoney. 1977. Preliminary hazard rating of grand fir stands for mortality by fir engraver. *For. Sci. 23*:103–110.

Schimitschek, E. 1974. Grundsatzliche Betrachungen zur Frage der öcologischen Regelung. *Z. ang. Entomol. 54*:22–48.

Shepherd, R. F. 1965. Distribution of attacks by *Dendroctonus ponderosae* Hopk. on *Pinus contorta* Dougl. var. *latifolia* Engelm. *Can. Entomol. 97*:207–215.

Stark, R. W. 1973a. Systems analysis of insect populations. *Ann. N.Y. Acad. Sci. 217*:50–57.

●Stark, R. W. 1973b. The systems approach to insect pest management—a developing programme in the United States of America. The pine bark beetles. *Mem. Ecol. Soc. Aust. 1*:265–273.

Waters, W. E. 1977. Evaluation of insect impacts on forest productivity and value, in D. L. Kibbee (ed.), *The Fundamental Bases for Integrated Pest Management*. Proc. Group 6, XVI World IUFRO Congr., Oslo, Norway (June 1976). College of Forestry, Wildlife and Range Science, University of Idaho, Moscow, pp. 15–18.

White, G. K. 1976. The impact of the pine bark beetle on recreational values. Ph.D. thesis, Washington State University, Pullman.

U.S. Forest Service

Amman, G. D. 1972. Mountain pine beetle brood production in relation to thickness of lodgepole pine phloem. *J. Econ. Entomol. 65*:138–140.

Amman, G. D. 1975. Abandoned mountain pine beetle galleries in lodgepole pine. *USDA For. Serv. Res. Note* INT-197, Ogden, Utah.

Amman, G. D., and B. H. Baker. 1972. Mountain pine beetle influence on lodgepole pine stand structure. *J. For. 70*(4):204–209.

Amman, G. D., M. D. McGregor, D. B. Cahill, and W. H. Klein. 1977. Guidelines for reducing losses of lodgepole pine to the mountain pine beetle in unmanaged stands in the Rock Mountains. *USDA For. Serv. Tech. Rep.* INT-36.

Cole, W. E. 1973. Interaction between mountain pine beetle and dynamics of lodgepole pine stands, *USDA For. Serv. Res. Note* INT-170.

Cole, W. E. 1975. Interpreting some mortality factor interactions within mountain pine beetle broods. *Environ. Entomol. 4*:97–102.

Cole, W. E., G. D. Amman, and C. E. Jensen. 1976. Mathematical models for the mountain pine beetle-lodgepole pine interaction. *Environ. Entomol. 5*:11–19.

Cole, W. E., and D. B. Cahill. 1976. Cutting strategies can reduce probabilities of mountain pine beetle epidemics in lodgepole pine. *J. For. 74*:294–297.

Roe, A. L., and G. A. Amman. 1970. The mountain pine beetle in lodgepole pine forests. *USDA For. Ser. Res. Paper* INT-71.

Sartwell, C. 1975. Thinning ponderosa pine to prevent outbreaks of mountain pine beetle, in D. M. Baumgartner (ed.), in *Management of Lodgepole Pine Ecosystems*. Vol. I. Symposium proceedings, Washington State University Cooperative Extension Service, Pullman, pp. 41–52.

Sartwell, C., and R. E. Stevens. 1975. Mountain pine beetle in ponderosa pine: prospects for silvicultural control in second-growth stands. *J. For. 73*(3):136–140.

Stage, A. R. 1973. Prognosis model for stand development. *USDA For. Serv. Res. Paper* INT-137.

Stage, A. R. 1975. Forest stand prognosis in the presence of pests, in D. M. Baumgartner (ed.), *Management of Lodgepole Pine Ecosystems*. Vol. I. Symposium proceedings, Washington State University Cooperative Extension Service, Pullman, pp. 406–428.

13

APPROACH TO RESEARCH AND FOREST MANAGEMENT FOR WESTERN PINE BEETLE CONTROL

D. L. Wood

Collator
Division of Entomology and Parasitology
The University of California
Berkeley, California

INTRODUCTION*

The objective of the pine bark beetles subproject has been stated in Chapter 12, "to obtain the necessary understanding of the role of bark beetles in forest ecosystems in order to develop strategies and tactics which forest managers can use for minimizing the adverse effects of these pests on the various forest values, with minimal disruption of the ecosystem." The organization and management of this subproject was divided into three separate but related programs, focusing on three distinct host/pest complexes. This separation was based upon the presumed economic significance of the major bark beetle species involved and the diversity of the forest habitats in which they occur. Each program has as its focus a different species of bark beetle: the mountain pine beetle, *Dendroctonus ponderosae* Hopkins; the western pine beetle, *Dendroctonus brevicomis* LeConte; and the southern pine beetle, *Dendroctonus frontalis* Zimmerman. The subproject coordinator was, until 1974, R. W. Stark (Univ. of Idaho), and, later, W. E. Waters (Univ. of Calif.). Each of the bark beetle programs had a leader or co-

*Authored by D. L. Wood.

ordinator. D. L. Wood (Univ. of Calif.) was the project leader for the work on the western pine beetle discussed in this chapter.

The western pine beetle program was organized to meet the following subobjectives.

1. To obtain the information necessary to understand the population dynamics of the western pine beetle, and to develop predictive models of its dynamics as basic inputs to models of ponderosa pine stand dynamics, of impacts on resource values, and of western pine beetle treatment strategies and tactics for use by forest managers.
2. To obtain the information necessary to understand the dynamics of ponderosa pine in mixed conifer forest stands, and to develop models of its dynamics, including growth, age, size, and species distribution, and effects of beetle-caused tree mortality, in the context of other important agents affecting stand parameters.
3. To develop strategies and tactics for bark beetle management and models for predicting and evaluating their outcomes, including the various costs and benefits associated with treatments, the forest values saved or gained, and environmental effects.

In a program of this large size and great degree of complexity it is necessary to develop an explicit organizational structure for (1) the system under study, (2) the investigators, and (3) the relationships of these to each other and to

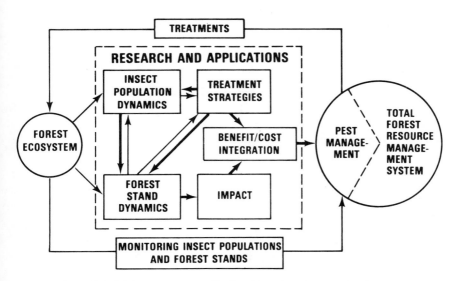

Figure 13.1. Major components of a forest insect pest management system.

the forest or forest pest manager. This systems structure is reported in the following section, while in the subsequent sections the biological and modeling studies performed under the above subobjectives are summarized to date.

It is important here to point out that pest management is only one aspect of the management of the total forest resource system, and that management of the western pine beetle is only one part of the pest management system (Fig. 13.1).*

THE INFORMATION SYSTEM FOR
RESEARCH AND PROJECT MANAGEMENT OF
THE WESTERN PINE BEETLE IPM PROJECT†

A major problem faced by project leaders of any complex or large scale eco-logical study is the development of a mechanism for securing and organizing the necessary data and achieving effective communication of information, that is, of data and ideas among the study participants. This includes informa-tion needed by project management to monitor and evaluate progress toward achieving program goals. In this section, we describe the role of a multi-faceted Information System concept for the western pine beetle (WPB) portion of the Integrated Pest Management (IPM) research program described in this book.

In such a project it is possible to simultaneously have a variety of distinct and helpful viewpoints or "images" of the project. (See, for example, the two sets of views of the larger but similar U.S.D.A. Gypsy Moth research and development program presented by Bean, 1973, and Campbell, 1973). In the WPB/IPM program, four major overviews evolved:

1. A discipline-oriented research and development viewpoint (Fig. 13.1)
2. A modeling viewpoint (Fig. 13.2)
3. A study system viewpoint (Fig. 13.3)
4. A data base management viewpoint (Fig. 13.4).

---

*Conceptualized by the Integrated Pest Management Coordinating Committee for the western pine beetle subprogram. The contributors were: F. W. Cobb, Jr., D. L. Dahlsten, B. Ewing, J. R. Parmeter, Jr., P. A. Rauch, W. E. Waters, and D. L. Wood, University of California; W. D. Bedard, C. J. DeMars, and B. H. Roettgering, U.S.D.A., Forest Service, Berkeley and San Francisco; and R. M. Bramson, Pacific Organization Development Associates, Oakland.

†Authored by P. A. Rauch, B. Ewing, D. L. Wood, and W. D. Bedard; from a paper presented by P.A.R. at the Annual Meeting of the Entomological Society of America, Honolulu, Hawaii, November 30, 1976.

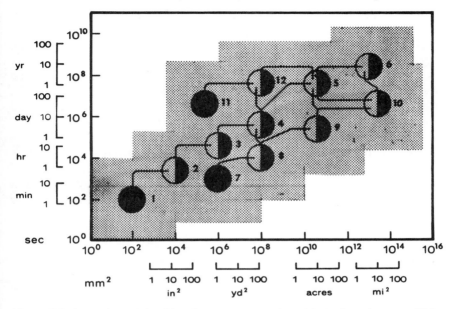

Figure 13.2. Space-time relationships of ecosystem network models (see legend on page 422 for explanation).

Each viewpoint is certainly a simplification of reality, but taken together they have been of enormous assistance in helping us to conceptualize the research goals and tasks, and thus to organize, mobilize, and communicate among ourselves, with greater economy of effort and fewer misunderstandings than previously (i.e., before these different viewpoints were developed and clarified). Each viewpoint complements the other, not by presenting greater or lesser detail, but in the way each relates to or views the entire project.

Before proceeding to detail these four viewpoints, we would first like to list several characteristics of this study (and undoubtedly of other large-scale ecological studies) which led us to develop these different views of the program.

Problems of scale arise from the following: the studies are multi-disciplinary; they are multi-agency in funding and accountability; the component studies themselves are subject to redefinition, as more is learned about the system; the need to mobilize a large number of diverse talents and operational resources gives rise to logistical complexity.

The forest ecosystem itself, like other living systems, is of course dynamic. The components are interactive and behave in nonintuitive ways. Also, the structure, processes, and interrelationships of forest ecosystems are inadequately understood. The forest ecosystem, unlike many other crops, encompasses a wider range of dominant and associated organisms, ecological conditions, socioeconomic factors, and space–time dimensions. The variety of personal expectations, skills, other commitments, understanding, and responsibilities of the participants add additional dimensions to the study. For example, interdisciplinary work brings in different vocabularies and concepts as communication barriers. Since it is essential for individual investigators to participate in group interactions, personality differences are certain to present a problem. Furthermore, such a group of investigators is commonly expected to function as a problem- and policy-formulating and decision-making group, rather than simply as contractual investigators of a single component problem. Its members should be selected not only for their

---

Explanation of Symbols:
1. Circles represent approximate space-time center of models.
   a. Shaded—descriptive part of model, empirical or statistical in nature, a "black box."
   b. Unshaded—explanative part of model, constructed from submodels and more basic concepts and principles, a "white box."
2. Connecting lines—indicate information flow between models, show strong biological interrelationships that are explicitly represented in the model network.
3. Dotted area—approximate space-time extent of model, about 10^4 in area and time; overlap is not indicated in the diagram.
4. Numbers—model numbers which are referred to in the section below:

Scope of Simulation Models:
  I. Western pine beetle-attacked ponderosa pine
    A. Western pine beetle brood development phase
      #1—within a gallery
      #2—between galleries
      #3—within a tree
    B. Western pine beetle dispersal and aggregation
      #4—between trees
      #5—between clusters of trees
      #6—between stands
 II. Flying western pine beetle
      #7—near trees and traps
      #8—between trees
      #9—between clusters
      #10—between stands
III. *Verticicladiella wagenerii*-diseased ponderosa pine
      #11—within a tree
      #12—between trees
      #5—between clusters
      #6—between stands
IV.   #13 (not shown)—management decision model, same space-time location as #6.

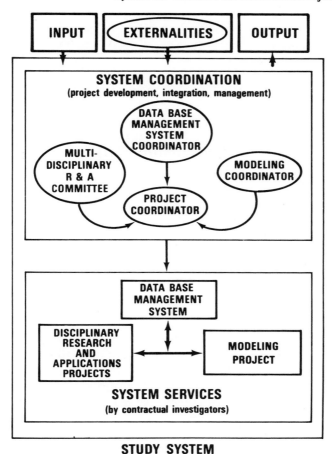

Figure 13.3. Major components and structural relationships for the WPB/IPM study system.

technical competence but also for their ability or potential to work effectively in an atmosphere with complicated and diverse lines of communication and authority. When competing interests or priorities occur or individual expectations are not met or shared, interpersonal problems of disappointment, frustration, and distrust can arise. Potential misunderstandings and disagreements can be minimized if the participants have experience with, and an understanding of, group dynamics under such circumstances. These people-related problems must be recognized and resolved (Stech, 1977; Bella & Williamson, 1976).

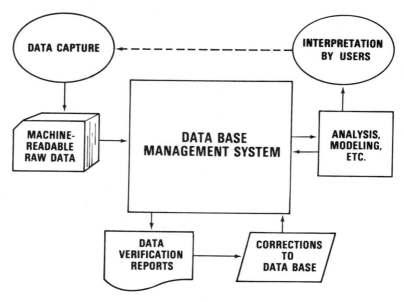

Figure 13.4. Basic relationships of a data base management system to other data processing activities.

If one can accept these observations as part of the reality of large ecological studies, it is clear why Patton (1975) said of the "Biome" programs that "The difficulty of bringing modelers and investigators together in meaningful productive interaction" was the "'system problem' which IBP [the International Biological Program] never solved." (See Duke et al., 1975; Krammer, 1975; Auerbach et al., 1977; Blair, 1977; Boffey, 1976; Downhower & Mayer, 1977; Gibson, 1977; and Mitchell et al., 1976, for detailed commentary on that program.)

The Information System

In *each* of the four viewpoints that follow, the principal features of an information system can be identified. There are distinct entities or components in each viewpoint and each component has its individual attributes, interrelationships, and dynamics. The principal function of the information system is to process and transfer the data or information as economically and accurately as required by the project. The major problem to be solved is to determine the system requirements and configuration that will achieve this function.

The Discipline-Oriented
Research and Development Viewpoint

Figure 13.1 presents the overall view of the major disciplinary components in our research and development program. It provides a common, goal-oriented focus for the biologists, modelers, and data management specialists engaged in the project.

The major components include:

1. The forest resource management system.
2. The forest itself.
3. The WPB pest management research program.

Two major activities in a pest management system are: (1) information accumulation and transfer, and (2) management actions, or in this case manipulation of the forest ecosystem (Waters, 1974). Each "box" in Figure 13.1 represents a complex of closely related research studies (Wood, 1974). These studies are selected and given a priority sequence. They are designed to produce the basic information needed as input to the other boxes. In this project, the ultimate research goal is to determine the effects of selected treatment strategies (including their various methodologies) on specified forest parameters and to use this information in models for predicting benefits/cost outcomes. The heavy arrows in Figure 13.1 emphasize the major flow of information toward this goal.

In order to explicitly identify the information to be obtained in each box and the nature of the models that would provide the mechanism of information flow, a logical sequence of goal-oriented questions was developed, and the resources required to support the studies were specified. This process is similar to the convergence technique (Carrese & Baker, 1967) used in developing the United States Dept. of Agriculture's Expanded Research and Applications Programs on the Southern Pine Beetle, the Douglas Fir Tussock Moth and the Gypsy Moth (Klassen, 1975). The system components, or boxes, were functionally defined in order to organize the research and coordinate related studies. Each element brings its own particular technology and perspective into the study, and produces a wide variety of data. The organization and integration of these diverse elements represent the basic step to develop an understanding about the pest management system itself.

The Modeling Viewpoint

Ecosystems such as the western pine beetle–ponderosa pine complex include an intricate network of strong interactions of processes that are generally nonlinear, and exhibit thresholds, saturations, and state-dependent time delays. Important interactions are discrete and stochastic, and many of the processes are sequential in the sense that past history strongly affects present and future behavior (Ewing et al., 1974, 1975a,b). Such systems may run counter to intuition in their response to direct manipulation (Pimentel, 1966). Also, such systems may have multiple equilibrium points, or under varying input no equilibrium at all (Holling, 1973). In this sense, this subproject of the IPM program is more similar to the IBP Biome studies than to the studies of the agricultural crop systems in this IPM program.

The overall WPB ecosystem under study spans a range of scale of about 10^{14} in area and 10^{10} in time (Fig. 13.2). Considering these problems and conditions, this ecosystem is clearly far too complex and extensive to either comprehend or implement as a single simulation model. The modeling effort is thus being designed to permit the decomposition of this large complex ecosystem into a space–time network of simpler, smaller ecosystem units. The resulting conceptual system consists of a group of models overlapping in time and space, each spanning about 10^4 in area and time, and communicating with adjacent models through the region of space–time overlap (Wood, 1973). The commentary by Botkin (1977) explains the problem of scale in ecosystem model design. For example, region 4 in Figure 13.2 represents root disease spread among ponderosa pine trees, while region 3 represents WPB brood development within a tree. This network of intercommunicating models results in another view of the study system organization that facilitates experimental design and the assignment of research priorities.

The Study System Viewpoint

Project management has the responsibility to establish a systematic, effective method for setting priorities, allocating resources, setting expectations for performance, and, of course, assuring that the necessary disciplinary and interdisciplinary efforts are developed.

Figure 13.3 depicts the study system viewpoint, with the major components and structural relationships for the WPB/IPM program. The primary emphasis is on information transfer and the functional separation of system coordination from system services. In coordination, the principal activity is to assure that the project coordinator receives useful (e.g., accurate, complete, timely) information with which he can develop plans along with the principal investigators, and then evaluate and maintain progress toward attaining the project's goals. In system services, the emphasis is on an exchange of information between the disciplinary research and applications projects and the modeling project, via the data base management system. The arrows emphasize the *principal* direction of information flow. Two features to note are: (1) the emphasis on supplying the project co-ordinator with necessary information and (2) the emphasis on system co-ordination driving the system services, even though system services are expected to produce and supply a substantial volume of information related to study goals.

It is necessary to connect the core of the study system to the rest of the world. Data and other information are required by the study from *input*. The study expects to produce new information, the *output*, to be used by others. *Externalities* represent uncontrollable forces that exert influences upon the study which must be taken into account, such as unpredictable weather conditions, vehicle or instrumentation breakdowns, and exercise of administrative prerogatives by funding agencies, deans, departmental chairmen, and others.

The necessity of providing a formally recognized *mechanism* for coordina-tion of a large multidisciplinary study is obvious. No one person is able to assume sole *technical* direction for the project because the scope is usually too broad for one person's technical competence. In addition to the working groups described earlier, we formed a coordinating committee to help define the major components of the WPB/IPM system (Fig. 13.1), the flow of information, the research areas to be emphasized and their priorities, and to help coordinate the work to be accomplished. The members (nine to eleven) met frequently. Many of the people-related problems mentioned earlier became apparent, as might easily be predicted (Congdon et al., 1971, 1975; Holling, 1972). Therefore, during part of the project an applied human behaviorist, whose interest was in the group dynamics of peer scientists participating in interdisciplinary research, contributed to the development of effective communication among program participants.

A structure like the one depicted in Figure 13.3 can be implemented in a variety of ways, from strong centralized management and control, to distri-bution of responsibility and control to the individual research projects. Which of these arrangements places the program in the best position to successfully achieve its goal depends in large part on how the various people-oriented problems mentioned earlier manifest themselves.

The Data Base Management Viewpoint

In a project encompassing both ecological and socioeconomic factors such as this one, a large number of variables are measured and many data points are collected. The tools and techniques used in analysis of the data strongly reflect the varied experiences and biases of the individual investigators. Their ideas must continually be tested against the available data. A large variety of analytical, numerical, computational, and modeling techniques were anticipated to be necessary to develop and analyse the behavior of forest pest management systems. It was therefore important to provide a great degree of data manipulation and computational flexibility in the data management and analysis segments of the project.

Basically, the implementation of an information system requires, among other things, that the various diagrammatic information structures (Figs. 13.1 to 13.3) be effectively mapped into data structures to make information available for machine manipulation and computation (Fig. 13.4). Much of this mapping activity involves the human/machine interface and the design of systems that reduce or avoid the problems of high cost, time loss, and proneness to error common to this interface (Holt & Stevenson, 1977; Cranmer et al., 1976). Of major usefulness to studies like ours is the concept of the data base management system (DBMS). Sophisticated DBMSs have only recently become widely available, somewhat behind the demand created by the large scale ecological projects initiated during the past few years. With careful planning and an understanding of the goals and activities of projects of this type, the inherently highly structured nature of their data base can be put into a form that makes it readily accessible for unanticipated studies and analyses, and to other investigators who may initiate complementary studies. Perhaps the most important contribution of a DBMS is to enhance the communications network among the various investigators and agencies. This is possible because the DBMS and its associated data dictionary require formal descriptions of all data collecting and processing activities, and in great detail. Furthermore, since data acquisition activities start early and tend to proceed throughout the duration of a project, a responsive DBMS can assist in uncovering many problems as they arise by providing program participants timely information of what *was* done relative to what *was supposed* to be done.

Results of Information-System Development

Our WPB-IPM "study system" has been organized to assist us to observe, guide and evaluate progress toward the specific program goals:

1. It has provided a mechanism for *coordinating the activities* of various scientists and pest control specialists.
2. It has provided a *structure* for the flow of data from the field and laboratory through to final analysis and synthesis.
3. It has permitted more rational ongoing *modifications* in study specifications, and the establishment of *priorities*, by providing more timely access to data.

PRINCIPAL RESEARCH RESULTS*

This summary consists of information obtained from published papers, dissertations, drafts of manuscripts, and reports of work in progress. The appropriate authors and the primary technical papers already published or to appear should be consulted prior to use of any information presented here.

Population Dynamics of the Western Pine Beetle

The first objective is to obtain the information necessary to understand the population dynamics of the western pine beetle, and to develop predictive models (Cobb & Parmeter, 1976; Waters & Ewing, 1976) of its dynamics as basic inputs to models of ponderosa pine stand dynamics, of impacts on resource values, and of western pine beetle treatment strategies and tactics for use by forest managers.

Between Tree Dynamics

In order to survive, the western pine beetle must invade living trees. Only after the death of the tree is assured does food (phloem and outer bark) become available for progeny survival. This process of host colonization can be described and analyzed in four phases: dispersal, selection, concentration, and establishment (Wood, 1972).

Dispersal/Selection. Following dispersal from the brood tree, the adult females must find individual ponderosa pine trees, often occurring in a mixture with several other tree species. Further, we know that the western pine beetle is especially successful in killing trees weakened by such agents as photochemical air pollutants (Stark et al., 1968), root disease (Cobb et al., 1973), and a variety of other agents, such as lightning and flooding. The mechanisms by which the western pine beetle detects weakened or "stressed" trees, and the subsequent relationship between density of attacking beetles and successful colonization of the living tree, are a critical part of the structure of the type of population model (Waters & Ewing, 1976) needed to reliably predict tree mortality. Colonization results in either death of the attacking beetles or death of the tree, as a result of the complex interactions between the beetle (Fig. 13.5) and its predators, the beetle-vectored phytopathogenic fungi (Fig. 13.6), weather, and the physico-chemical properties

*Authored by D. L. Wood and P. A. Rauch; from a paper presented by D. L. W. at the annual meeting of the Entomological Society of America, Honolulu, Hawaii, November 30, 1976.

of the host (Wood, 1972). A working hypothesis states that in stressed host plants fewer beetles, and consequently less inoculum, are required than in less stressed (so-called healthy) trees to reach a threshold density where death of the tree occurs. Therefore, in predicting the occurrence of tree mortality, predisposition through stress or injury must be considered.

The landing rate of western pine beetle and other bark beetles on "nonstressed " and " stressed " hosts was investigated in field experiments utilizing: (1) ponderosa pine bolts subjected to conditions favorable for anaerobic fermentation (see Wood, 1972, and citations therein, pp. 104–105); (2) standing ponderosa pines artificially stressed by injections of the herbicide cacodylic acid, and by lower stem freezing with dry ice; and (3) standing ponderosa pines naturally predisposed to bark beetle infestation by severe infections of the root pathogen, *Verticicladiella wagenerii* Kendrick (Moeck & Wood, 1977) (Fig. 13.7). In summary, these experiments showed that the bark beetle species trapped do not appear to orient to stressed host trees. The western pine beetle, the mountain pine beetle, *Ips paraconfusus* (Lanier), *I. latidens* (LeConte), *Gnathotrichus retusus* (LeC.), and *Hylurgops subcostulatus* (Mannerheim), all Scolytidae, landed, apparently indiscriminately, on healthy and stressed hosts. Therefore, the higher incidence of western pine beetle concentration on trees stressed either artificially or naturally is probably a result of increased feeding stimulation and/or decreased host resistance mechanisms (e.g., oleoresin exudation pressure, rate of oleoresin flow, etc.) in the stressed trees.

Concentration.   The concentration phase is initiated by the feeding and boring activity of female beetles in the outer bark and phloem. A complex pheromone communication system is initiated that results in the timely arrival of both male and female western pine beetles so that the fungal inoculation process will succeed and brood production can thus proceed.

Wood and Bedard (1977) have recently summarized the role of pheromones in the population dynamics of the western pine beetle. The concentration phase is initiated following release of the attractive pheromone, *exo*-brevicomin, and a host oleoresin constituent, myrcene, by females feeding in the phloem. Increased attraction occurs when frontalin is released by the male at a time in the attack process when female galleries are less than 2 cm long and males have joined about one-half of the female galleries. Mixtures containing (1*R*, 5*S*, 7*R*)-(+)-*exo*-brevicomin and (1*S*, 5*R*)-(−)-frontalin are much more attractive than mixtures containing their antipodes. Only (+)-*exo*-brevicomin and (−)-frontalin are produced by the beetle.

At this time predators, competitors, and many other species of insects begin to arrive. These arriving insects are not just those associated with the western pine beetle; many are exploiters of other resources now made

Figure 13.5 Female western pine beetle on a resin tube which forms when xylem resin ducts are severed during entrance tunnel excavations in the phloem. Females release the attractive pheromone *exo*-brevicomin and the host terpene myrcene at this time [courtesy of C. Willson].

Figure 13.7 Vertical black-stained streaks in the xylem caused by infection of the root-pathogen, *Verticicladiella wagenerii,* following invasion of the root system [courtesy of D.J. Goheen and F.W. Cobb, Jr.].

Figure 13.6 Blue-stain fungi form lenses of discolored tissues in the xylem around each female entrance tunnel. At this time the female beetle introduces three nonstaining (hyphae) fungi which are believed to cause the death of the tree by interrupting water transport to the crown.

available in the dying tree. *I. paraconfusus* has been observed to colonize the upper bole and larger branches of trees killed by the western pine beetle. Its developmental stages appear to be closely synchronized with those of the western pine beetle, which suggests that the two species may compete for phloem tissue in the upper bole. In a recent study (Byers & Wood, 1979), it was discovered that each of these species interrupts the other's response to its own aggregation pheromone. This behavior is elicited by volatile compounds associated with each species boring in ponderosa pine. These findings suggest that pheromones (or other beetle-produced compounds) may play a critical role in the process of interspecific competition during host colonization.

In another study of inter-species interactions, continuous trapping on the bark surface of trees under attack by the western pine beetle was carried out during six beetle generations over a $2\frac{1}{2}$ year period (Stephen & Dahlsten, 1976a,b). A significant correlation was found between the density of the western pine beetle trapped on the surface of attacked trees and the initial within-tree beetle densities. In the late spring and early summer generation, mass arrival (the period over which 70 percent of the cumulative catch occurred) was rapid (occurring over a period of about nine days) and intense (mean catch = 43.2 beetles/dm^2 of trapping surface per tree*). This pattern was quite consistent among trees. During the late summer generation, mass arrival occurred over a period of about three weeks and fewer beetles were trapped than in the earlier generation (mean catch = 24.5* beetles/dm^2 of trapping surface per tree*). In addition, this pattern was more variable among trees than in the early season generation.

The four principal insectan predators of western pine beetle adults and larvae were among the first of some 100 species to arrive following initiation of the concentration phase. These were: *Aulonium longum* LeConte (Coleoptera: Colydiidae), *Enoclerus lecontei* (Wolcott) (Coleoptera: Cleridae), *Medetera aldrichii* Wheeler (Diptera: Dolichopodidae), and *Temnochila chlorodia* (Mannerheim) (Coleoptera: Trogositidae). The two principal parasitoids, *Dinotiscus burkei* (Crawford) (Hymenoptera: Pteromalidae) and *Roptrocerus xylophagorum* (Ratzburg) (Hymenoptera: Torymidae), arrived when late instar larvae were available as hosts. The density of most species trapped on the bole was greater in the spring than in the fall generation. Differences between the vertical distributions of the species were common and often varied with the beetle generation. A highly significant increase in species diversity occurred on the traps between the time of the beetles' mass arrival and emergence of the new brood. The arrival sequence among the

---

*Trapping surface area incorrectly reported in the abstract from Stephen, F. M. and D. L. Dahlsten, 1976a; should be 27 dm^2/tree. Personal communication with H. A. Moeck and F. M. Stephen.

principal parasites and predators of the western pine beetle was independent of the total number of these natural enemies trapped on the trees. Further, the trend in density of some species trapped, such as *Lechriops californicus* (LeC.) (Coleoptera: Curculionidae), may indicate a pheromone response. In this example, the rate of increase in the number trapped appears to be sigmoid when compared to *Avetianella dahlsteni* Trjapitzin (Hymenoptera: Encyrtidae), where the rate appears to be constant.

Because of the difficulty in locating living trees early in the concentration phase, trap counts were made on trees baited either with female-infested bolts or with frontalin (Stephen & Dahlsten, 1976a,b). These baits were left on the trees for periods of 1 to 3 weeks. Trap counts on naturally attacked trees were also taken. In this instance, in the spring generation the number of western pine beetles trapped on the two trees baited with female-infested bolts was greater than on the naturally attacked tree and on the four trees baited with frontalin. Also, in this generation higher numbers of *R. xylophagorum* and *M. aldrichii* were trapped on naturally attacked trees (four trees were used in analysis of arrival of arthropod complex) than on the two classes of baited trees. The density of all species combined, except for the western pine beetle, that were trapped during the fall generation was greater on the naturally attacked tree (only one fall generation tree was observed) than on either type of baited trees. Of all the associate species analyzed, only the density of *M. aldrichii* was greater on the trees baited with female-infested bolts than on trees baited with frontalin. The above results suggest that this variation in density of trapped insects among baited and naturally attacked trees may be a result of the baiting technique used to bring the tree under attack. However, in two of the above comparisons only one tree was observed. Therefore, the possible effect of baiting on the natural arrival patterns must await further study.

In other experiments it has been shown that a high release rate of the synthetic attractive pheromones (*exo*-brevicomin, frontalin, and myrcene) will cause a reduced catch of the western pine beetle at the pheromone source (Tilden, 1976). In the above study (Stephen & Dahlsten, 1976a) frontalin was released from trees at rates of 8 to 30 mg/hr/tree. However, the release rate of frontalin from trees under natural attack was found to be much less than this amount (Browne et al., 1979). Also, bolts infested with females release *exo*-brevicomin and myrcene, which together are weak attractants for the western pine beetle (Bedard et al., 1968). Similarly, frontalin, which is the male component of the western pine beetle pheromone, is also a weak attractant (Bedard et al., 1970). However, all three components together produce a very attractive mixture (Bedard et al., 1970). Recently, *exo*-brevicomin, frontalin, and myrcene have been shown to be produced in trees under natural attack (Browne et al., 1979).

As with the arrival pattern of the western pine beetle on trees during the concentration phase, higher attack rates were observed in the late spring and early summer generation than in the mid- to late-summer generation (Dahlsten & coworkers, unpublished data). The concentration phase probably continues beyond the attack density required to kill the tree, and is likely caused by the continued production of attractive pheromones. Wood and Bedard (1977) have defined the termination of the concentration phase to be that point when no further pheromone is produced. Pheromone release is dependent upon sustained feeding and boring activity by the female beetle. An interruptant system may also terminate the concentration phase by preventing further landing on the host tree. This has been hypothesized by Renwick and Vité (1970).

Establishment. The establishment phase is initiated at some threshold attack rate and density where sufficient fungal inoculum and development (Whitney & Cobb, 1972) has occurred in the xylem to overcome tree resistance (Fig. 13.6) so that mating, gallery elongation, and oviposition can proceed (Wood, 1972; Wood & Bedard, 1977). This phase has been terminated when elongation of egg galleries and oviposition cease.

Within-Tree Dynamics

All western pine beetle mortality from egg to adult emergence from the tree occurs beneath the bark of the dead tree. In addition, about one-half of the parent adults have been observed to emerge following oviposition and to attack new trees (Miller & Keen, 1960, p. 19). Also, Bongberg (in Miller & Keen, 1960, p. 19) observed a second re-emergence of parent adults and showed that they were capable of producing brood in caged, cut logs.

The complexity of the relationships among the bole-inhabiting organisms following termination of the bark beetle concentration phase has been investigated in a mixed conifer forest in the central Sierra Nevada of California (Stark & Dahlsten, 1970; Dahlsten, 1976; Dahlsten & coworkers, unpublished data). Relative and seasonal abundance were obtained for about 70 insectan associates of the western pine beetle over a 10 year period. Preliminary analyses showed that: (1) within-tree densities of late instar larvae, pupae, and adults were higher in the late spring and early summer generation than in the fall generation, and the densities of these life stages varied 15-fold over ten generations; (2) the highest density of beetles observed during the study occurred in the generation following the one having the lowest density of western pine beetle parasitoids and predators; and (3) predation by woodpeckers was about ten times greater in the late summer or fall generation than in the late spring and early summer generations, as estimated by average percentage of woodpecker activity (i.e., chipping off the outer bark) on samples.

Ponderosa Pine Stand Dynamics, Including Bark Beetle/Root Pathogen Caused Tree Mortality

The second objective is to obtain the information necessary to understand the dynamics of ponderosa pine in mixed conifer forest stands and to develop models of its dynamics, including growth, age, size, and species distribution, and effects of beetle-caused tree mortality, in the context of other important agents affecting stand parameters.

Cobb et al. (1973) discovered that western pine beetle, mountain pine beetle, or both species together killed ponderosa pines infected by the root pathogen, *V. wagenerii*, at a much higher rate than symptomless trees in the same stand. Mortality occurred after the fungus had reached the root collar (Fig. 13.7). They (Cobb et al., 1973) concluded that the tree was predisposed to bark beetle invasion either because the beetles could detect a diseased tree or because *V. wagenerii* alters the physico-chemical status of the tree, leading to successful invasion of the host tissues by the beetles and their vectored fungi [*Ceratocystis nigrocarpa* Davidson, *C. minor* (Hedgc.) Hunt], an unidentified hyphomycete and an unidentified basidiomycete (Whitney & Cobb, 1972).

Extent of Association

The association of *V. wagenerii*, western pine beetle, and ponderosa pine mortality was studied further (Goheen, 1976). Over three consecutive seasons, 265 trees in the following categories were monitored: 136 healthy, 60 moderately diseased, and 60 severely diseased. Disease ratings were updated by periodic examination (of samples extracted with a 6 mm dia. punch driven into the sapwood) of the root collar for the black stain in the sapwood that is associated with the fungus (Fig. 13.7). During the course of the study, 96 trees died, 42 of which were infested by western pine beetle alone, 20 by *D. ponderosae* alone, 18 by both species concurrently, five by buprestids alone, and one by *V. wagenerii* alone. In all, 62 of the trees that became infested with bark beetles were severely diseased, 25 moderately diseased, and only three lacked disease symptoms. Thus, bark beetles were much more likely to infest infected than uninfected trees. Among the diseased trees infested with bark beetles, the mortality rate was greater in those that exhibited advanced symptoms (50 percent or more of the root collar samples were discolored by the fungus). There was evidence that buprestids, especially *Melanophila* spp., and possibly *Ips* spp., attacked diseased trees before *Dendroctonus* spp. did. At the end of the study only 3 percent of the nondiseased trees, respectively, had died.

Mechanism of Spread

Based upon this very close relationship between *V. wagenerii* and infestation by bark beetles, the mechanisms of spread of the root-infecting fungus were investigated (Goheen, 1976). The colonization patterns of the fungus in the root systems of 126 infected trees occurring in 19 infection centers were examined by total root excavation. The study indicated that the fungus occasionally spreads from tree to tree across grafts and major root contacts, but that it much more commonly entered roots through small rootlets. The small rootlets were rarely in contact with other infected roots or rootlets. Thus, the major mechanism of spread of infection is still unknown.

For the first time, the perfect and imperfect reproductive stages of this root-infecting fungus were found in the galleries of several root colonizing beetles. Conidia were found in one gallery each of a buprestid, a cerambycid, and an unidentified insect, in two galleries of *D. valens* LeConte (Scolytidae) and in 96 galleries of *Hylastes macer* (LeConte) (Scolytidae). This represents 24 percent of the *H. macer* galleries and only 1 percent of the other insect galleries examined. The perithecia of *V. wagenerii* were discovered in 19 galleries of *H. macer* and described (Goheen, 1976). *V. wagenerii* classification is based upon the imperfect stage (conidium) because, until now, the perfect stage (perithecium) had not been known. Thus the species has been removed from the Fungi Imperfecti and described as an Ascomycetes, *Ceratocystis wagenerii sp. nov.* (Goheen and Cobb, 1979).

Forest-wide Distribution

In order to determine the pattern and extent of tree mortality associated with this root pathogen/insect complex (*V. wagenerii* and *D. brevicomis/D. ponderosae*), a large scale evaluation was undertaken (Cobb et al., unpublished data). The rate of enlargement of existing *V. wagenerii* infection centers and the rate of generation of new infection centers in a mixed conifer forest in the central Sierra Nevada of California were determined on approximately 2200 hectares. Aerial photographs taken periodically from 1963 to 1974 were interpreted for tree mortality, and then ground-checked to establish the causes of death. The rate of center enlargement was calculated from a sketch map of each center made to scale. The change in the perimeter of centers was plotted for each year in each of eight sectors of the center. The new photo-detected dead trees indicated the extent of the enlargement of the center. Ponderosa pine mortality within the 54 infection centers studied averaged 1.4 trees/hectare/year. The total area of increase in mortality during the 11 year period was 22 hectares. The average annual radial increase in area of mortality in the 54 centers studied was 1.2 meters. New infection centers were generated at the rate of 2.7/year for the entire study area. The largest single cause of mortality was the *V. wagenerii*-bark beetle complex. This mortality occurred in relatively large aggregates, almost entirely in dense, pure (i.e., largely one species in the overstory) ponderosa pine stands. The western pine beetle was the most common insect associated with tree mortality in pure ponderosa pine stands, while *D. ponderosae* was the most common insect associated with ponderosa pine mortality in the areas with mixed conifer species in the overstory.

Aerial photography of an 11,000 hectare area in the same forest was used to identify centers of ponderosa pine mortality and forest types (Cobb et al., unpublished data). A sample of the photo-detected mortality centers was visited and the area, number, and characteristics of affected trees, the bark beetle/root pathogen complex associated with the trees, and the terrain characteristics were recorded. Based on the changes in foliage color and other manifestations of the deterioration of the trees observed, the area of new tree mortality was determined to be increasing at a rate of 14 hectares/year, equivalent to 4,000 cubic meters of wood/year. This estimate of the rate of mortality was based on the four years immediately preceding the ground visit because the year of death and associated insect/root pathogen complex could not be determined reliably beyond that period. The species complex *V. wagenerii*, *D. brevicomis*, and *D. ponderosae* accounted for 46 percent of the total area in the mortality centers. *D. ponderosae* and *D. brevicomis* alone, without (or with undetectable) disease, accounted for an additional 44 percent of the area in the mortality centers, while other causes, mostly flat-headed borers (Buprestidae), accounted for about 10 percent of the area.

Prediction of Occurrence

In the central Sierra Nevada where ponderosa pine is the principal host affected by *V. wagenerii*, many of the largest and most active infection centers have been found in low lying sites. In such locations, soil moisture content is generally high, especially in the winter and spring when observations (Cobb & Goheen, unpublished data) indicate that the fungus is most actively infecting and colonizing new hosts. A study was initiated to test the hypothesis that high soil moisture may favor fungus establishment and disease development (Goheen et al., 1978). Year-old ponderosa pine seedlings were planted in equal amounts of soil in containers. Seedlings were inoculated by taping a block of inoculum 3 cm long to the roots 3 cm below the root crown. An irrigation technique was used to generate the various soil moisture treatments. Soil moisture in all treatments was started at field capacity, and the plants were allowed to extract water from the soil until preselected stresses below field capacity were reached. When these points were attained, all containers were watered again to field capacity.

A total of 31 of 60 ponderosa pine seedlings inoculated in 1972 and 17 of

100 inoculated in 1973 exhibited stained sapwood when examined. *V. wagenerii* was isolated easily from these stained seedlings. The largest percentages of infected seedlings were from those in the wettest soil treatments, with the percentages decreasing rapidly as the soil treatments became drier. Though the extent of *V. wagenerii* colonization differed very little between wet and dry treatments, the significant difference in number of infections strongly suggests that infectivity of *V. wagenerii* is favored by high soil moisture.

The implications of a relationship between high soil moisture and successful infection by *V. wagenerii* may be far reaching to forest managers. Possibly, the disease may be limited or controlled by manipulating forest stands on poorly drained sites.

Treatments for Western Pine Beetle Management

The third objective is to develop strategies and tactics for bark beetle management and models for predicting and evaluating their outcomes, including the various costs and benefits associated with treatments, the forest values saved or gained, and environmental effects.

Currently, the only direct method to lower western pine beetle populations (purportedly to reduce the rate of beetle-induced tree mortality) involves locating and felling infested trees and killing the beetles with insecticides (BHC and Lindane, primarily), or by burning, burying, and peeling, or by removing the infested trees from the forest (Miller & Keen, 1960). The use of insecticidal chemicals is controversial on the ground of its unproven efficacy and its environmental safety (Koerber, 1976). Further, the value of the trees varies greatly, from very high for trees around homes, to very low for those in remote areas. Therefore, the benefit-cost relationships are also placed in question.

Employment of the behavior-modifying chemicals used by the beetles to coordinate and regulate their natural attacks on living trees (see above section on population dynamics of the western pine beetle) offers promise for detection, monitoring, and manipulation of beetle populations and for individual tree protection (Wood, 1977; Bedard et al., 1979c). We have conducted a variety of experiments to determine the effects of treatment variables and environmental factors on the catch of the western pine beetle on sticky traps baited with the mixture of myrcene and synthetic, racemic *exo*-brevicomin and frontalin.

Trap Size, Release Rate of Attractants, and Tree Silhouette

Tilden et al. (1979) studied the behavior of the western pine beetle at and near sources of attractive pheromones in nature, with two principal results. In one experiment beetles flew into unbaited sticky traps placed at various distances and heights around a single source of the attractants. Doubling the area of unbaited traps significantly increased the number of beetles caught 1.5 m, but not 4.5 m, away from the source of attractants. Also more beetles were caught at a height of 3 m than at 1.5 m and the catch at the source was significantly decreased. In other experiments (Tilden, 1976), as the attractant release rate was increased, beetles were caught at greater distances from the pheromone source, and a greater proportion of the beetles was trapped farther from the source. Further, the presence of a tree trunk-simulating silhouette increased the catch.

The findings from these earlier studies indicated that use of a very large trap at the pheromone source would probably maximize the catch of beetles flying in the vicinity of the baited trap. By using the large trap and by not placing traps near ponderosa pines it was concluded that tree killing around the traps themselves could be minimized. Based on these findings large scale "trap-out" experiments were designed and implemented.

Trap Site Features

A study by Bedard et al. (1978a) of the site features associated with the catch of *D. brevicomis* on traps baited with attractants showed the following:

1. Total catch was directly related to current tree mortality in the 65 km^2 study area.
2. Catch was not significantly correlated with local *D. brevicomis* abundance as estimated from proximity of the trap to nearby *D. brevicomis*-infested trees.
3. Three peaks of abundance were revealed during the season, and these peaks were correlated with three distinct generations as determined by examination of within-tree *D. brevicomis* development throughout the year (DeMars et al., 1979).
4. Catch was negatively correlated with slope and distance to the nearest ponderosa pine.
5. The relationship of the number of beetles caught on all traps to those caught on individual traps was consistent through time.

Trap-out Method

In two large scale field experiments an attempt was made to lower the western pine beetle population density, and thus subsequent tree mortality, by attracting the beetles to sticky traps deployed in various configurations. A trapping system utilizing attractive pheromones (myrcene and racemic *exo*-brevicomin and frontalin) (see Wood & Bedard, 1977) was developed for the suppression and monitoring of beetle populations over large forested areas and for extended periods of time (Bedard & Wood, 1974). A portable, reusable sticky trap was developed and fabricated (Browne, 1978). This is a large trap designed to catch large numbers of beetles for population suppression. In order to efficiently estimate suppression trap catch (many thousands of beetles per trap were anticipated), an unusual method of counting by sample splitting was developed to be used with a nearly unbiased estimator to derive the total trap catch (Browne, 1978; Lindahl, 1979). The precision of the estimator, or the value of the threshold count (beyond which no further counting is necessary in order to estimate the total trap catch with the given precision) can be specified to suit the users' needs. The estimated catch and variation of individual trap counts can be pooled by simple addition to determine the effects of groups of traps in a specified area. Also, estimates were made of the catch of two predators of *D. brevicomis*: *E. lecontei* and *T. chlorodia*.

In one area (at Bass Lake in the Sierra Nevada Mountains), traps were deployed in two configurations over 65 km^2 for one season (Bedard & Wood, 1974). Nearly one million western pine beetles were trapped. Tree mortality declined throughout the area. Tree mortality attributed to the western pine beetle and its vectored pathogens in the generation preceding treatment was 283 $\pm$ 46 trees (overwintering in 1969–1970) (DeMars et al., 1979). Following the treatment period, *D. brevicomis* was found infesting 90 $\pm$ 16 dead trees (overwintering in 1970–1971). Tree mortality then remained low for about 4 years. A second test was performed at McCloud Flats in the southern Cascade Mountains of northern California with a more intensive monitoring system. Results of this experiment are currently being analyzed.

Evidence from these two experiments indicates that western pine beetle populations can be manipulated over a large area. This strengthens our belief that effective control strategies can be developed using methods similar to those employed in these experiments (Bedard et al., 1979c). Because chemicals used in this manner are considered pesticides by the Environmental Protection Agency (EPA) they are subject to the same scrutiny as insect toxicants with regard to their efficacy, toxicology, and production and environmental chemistry, so as to insure that any recommended use of a commercially registered material will be safe and effective (Phillips, 1976). The following points relate to the compounds used in these field tests: (1) they were non-toxic in our panel of tests (Bedard, unpublished data); (2) they occur in nature; (3) they would be applied only in their normal environment; (4) they would probably be applied at rates that do not exceed those found in nature (Browne et al., 1979); and (5) they are not introduced into plant, animal, soil, or aquatic systems (the trapping method employs bait stations). We feel therefore that there is a high likelihood that these compounds can be registered under EPA requirements.

Interruption Method

With this method, attractants or "interruptants" are either broadcast over a large area or are placed on individual trees in order to interrupt the normal behavior that results in beetle aggregation on its host plant.

Attractive Pheromones. We have demonstrated that the release of pheromone components (myrcene and racemic *exo*-brevicomin and frontalin) from many points in an 0.81 hectare area greatly lowers the catch of western pine beetles on traps baited with the same compounds in the center of the treated area (Tilden et al., manuscript in preparation).

Interruptants. When we baited paired ponderosa pines with the above pheromone components and released verbenone between them, only a few beetles were trapped on these trees and they were not mass-attacked (Bedard et al., 1979). However, paired comparison trees baited with pheromones at the same time but not with verbenone were attacked and killed. Further experiments are required as this result could not be repeated.

Population Monitoring

As part of the field experiments designed to evaluate a population suppression technique by trapping beetles, an attempt was made to develop a method to monitor the in-flight western pine beetle population through time and space (Gustafson et al., 1971). If a pattern of catch using such a survey or monitoring system can be detected, then it may be possible to exploit the method for measurement and prediction of population movement and trend in tree mortality. A method of population monitoring that utilizes attractant-baited traps should complement within-tree population sampling, thus reducing the difficulty and cost of estimating population trend over large areas (DeMars et al., unpublished data).

In a population suppression experiment conducted at McCloud Flats a total of 219 small survey traps were located on an 0.8 km grid during each of two years, 1971 and 1972. The traps were in place from the early part to the end of the flight season (22 and 27 weeks, respectively) of the western pine beetle. Counts of the western pine beetle and several major predators were made at approximately 7-day intervals. In addition, the catch of all insect species was counted on 5 percent of the traps selected randomly each week. Ten unbaited survey traps were placed nearby and in similar sites to the survey traps. These traps were designated for complete enumeration of all species to aid in the interpretation of trap selectivity.

Preliminary analyses (Wood & coworkers, unpublished data) of the survey trap catch data indicate a distinct pattern of spatial and numerical distributions. Certain traps, whose site features have not been analyzed yet, consistently had much higher catches than other traps in each of the two trapping seasons. The variability in total catch between seasons at a given trap site appears in many cases to be low, and does not appear to be correlated with magnitude of catch (e.g., traps with low catch, 98 and 100 *D. brevicomis* for the two seasons; moderate catch, 1716 and 1714; high catch, 3711 and 3981). The relative proportions of the total seasonal catch of *D. brevicomis* and two of its major predators, *T. chlorodia* and *E. lecontei*, were also in the same order of magnitude between years (412:41:1 in 1971 and 632:27:1 in 1972). The total survey trap catch for the two seasons is also comparable by species. The attractivity of the baited compared to unbaited survey traps is significantly different, as expected, with a weekly average of 58.8 versus 0.37 *D. brevicomis* per trap, respectively.

There is an abundance of information yet to be extracted from the survey trap catch and photo-detected tree mortality data bases. Part of the analyses must rely on efficient photointerpretation and related data recording techniques presently under development (DeMars & Aldrich, 1978*). Aerial photographic estimates of tree mortality by generation (DeMars et al., 1973, DeMars et al., 1978) and the estimation of within-tree beetle populations furnish an indirect measure of population movement and numerical change. At each sampling time these measurements show the size and location of the beetle population but not how their distribution came about. The survey trap network gives a more direct measurement of the movement of western pine beetles because of the greater frequency of observations that are made and because the dispersing adults are directly sampled.

In the McCloud Flats study, we expect aerial photography to help us understand the observed patterns of survey trap catch. By use of a combined photographic stereo viewing and computerized digitizing system the forest cover type and tree mortality in the immediate vicinity of the survey traps can be conveniently observed and recorded. Forest stand species composition and tree size distributions are recorded, and distances between trees and traps can be computed from the digitized locational data. In this way, trap catch can be more thoroughly analyzed with respect to these variables and through time. By use of the photo-digitization techniques developed thus

*This work is also being supported in part by the *Expanded Southern Pine Beetle Research and Applications Program*, U.S. Department of Agriculture.

far, accuracy of mapping point locations is on the order of 25 to 50 feet (DeMars & Aldrich, 1978).

BENEFITS-COST ANALYSIS*

An essential area of bark beetle management is concerned with benefit-cost analysis of tree mortality reduction made possible from employment of specific treatments. The impact of bark beetles on ponderosa pine forests has consequences other than loss in growth or loss of products. The effects of bark beetle-caused tree mortality on recreation and watershed values are examples, but these consequences are not known and are also difficult to quantify. Recently work has been undertaken to model the possible benefits relative to timber production of western pine beetle management tactics or strategies that utilize attractive pheromones to: (1) lower the rate of tree mortality; (2) aggregate tree mortality in locations convenient for logging; or (3) achieve a combination of (1) and (2).

The large amount of tree mortality, both from insects and other causes, has attracted considerable national attention. Two recent forest resource reports (Outlook for Timber in U.S., 1973; Report of the President's Advisory Panel on Timber and the Environment, 1973) stress the need to salvage dead timber, as does a special report of the Comptroller General (1973). Because of the large volume of timber involved and the perceived ability to save or salvage some of it, both the new National Forest Management Act of 1976 and U.S. Forest Service regulations (U.S. Forest Service Manual, 1976) issued pursuant to it specifically deal with pest control programs. The National Forest Management Act exempts pest control projects or salvage cutting from the long-term planning requirements of the act. The United States Forest Service Manual, chapter 5280, paragraph 5281.4, sets forth new selection criteria which identify the United States Forest Service standards for approving insect and disease prevention suppression projects. The principal criteria—biological effectiveness, environmental acceptability, and economic efficiency—are each supported by specific evaluation guidelines. Economic efficiency is very difficult to demonstrate with current conceptual and modeling technology (Berck & Hanemann, unpublished). Also, there is a demonstrated need for a method to evaluate control strategies for forest pests, particularly tree-killing bark beetles, that can take into account multiple resource values.

*Authored by P. Berck and W. M. Hanemann. From: Economic evaluation of forest pest management strategies. Project description—1977. Department of Agricultural and Resource Economics, California Agricultural Experiment Station and the Giannini Foundation, University of California, Berkeley.

EXPECTED NEW TECHNOLOGY AND ITS PROBABLE VALUE*

The most significant new technologies at hand (1, 2) and expected (3, 4) for management of the western pine beetle/ponderosa pine system are:

1. A formalized pest management research and implementation system for the western pine beetle within which current and new work can be evaluated and guided.
2. A conceptual framework for a network of models that describe the important aspects of the western pine beetle ecosystem. A spatial and temporal framework of simulation models, and a modeling strategy involving a network of thirteen models has been described. However, implementation of these models is extraordinarily difficult, due principally to the following characteristics of the forest resource management system under consideration:
   a. High dimensionality (e.g., time, space, number of individuals and species).
   b. Complex structure (e.g., spatial heterogeneity, population age structures, multiple species interactions).
   Also, the lack of, and high cost of acquiring, data for many segments of the models, both to efficiently postulate appropriate model elements as well as for model testing and validation, is a major obstacle to successful implementation of these methods.
3. A treatment method utilizing attractive pheromones where the density of in-flight beetles is manipulated so as to lower the rate of tree mortality caused by the western pine beetle.
4. An economic model to explore the benefits and costs of tree mortality reduction in ponderosa pine stands by manipulation of beetle populations with attractive pheromones.

ACKNOWLEDGMENTS

The authors are grateful to the many reviewers of this chapter, especially C. B. Huffaker and W. E. Waters.

These studies were supported in part by the Rockefeller Foundation; U.S.D.A. Forest Service; and Cooperative State Research Services (2598-RRF, W-110); McIntire-Stennis (Project No. 2800); and the National Science Foundation and Environmental Protection Agency through a grant (NSF GB-34719/BMS 75-04223) to the University of California. The

*Authored by D. L. Wood.

findings, opinions, and recommendations are not necessarily those of the University of California or the funding agencies. The contributors to this project are: J. F. Barbieri, P. Berck; L. E. Browne, J. A. Byers, F. W. Cobb, Jr., W. A. Copper, D. L. Dahlsten, J. S. Elkinton, B. Ewing, P. A. Felch, W. M. Hanemann, K. Q. Lindahl, J. R. Parmeter, Jr., P. A. Rauch, D. L. Rowney, M. E. Schultz, G. W. Slaughter, W. E. Waters, D. L. Wood, and B. S. Yandell, University of California, Berkeley; R. M. Silverstein, State University of New York, Syracuse; F. M. Stephen, University of Arkansas, Fayetteville; L. J. Edson, Texas A & M University, College Station; W. D. Bedard, C. J. DeMars, Jr., M. M. Kramer, N. X. Norick, and P. E. Tilden, U.S.D.A. Forest Service, Berkeley: J. A. Caylor, R. W. Gustafson, B. H. Roettgering, U.S.D.A.F.S., San Francisco; D. J. Goheen, U.S.D.A.F.S., Portland, Oregon; H. A. Moeck, Canada Department of Environment, Pacific Forest Research Centre, Victoria, B.C.; and R. M. Bramson, Pacific Organization Development Associates, Oakland, California.

LITERATURE CITED

Auerbach, S. I., R. L. Burgess, and R. V. O'Neill. 1977. The biome programs: evaluating an experiment. *Science 105*:902–904.

Bean, J. L. 1973. The USDA forest service gypsy moth research and development program. Proc. 24th Annual Western Forest Insect Work Conference, March 6–8, 1973, Rocky Mountain Forest and Range Experiment Station, USDA, pp. 5–24.

Bedard, W. D., and D. L. Wood. 1974. Program utilizing pheromones in survey or control: Bark beetles—the western pine beetle, in M. C. Birch (ed.), *Pheromones*, North-Holland, New York, pp. 441–449.

Bedard, W. D., R. M. Silverstein, and D. L. Wood. 1970. Bark beetle pheromones. *Science 167*:1638–1639.

Bedard, W. D., J. E. Coster, and M. M. Kramer. 1979a. Attractive pheromones for suppression and survey of western pine beetle populations: exploration of the survey potential of baited traps. (In manuscript.)

Bedard, W. D., P. E. Tilden, D. L. Wood, K. Q. Lindahl, and P. A. Rauch. 1979b. Effects of the anti-attractant verbenone on the response of *Dendroctonus brevicomis* to natural and synthetic attractant in the field. (In manuscript.)

Bedard, W. D., D. L. Wood, and P. L. Tilden. 1979c. Using behavior modifying chemicals to reduce western pine beetle-caused tree mortality and protect trees, in W. E. Waters (ed.), *Current Topics in Forest Entomology*. USDA Forest Service Gen. Tech. Rep. WO-8, pp. 159–163.

Bedard, W. D., P. E. Tilden, D. L. Wood, R. M. Silverstein, R. G. Brownlee, J. O. Rodin. 1969. Western pine beetle: field response to its sex pheromone and a synergistic host terpene, myrcene. *Science 164*:1285–1286.

Bella, D. A., and K. J. Williamson. 1976–1977. Conflicts in interdisciplinary research. *J. Environmental Systems 6(2)*:105–124.

Blair, W. H. 1977. Letter. *Science 195*:823.

Boffey, P. M. 1976. International biological program: was it worth the cost and effort? *Science 193*:866–868.

Botkin, D. B. 1977. Bits, bytes, and IBP (editorial). *BioScience 27*:385.

Browne, L. E. 1978. A trapping system for the western pine beetle using attractive pheromones. *J. Chem. Ecol. 4*:261–275.

Browne, L. E., D. L. Wood, W. D. Bedard, R. M. Silverstein, and J. R. West. 1979. Quantitative estimates of the attractive pheromone components, *exo*-brevicomin, frontalin and myrcene, of the western pine beetle in nature. *J. Chem. Ecol. 5*: 397–414.

Byers, J. A., and D. L. Wood. 1979. Interspecific inhibition of the response of the bark beetles, *Dendroctonus brevicomis* Le Conte and *Ips paraconfusus* Lanier, to their pheromones in the field. *J. Chem. Ecol.* (In press.)

Campbell, R. W. 1973. The conceptual organization of research and development necessary for future pest management, in R. W. Stark and A. R. Gittins (ed.), Pest management for the 21st century. Idaho Research Foundation, Inc., National Resource Series No. 2, pp. 23–38.

Carrese, L. M., and C. G. Baker. 1967. The convergence technique: a method for the planning and programming of research efforts. *Management Science 13*:B-420–B-438.

Cobb, F. W., Jr., and J. R. Parmeter, Jr. 1976. Modeling of plant pathogen populations in the forest ecosystem, in R. L. Tummala, D. L. Haynes, and B. A. Croft (eds.), Modeling for pest management: concepts, techniques, and applications. Michigan State Univ., East Lansing, Mich., pp. 165–167.

Cobb, F. W., Jr., J. R. Parmeter, D. L. Wood, and R. W. Stark. 1973. Root pathogens as agents predisposing ponderosa pine and white fir to bark beetles. Proc. Fourth International Conference on *Fomes annosus*. Athens, Georgia. International Union of Forest Research Organizations, Section 24: Forest Protection, pp. 1–8.

Comptroller General, U.S. General Accounting Office Report B-125053, October 5, 1973; Report to the Congress: " More usable dead or damaged trees should be salvaged to help meet timber demand," pp. 1–35.

Congdon, C. C., A. I. Chernoff, R. D. Lange, D. A. Lingwood, F. J. Miller, and W. C. Morris. 1975. Management and the achievement of research goals. Progress Report: Phases I and II. Univ. Tenn. Memorial Research Center Bull. 17, pp. 1–50.

Congdon, C. C., and D. G. Doherty. 1971. Science-based management for scientists. *The Chemist 48*:299–302.

Cranmer, M. F., L. R. Lawrence, A. K. Konvicka, D. W. Taylor, and S. S. Herrick. 1976. Research data integrity: a result of an integrated information system. *J. Toxicology and Environmental Health 2*:285–299.

Dahlsten, D. L. 1976. *D. brevicomis*, integrated control potential using biotic factors. XVI International Union of Forest Research Organizations World Congress. Proc. Div. II, pp. 447–448.

DeMars, C. J., Jr., and R. Aldrich. 1978. Development of an on-line, interactive, digitizer-computer aided system for tabulating, mapping, and comparing tree mortality detected on successive stereo aerial photographs. Final Report—Expanded

Southern Pine Beetle Research and Application Program. U.S. Dept. of Agric. pp. 1–5, with appendices.

DeMars, C. J., J. Caylor, B. Ewing, and P. Rauch. 1973. Estimating cause and pattern of insect-killed trees from aerial photographs. Proc. of the 4th Conf. of the Advisory Group of Forest Statisticians. Int'l Union of For. Res. Organizations, pp. 37–47.

DeMars, C. J., W. D. Bedard, N. Norick, B. Roettgering, and G. W. Slaughter. 1979. Attractive pheromones for suppression and survey of western pine beetle populations: estimation of beetle-caused tree mortality. (In manuscript.)

Downhower, J., and R. Mayer. 1977. Letter. *Science 195*:823.

Duke, K. S. Globe, W. Martin, R. Mayer, J. McGinnis, and R. Mitchell. 1975. Evaluation of three of the biome studies funded under the Foundation's international biological program (IBP). Battelle Columbus Laboratories, pp. 1–332.

Ewing, B., P. A. Rauch, and J. F. Barbieri. 1974. Simulating the dynamics and structure of populations. Progress report. Univ. of California Lawrence Livermore Lab. UCRL-760-46 (Rev. 1), pp. 1–59.

Ewing, B., J. F. Barbieri, and P. A. Rauch. 1975a. Simulating the dynamics and structure of populations. Progress report. Suppl. to Univ. Calif. Lawrence Livermore Lab. Publ. UCRL-76046 (Rev. 1), pp. 1–59.

Ewing, B., J. F. Barbieri, and P. A. Rauch. 1975b. The principles, strategies, and tactics of pest population regulation and control in major crop ecosystems. Progress report. Vol II. Intern. Center for Integrated and Biological Control, Univ. of California, Berkeley, pp. 262–295.

Gibson, J. H. 1977. Letter. *Science 195*:822–823.

Goheen, D. J. 1976. *Verticicladiella wagenerii* on *Pinus ponderosa*: epidemiology and interrelationships with insects. Ph.D. dissertation, Univ. of California, Berkeley, pp. 1–118.

Goheen, D. J., F. W. Cobb, Jr., and G. N. McKibbin. 1978. Influence of soil moisture on infection of *Pinus ponderosa* by *Verticicladiella wagenerii*. *Phytopathology. 68*: 913–916.

Goheen, D. J., and F. W. Cobb, Jr. 1978. Occurrence of *Verticicladiella wagenerii* and its perfect state, *Ceratocystis wagenerii sp. nov.*, in insect galleries. *Phytopathology 68*:1192–1195.

Gustafson, R. W., W. D. Bedard, and D. L. Wood. 1971. Field evaluation of synthetic pheromones for the suppression and survey of the western pine beetle, McCloud Flats, Shasta-Trinity National Forest, California. U.S. Forest Service, pp. 1–20.

Holling, C. S. 1973. Resilience and stability of ecological systems. *Annu. Rev. Ecology and Systematics 4*:1–23.

Holling, C. S. 1972. Resource Science Center, U. of British Columbia in J. L. Aldrich and E. J. Kormondy (eds.), *Environmental Education: Academia's Response*. Publ. No. 35. The Commission on Undergraduate Education in the Biological Sciences, A.I.B.S. Washington, D.C., pp. 26–30.

Holt, H. O., and F. L. Stevenson. 1977. Human performance considerations in complex systems. *Science 195*:1205–1209.

Klassen, W. 1975. Pest management: organization and resources for implementation, in D. Pimentel (ed.), *Insects, Science, and Society*. Academic Press, New York, pp. 227–256.

Koerber, T. W., compiler. 1976. Lindane in forestry. . . a continuing controversy. U.S.D.A. Forest Serv. Gen. Tech. Rep. PSW-14, illus. Pacific Southwest Forest and Range Exp. Stn., Berkeley, Calif., pp. 1–30.

Krammer, P. J., Chairman of Study Committee. 1975. An evaluation of the international biological program. National Research Council, Washington, D.C., pp. 1–81.

LeMaster, D. C. 1976. The Resources Planning Act as amended by the National Forest Management Act. *J. Forestry 74*:798.

Lindahl, K. Q. 1979. Reverse sequential estimation of insect trap catches from geometrically divided subsamples. Report to Southwest Forest and Range Expt. Sta. U.S. Forest Service, Berkeley, California.

Miller, J. M., and F. P. Keen. 1960. Biology and control of the western pine beetle. USDA Misc. Publ. 800. 381 pp.

Mitchell, R., R. A. Mayer, and J. Downhower. 1976. An evaluation of three biome programs. *Science 192*:859–865.

Moeck, H. A., D. L. Wood, and K. Q. Lindahl. 1979. Host selection behavior of bark beetles (Coleoptera: Scolytidae) attacking *Pinus ponderosa*, with special emphasis on *Dendroctonus brevicomis*. *J. Chem. Ecol.* (Submitted.)

Patten, B. C., ed. 1975. *Systems Analysis and Simulation in Ecology*. Vol. III. Academic Press, New York.

Phillips, W. G. 1976. EPA's registration requirements for insect behavior controlling chemicals—philosophy and mandates, in M. Beroza (ed.), *Pest Management with Insect Sex Attractants and other Behavior-Controlling Chemicals*. Am. Chem. Soc. Symp. Series 23, pp. 135–144.

Pimentel, D. 1966. Complexity of ecological systems and problems in their study and management, in K. E. F. Watt (ed.), *Systems Analysis in Ecology*. Academic Press, New York, pp. 15–35.

Renwick, J. A. A., and J. P. Vité. 1970. Systems of chemical communication in *Dendroctonus*. *Contribs. Boyce Thompson Inst. 24*:283–292.

Report of the President's Advisory Panel on Timber and the Environment. April, 1973, pp. 1–541.

Stark, R. W., and D. L. Dahlsten. 1970. Studies on the population dynamics of the western pine beetle, *Dendroctonus brevicomis* LeConte (Coleoptera: Scolytidae). Div. of Agric. Sci., Univ. of Calif., pp. 1–174 (see especially Sections 7, 8, 9, 10, 11, and 16).

Stark, R. W., P. R. Miller, F. W. Cobb, Jr., D. L. Wood, and J. R. Parmeter, Jr. 1968. Photochemical oxidant injury and bark beetle (Coleoptera: Scolytidae) infestation of ponderosa pine. I. Incidence of bark beetle infestation in injured trees. *Hilgardia 39*:121–126.

Stech, F. J. 1977. Scientific rivalries: a sign of vitality. *Science 196*:830–831.

Stephen, F. M., and D. L. Dahlsten. 1976a. The arrival sequence of the arthropod complex following attack of *Dendroctonus brevicomis* (Coleoptera: Scolytidae) in ponderosa pine. *Can. Entomol. 108*:282–304.

Stephen, F. M., and D. L. Dahlsten. 1976b. The temporal and spatial arrival pattern of *Dendroctonus brevicomis* in ponderosa pine. *Can. Entomol. 108*:271–282.

Tilden, P. E. 1976. Behavior of *Dendroctonus brevicomis* near sources of synthetic pheromones in the field. M.S. thesis, Univ. of California, Berkeley, pp. 1–66.

Tilden, P. E., W. D. Bedard, D. L. Wood, K. Q. Lindahl, and P. A. Rauch. 1979. Trapping the western pine beetle at and near a source of synthetic attractive pheromone: Effects of trap size and position. *J. Chem. Ecol. 5*:519–531.

U.S. Forest Service, Forest Service Manual, Interim Directive No. 2 (December 1, 1976). Chapter 5280—Forest Insect and Disease Control Administration.

Waters, W. E. 1974. Systems approach to managing pine bark beetles, in T. L. Payne, R. N. Coulson, and R. C. Thatcher (eds.), Southern pine beetle symposium, March 7–8, 1974. Texas A&M University and U.S. Forest Service, pp. 12–14.

Waters, W. E., and B. Ewing. 1976. Development and role of predictive modeling in pest management systems—insects, in R. L. Tummala, D. L. Haynes, and B. A. Croft (eds.), *Modeling for Pest Management: Concepts, Techniques, and Applications.* Michigan State Univ., East Lansing, Mich., pp. 19–28.

Whitney, H. S., and F. W. Cobb, Jr. 1972. Non-staining fungi associated with the bark beetle *Dendroctonus brevicomis* (Coleoptera: Scolytidae) on *Pinus ponderosa. Can. J. Bot. 50*:1943–1945.

Wood, D. L. 1972. Selection and colonization of ponderosa pine by bark beetles, in H. F. Van Emden (ed.), *Insect/Plant Relationships.* Blackwell Sci. Publ., Oxford, pp. 101–117.

Wood, D. L. 1973. The principles, strategies, and tactics of pest population regulation and control in major crop ecosystems, Progress Report, Vol. II, Intern. Center for Integrated and Biological Control, Univ. of California, Berkeley, pp. 446–468.

Wood, D. L. 1974. The principles, strategies, and tactics of pest population regulation and control in major crop ecosystems, Progress Report, Vol. II, Intern. Center for Integrated and Biological Control, Univ. of California, Berkeley, p. 692–697.

Wood, D. L. 1977. Manipulation of forest insect pests in H. H. Shorey, and J. J. McKelvey, Jr. (eds.), *Chemical Control of Insect Behavior: Theory and Application.* Wiley, New York, pp. 369–384.

Wood, D. L., and W. D. Bedard. 1977. The role of pheromones in the population dynamics of the western pine beetle. Proc. XV Intern. Congress of Entomol., Washington, D.C., 1976, pp. 643–652.

14

APPROACH TO RESEARCH AND FOREST MANAGEMENT FOR SOUTHERN PINE BEETLE CONTROL

R. N. Coulson

Department of Entomology
Texas A & M University, College Station, Texas

W. A. Leuschner

Department of Forestry and Forest Products
Virginia Polytechnic Institute and State University,
Blacksburg, Virginia

J. L. Foltz

Department of Entomology
Texas A & M University, College Station, Texas

P. E. Pulley

Data Processing Center
Texas A & M University, College Station, Texas

F. P. Hain

Department of Entomology
North Carolina State University, Raleigh, North Carolina

T. L. Payne

Department of Entomology
Texas A & M University, College Station, Texas

INTRODUCTION

Pine forests were selected as one of the original crop systems to be investigated in the Integrated Pest Management Program (IPM). Emphasis was directed to bark beetles, as these insects are often important mortality agents in mature forests. Three pine forest types and the predominant bark beetle species associated with each were identified for study and included the western pine beetle (WPB), *Dendroctonus brevicomis* LeConte, occurring in ponderosa pine, *Pinus ponderosa* Lawson; the mountain pine beetle (MPB), *D. ponderosae* Hopkins, occurring in lodgepole pine, *P. contorta* Douglas, var. *latifolia* Englemann; and the southern pine beetle (SPB), *D. frontalis* Zimmerman occurring in loblolly pine, *P. taeda* L.

A similar format was developed and adopted for investigating each of the three host-insect systems, but the specific approaches taken and subject areas of initial emphasis varied as a function of existing background information and the resources available.

The objectives of this chapter are (1) to trace briefly the history and structure of the southern pine beetle program and (2) to focus on the approaches taken and the accomplishments realized in the component areas of impact definition and populations dynamics and sampling.

HISTORY AND STRUCTURE OF
THE INTEGRATED PEST MANAGEMENT PROGRAM
FOR SOUTHERN PINE BEETLE

The basic management systems structure incorporating forest stand dynamics, pest population dynamics, impact, and treatment components was adopted for use in each of the three bark beetle subprojects (WPB, MPB, and SPB) of the International Biological Program (IBP)-initiated project funded by NSF and EPA. This systems structure was first reported in the literature by Waters (1974) and was later illustrated and discussed in detail by Stark (1975) and, more briefly, in Chapter 12 herein.

Originally the Southern Pine Beetle project focused on only two components of the system: impact and population dynamics. The impact analysis was initially undertaken at Duke University and later transferred to personnel of Virginia Polytechnic Institute and State University. Emphasis of the study was directed to impact in the areas where the population dynamics studies were being conducted in east Texas. The population dynamics effort was centered at Texas A&M University and had two principal components: one dealing with quantitative aspects of within- and between-tree population dynamics and the development of quantitative sampling plans for $D. frontalis$, and the other directed to the pheromone behavior of the beetle. The first effort was supported through this NSF-EPA program and the second through a cooperative aid agreement with the United States Forest Service, Texas A&M University, and Stephen F. Austin State University.

In 1974 the United States Department of Agriculture " Expanded Southern Pine Beetle Research and Application Program " (ESPBRAP) was funded and the International Biological Program (IBP) initiated, and the NSF-EPA supported program of research on integrated pest management (IPM) was absorbed into the program. The ESPBRAP adopted the basic management system structure, developed in the IBP-IPM program, as their basic organizational format. The ESPBRAP originally funded 48 proposals representing efforts in twelve universities, four United States Forest Service experiment stations, and two state forestry organizations.* With the influx of new funding and scientific expertise resulting from the ESPBRAP, investigations of the remaining components in the management systems structure not covered in the original IBP-IPM program, that is, treatments and stand dynamics, were initiated. The program goals of the ESPBRAP were expressed in terms of "application" needs and included the following

*Information regarding the Expanded Southern Pine Beetle Research and Applications Program (ESPBRAP) may be obtained from Dr. Robert C. Thatcher, Program Director, Southern Forest Experiment Station, 2500 Shreveport Hwy., Pineville, LA 71360.

categories: (1) detection, (2) evaluation, (3) prediction, (4) suppression, and (5) prevention. In the discussion that follows emphasis has been placed on the approaches taken and accomplishments realized in the studies of impact and population dynamics initiated in the IBP-IPM program.

IMPACT

The concept of insect impact on forest ecosystems evolved throughout the IBP (NSF-EPA) program of integrated pest management, from one which was primarily timber oriented to one which contains broad ecological, economic, and philosophical implications. These views have been discussed by R. W. Stark and W. A. Leuschner in presentations at the XV International Congress of Entomology, and will not be pursued here. Rather we will trace some of this evolution as it occurred under the IBP-initiated effort and continued under the ESPBRAP.

IBP-IPM funding of *D. frontalis* impact studies began in 1973. Funds were limited, so activities were restricted to east Texas and the financial aspects of impact. East Texas has been chronically attacked by *D. frontalis* and presumably would have the best impact data base. The first study examined expenditures made to control *D. frontalis* during 1971 and 1972, generally considered to be endemic and epidemic years, respectively. The study found that about $185,000 were spent on detection and control in 1971, whereas about $391,000 were spent in 1972 (Leuschner, Newton, & Neal, 1974). About 90 percent of these funds came from federal and state sources in 1971 but only two-thirds from these sources in 1972, thereby indicating the private sector's willingness and flexibility to meet extreme conditions. Timber losses were estimated also, but because differences of up to 25 percent between two independent data sources were obtained, no further analyses were attempted. About $871,000 were spent on *D. frontalis* research during these two years by organizations located in east Texas. Over 90 percent of these funds were spent by the Boyce Thompson Institute and the Texas Forest Service and over 75 percent of the funds expended were provided by the Southern Forest Research Institute, the Texas Legislature, and the National Science Foundation.

Concurrently, a conceptual framework was being constructed to define the various benefits from forest insect control in general. Plans were to then examine the applicability of these various benefits to *D. frontalis* control. A benefit-cost framework was chosen because pest management programs are usually undertaken by public agencies, or at least with public funding, and many of the program benefits accrue to the public at large, irrespective of land ownership. The benefits of control in this framework are the losses which a management program prevents; therefore the definition of benefits also defines specific elements of impact. In general, the following categories were identified: (1) timber, (2) recreation, (3) esthetic, (4) watershed, (5) soils, (6) wildlife, and (7) grazing (Leuschner & Newton, 1974).

With the funding of ESPBRAP and a year of bark beetle experience the study was broadened to include not only a benefit-cost approach but also examination of impacts on income and wealth distribution and construction of spot and stand projection models to assess likely future impacts (Fig. 14.1). The spot projection model, FRONSIM, simulates timber mortality damages in future years over large geographical areas. FRONSIM estimates the number of southern pine beetle spots occurring in a year by summing monthly spot number estimates generated by either a lognormal distribution of random numbers having the same mean and variance as the historical spot numbers in that month or a regression model which includes number of spots in the preceding month, mean monthly temperature, and month of the year as independent variables (Leuschner, Matney, & Burkhart, 1977).

The number of spots per year are then multiplied by a frequency distribution of the number of trees per spot to estimate total number of trees, and the total number of trees are multiplied by a dbh frequency distribution to estimate number of trees in each dbh class. The volume of timber damaged is estimated using a local timber volume table, and a dollar value placed upon it. The procedure is repeated for any number of years and iterations and the

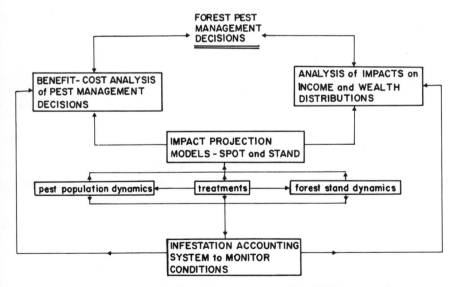

Figure 14.1. Interrelationships of the components of the SPB impact project.

present value of the mean damage in each future year is calculated and summed over all years. Damage reduction patterns for proposed management programs can also be included to calculate benefits for use in benefit-cost analyses.

Using and testing FRONSIM required several frequency distributions which were previously unavailable. Therefore, a descriptive study of host and site variables was made in east Texas which measured 477 spots and 4956 attacked trees (Leuschner et al., 1976). Frequency distributions of dbh, merchantable height, basal area, drainage-soil type, pine basal area ratio, trees per spot, and acres per spot were examined. It was found that pine trees attacked by *D. frontalis* occurred in denser, purer stands and on fairly moist sandy loam soils and had larger than average diameters.

A stand simulation model which describes the growth and competitive interaction of individual trees in managed loblolly pine plantations was also developed (Daniels & Burkhart, 1975). The simulator, PTAEDA, has an initial generation stage and a growth and stand dynamics stage. The initial generation stage allows several spacing options, applies juvenile mortality functions, and assigns dbh to live trees using a Wiebull distribution. Tree height and crown length are then predicted based on dbh.

Trees are grown annually as a function of their size, the site quality, and competition from their neighbors in the growth and stand dynamics stage of the simulator. Growth increments are adjusted by stochastic elements representing genetic and microsite variability. Mortality is generated stochastically through Bernoulli trails. Management subroutines are included to simulate tree and stand response to site preparation, thinning, and fertilization.

PTAEDA is currently being adapted and refined to simulate seeded stands and the probability of *D. frontalis* attack and spread as a function of site and stand parameters. This approach should provide more precise estimates of potential damages and the value of their prevention. However, it does require a great deal more information than the spot model technique.

In addition, we are currently working on a methodology for estimating the economic impact on reservoir recreation and have developed a Hotelling-Clawson demand model for east Texas reservoirs; an assessment of impact on landscape preferences on the Blue Ridge Parkway using the Thurstone Method of Paired Comparisons and semantic differential tests; and modeling *D. frontalis*'s hydrologic impact using existing hydrologic models.

Emphasis has been placed on the "benefits" side of benefit-cost analysis because this is where the potentially more difficult problems lie and because cost assessment must await development of control tactics by other parts of the program. However, we will provide tested methodologies for future analysts to use to assess the merits of their individual cases.

POPULATION DYNAMICS AND SAMPLING

There have been two basic goals of the population dynamics aspect of the project, which included (1) the development of a population dynamics model for *D. frontalis* and (2) the development of quantitative estimation procedures for the insect. This research has been conducted cooperatively with members of the Biosystems Research Division, Department of Industrial Engineering, Soil and Crop Sciences Department, and the Entomology Department at Texas A&M University. The basic approach taken is illustrated in Figure 14.2 and consisted of a blend of both descriptive and biophysical submodels. The following discussion is directed primarily to the activities associated with the center and right hand sections of Figure 14.2, which were initiated in the IBP-IPM program. The activities associated with the left hand portion of Figure 14.2 are being pursued in the ESPBRAP program in the TAMBEETLE project at Texas A&M University.

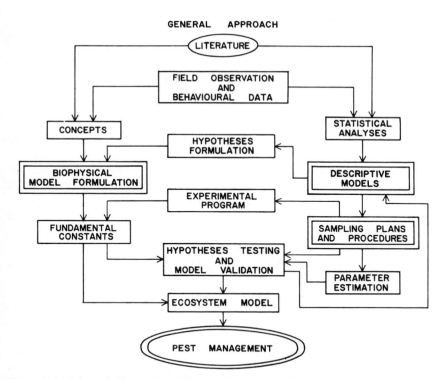

Figure 14.2. Schematic illustration of the general approach utilized in conducting research pursuant to the goal of developing statistical and biophysical models of *D. frontalis* populations.

At the onset of the IBP-IPM program a considerable body of information existed in the published literature on various aspects of the population ecology of *D. frontalis*. This literature has been reviewed by Thatcher (1960), Dixon and Osgood (1961), and Coulson et al. (1972), and has served as the basis for identification of pertinent processes in the population ecology of the insect needed for model development.

The general life history of the insect as excerpted from this literature begins with the selection of suitable host trees by adults through either random or directed behavior (the exact process is unknown). Colonization of trees is regulated by a blend of both insect-produced pheromones and host-produced attractants. Females initiate the construction of "egg galleries" by boring into the inner bark region where they are joined by males. Mating takes place within the galleries and eggs are oviposited in niches at intervals along the lateral walls of the galleries. Both the males and females re-emerge and are capable of attacking and colonizing new host trees. Eggs hatch shortly after oviposition and the ensuing larvae begin to excavate "larval galleries" at right angles away from the egg galleries. There are four larval instars. The first two remain in and feed in the phloem region and the last two migrate into the outer corky bark where pupation and adult emergence occur. Development can be completed in as few as 35 days, and six to eight generations per year occur in some regions of the South. Numerous parasites, predators, and associates have been identified as occurring in the tree with *D. frontalis*, but their effects on within-tree populations have not been quantitatively defined.

Infestation spots often enlarge dramatically in a short period of time, are characterized by an active front or head, and are comprised of multiple asynchronous generations of the insect developing concurrently. Both re-emerging and brood adults are involved in the colonization of trees. Various site and stand conditions have been associated with the occurrence of *D. frontalis*. Prominent correlates include soil type, stand density, and stand disturbances, species composition, radial growth, and stand size.

There were no studies reported in the literature dealing with the development of quantitative sampling methodologies for *D. frontalis*. Radiographic techniques have been tested and found suitable for identification of adults, late-stage larvae, and pupae, as well as several predators and associates.

Based on the reports in the literature, the following within-tree processes were identified to be in need of further detailed study: colonization, re-emergence, brood development and survivorship, and emergence. Concurrent with these studies would be the accumulation of data in sufficient quantity to develop and test various sampling plans which would subsequently be used to assess between-tree population dynamics.

Development of statistical models of the within-tree processes was oriented to provide a better understanding of the structure of within-tree populations of *D. frontalis* and also to provide density and distribution functions and rate constants needed for the development of biophysical models.

Emphasis in the remainder of this discussion is directed to the approach taken and the results realized in the development of statistical models of *D. frontalis* populations and the synthesis of quantitative sampling methodologies for the insect. Work on pheromone behavior, models of the host system, and macro- and micro-environmental needs is underway and is not treated here.

The general approach taken was to proceed in steps from the simple to the complex. There were several basic activities associated with this approach. These included (1) gathering the data on within-tree populations, (2) data file management, (3) functional descriptions of the *D. frontalis* life stages, (4) development of needed analytical methodologies, (5) development of within-tree and within-spot sampling methodologies, and (6) description of within-tree population processes. The components investigated within each of these six activities are illustrated in Figure 14.3 and each is discussed in turn below.

Data Base Accumulation

Methods and procedures for collecting data on within-tree populations of bark beetles have been researched by scientists at the University of California, Berkeley, and the United States Forest Service Pacific Southwest Forest and Range Experiment Station, and reported in detail by Stark and Dahlsten (1970). The format used for collecting within-tree data on *D. frontalis* was adapted from the basic procedure used for estimating within-tree populations of *D. brevicomis*. Although the procedures were modified somewhat during the course of the study, the basic routine consisted of extracting 100 cm^2 bark disks from standing trees. Four disks were removed at the NE, NW, SE, and SW aspects at 1.5 m intervals beginning at 2.0 m from the base of the tree and continuing to the top of the infested bole. Cups coated internally with Stickem Special® were placed on the tree prior to the beginning of adult emergence. The catch in the sticky cups was used together with the count obtained from the disk below the cup in order to obtain a measure of emergence. The disks were placed in rearing containers.

Each tree was sampled three times to characterize attacking adults, eggs, larvae, pupae-callow adults, and emerging adults. In addition basic tree parameters such as dbh, tree height, and infested bole height, were measured. Radiographic techniques were used to detect the various life stages of *D.*

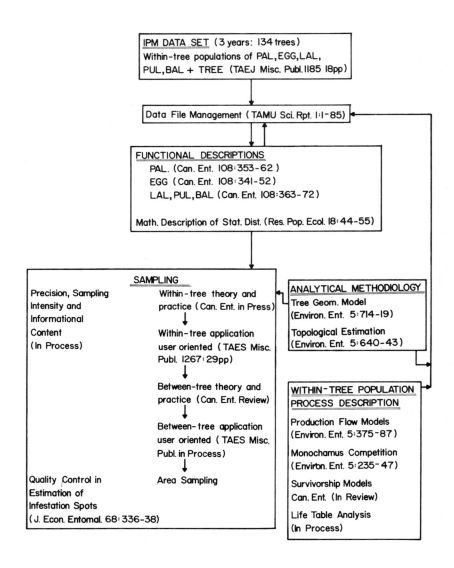

Figure 14.3. The basic components and accomplishments in the SPB population dynamics and sampling project.

frontalis. The data were recorded so as to permit machine processing. Details regarding the methods and procedures used in both the field and laboratory phases of the study have been described and illustrated (Coulson et al., 1975a).

As bark beetle populations in general are characterized by a tremendous inherent variation, the data collection phase of the study was extended over three summer field seasons. In all, 134 trees were sampled from 11 plots located in southeast Texas.

Data collected in the manner described above provide point estimates of peak density of the various life stages and are suitable for development of partial life tables, description of spatial distributions, and development of sampling methodologies. The data collection procedures for studying the processes of colonization, re-emergence, and emergence were modified to provide daily estimates of population density. For the colonization study, bark disks were removed daily along the infested bole for the entire period of colonization. For re-emergence and emergence, cage traps were placed on the trees and insects were collected daily. The same heights and aspects were sampled as described above. Data collected in this format provide estimates of density at various heights along the tree bole through time and is particularly useful in describing the distribution functions needed for model development.

Data File Management

Six basic data sets were accumulated for each of the 134 trees sampled using the first data collection format: $PAL (attacking adults), $EGG (eggs), $LAL (larvae), $PUL (pupae-callow adults), $BAL (brood adults), and $TREE (tree parameters). These data were keypunched, formatted in a manner convenient for perusal, printed, and verified. A rather cumbersome storage and retrieval system was initially developed which permitted either random or sequential access to the data. This system was later abandoned and a new system developed to accommodate α, β, and γ level files and to emphasize ease of access for users. Users' guides were prepared for both systems.

Functional Descriptions of *D. frontalis* Life Stage Distributions

With the data in a computer-accessible form, the first analytical activities involved the description of the density of the various life stages in relation to the infested portion of the tree. Two- and three-parameter models were selected for this purpose and nonlinear regression techniques were employed.

Attacking adults, larvae, pupae-callow adults, and brood adults were described by two-parameter models (Coulson et al., 1976a; Mayyasi et al., 1976a) and eggs by a three-parameter model (Foltz et al., 1976a). Figure 14.4 illustrates the general distribution pattern along the infested bole for the various life stages. The functional distributions were mathematically described by Mayyasi et al. (1976b). The analyses not only provided a view of the spatial distribution of the various life stages along the infested bole but also permitted estimation of within-tree population density. Furthermore, the results of these model fits were used (1) in defining fundamental features of the structure of the within-tree population systems, (2) in developing partial life tables, (3) as a basis for one of the quantitative estimation procedures described below, and (4) in evaluating potential within-tree treatment effects (Coulson et al., 1975b). An example of the type of information

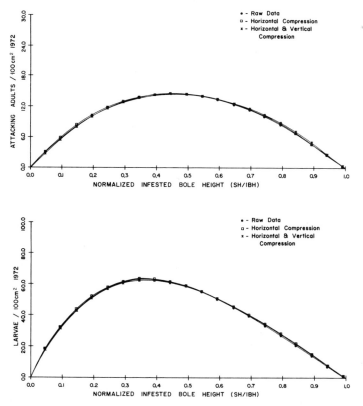

Figure 14.4. Functional distributions of *D. frontalis* life stages in relation to the normalized infested bole.

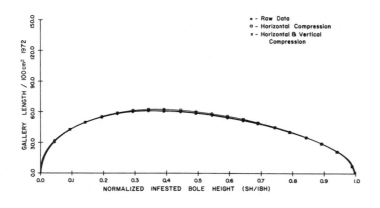

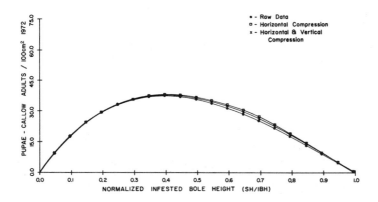

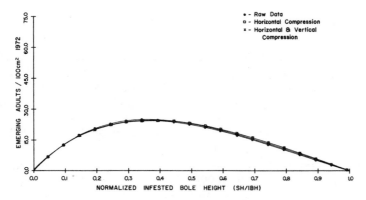

Figure 14.4. (continued)

461

Comparative Life Tables for Cacodylic Acid, Control, and Frontalure Treatments (1972).

Stage	Cacodylic Acid No./dm²	%Mortality	Control No./dm²	%Mortality	Frontalure No./dm²	%Mortality
Attacking Adults	3.84		5.94		7.38	
(Eggs / ♀)	(28.54)		(32.15)		(28.80)	
Eggs	54.80		95.49		106.26	
Larvae	8.99	84	32.29	66	34.42	68
Pupae-Callow Adults	2.07	77	21.69	33	20.28	41
Brood Adults	1.52	27	14.95	31	14.47	29
Generation		97.23		84.34		86.38
Increase Ratio	0.40		2.52		1.96	

Figure 14.5. Comparative partial life tables developed from the two and three parameter models of the various *D. frontalis* life stages for three classes of trees: cacodylic acid injected, frontalure baited, and untreated controls.

obtainable from the model fits for the various life stages is illustrated in Figure 14.5. These partial life tables were constructed to illustrate the morality occurring in within-tree populations resulting from injections of cacodylic acid relative to that occurring in untreated "control" trees and trees baited with attractants.

Analytical Procedures

One of the fundamental problems encountered in developing quantitative sampling plans and describing within-tree populations of bark beetles has been the absence of a procedure for summarizing population measurements without introducing errors of undefined magnitude (e.g., as with the regression procedures described above). Pulley et al. (1976) solved this problem by developing a topological mapping procedure for estimating total within-tree populations. The data requirements for using the procedure consist of the insect densities at measured intervals along the bole. The procedure minimizes errors resulting from smoothing, regression, or extrapolation, and provides estimates suitable for evaluating the bias and variability of other sampling plans and for use in population modeling.

Another common obstacle in developing sampling plans and studying population processes has been the absence of a simple procedure for estimating habitat area and volume. Most of the procedures available are data intensive and, therefore, of limited use. Foltz et al. (1976b) developed regression equations for predicting diameter outside bark and bark thickness as a function of height and diameter or bark thickness class. The two equations were combined and integrated over height to provide estimates of the phloem area and bark volume infested by *D. frontalis*. These models are extremely useful, as the only data requirements for use are diameter and bark thickness at 2.0 m and the height of the infested portion of the tree. Coulson et al. (1976b) provide tables of bark surface area for three bark thickness classes over a range of tree diameters and heights.

Both the topological estimation procedure and the tree geometry model have been used extensively in the development of sampling plans and description of population processes for *D. frontalis*.

Development of Within-Tree and Within-Spot Sampling Methodologies

With the data base on *D. frontalis* populations, the functional and mathematical descriptions of the life stage distribution, and the methodology described above, many options become available for the development of quantitative estimation procedures. The fundamental reasons for the development of sampling plans are to (1) evaluate treatment tactics, (2) survey populations, and (3) study and describe population processes. Obviously, the requirements for precision in sampling vary with the intended use of the estimate obtained. The approach taken in the development of both within-tree and within-spot estimation plans was to survey possible options and appraise the costs and benefits associated with each option. Accuracy and precision of the procedures were evaluated in relation to the data requirements. Based on this survey of possible plans, those procedures which provided suitable accuracy and precision for certain applications were selected and described in detail in a format oriented to entomologists or foresters.

For example, in surveying the possible options for estimating within-tree population, five tree surface area estimation procedures and five beetle estimation procedures were investigated. The different procedures required varying amounts of data and were classed as "small" and "large" sample plans. Accuracy, precision and bias for each of the plans were defined by comparison with topological estimates (Pulley et al., 1977a,b). Two of the "small" sample procedures were then described in detail for each life stage.

All tabular material needed for using the procedures, as well as step-by-step instructions have been published (Coulson et al., 1976b).

Sampling plans for within-spot estimation were developed using the same procedural format described for within-tree populations. The basic survey of potential procedures was described by Pulley et al. (1976c) and the instructional guide by Foltz et al. (1977).

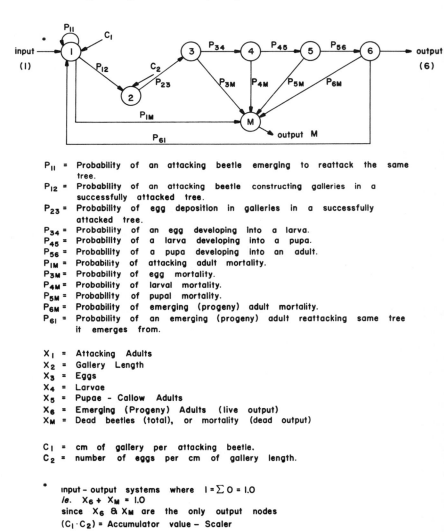

P_{11} = Probability of an attacking beetle emerging to reattack the same tree.

P_{12} = Probability of an attacking beetle constructing galleries in a successfully attacked tree.

P_{23} = Probability of egg deposition in galleries in a successfully attacked tree.

P_{34} = Probability of an egg developing into a larva.

P_{45} = Probability of a larva developing into a pupa.

P_{56} = Probability of a pupa developing into an adult.

P_{1M} = Probability of attacking adult mortality.

P_{3M} = Probability of egg mortality.

P_{4M} = Probability of larval mortality.

P_{5M} = Probability of pupal mortality.

P_{6M} = Probability of emerging (progeny) adult mortality.

P_{61} = Probability of an emerging (progeny) adult reattacking same tree it emerges from.

X_1 = Attacking Adults
X_2 = Gallery Length
X_3 = Eggs
X_4 = Larvae
X_5 = Pupae - Callow Adults
X_6 = Emerging (Progeny) Adults (live output)
X_M = Dead beetles (total), or mortality (dead output)

C_1 = cm of gallery per attacking beetle.
C_2 = number of eggs per cm of gallery length.

* input – output systems where $I = \sum O = 1.0$
ie. $X_6 + X_M = 1.0$
since X_6 & X_M are the only output nodes
$(C_1 \cdot C_2)$ = Accumulator value – Scaler

Figure 14.6. Production flow model structure for within-tree populations of *D. frontalis*.

Description of Within-Tree Population Processes

One of the main goals of the project was the development of statistical models of within- and between-tree population processes, and the work described above is part of the foundation on which these models will be based. The basic processes of major interest are host-tree colonization, re-emergence of parent adults, within-tree survivorship, and emergence and dispersal of new brood adults. To this point primary emphasis has been placed on within-tree survivorship; the remaining processes are being researched in the ESPBRAP.

The first phase in the investigation of within-tree populations was the construction of production flow models for the three years of population data. These models illustrate mathematically and graphically the mortality and transition probabilities for the various life stages (Coulson et al., 1976c). An example of the production flow model constructed for *D. frontalis* is illustrated in Figure 14.6. The initial flow models do not identify the causes of mortality but identify where in the life cycle of *D. frontalis* it is occurring. Figure 14.7 provides the transition probabilities between life stages for *D.*

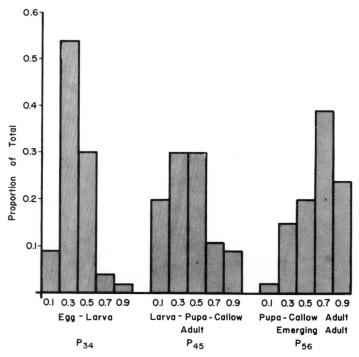

Figure 14.7. Transition probabilities between life stages of *D. frontalis* based on 46 flow models developed from trees selected from the three year study.

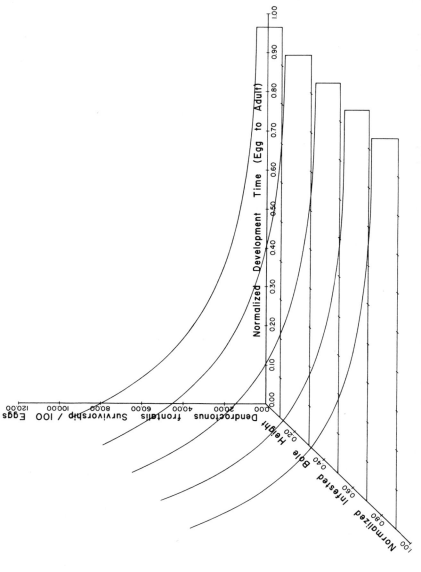

Figure 14.8. Survivorship curves for within-tree populations of *D. frontalis* through time at intervals along the infested tree bole

frontalis. These histograms were developed from 46 individual flow models selected from the three year study.

Another more detailed view of within-tree survival was obtained by modelling survivorship for various sections of the tree bole (Coulson et al., 1977). These statistical models provide insight into the basic structure of within-tree populations and were used to appraise the probable contribution of parasites and predators as suppressants or regulators of *D. frontalis* populations. Survivorship curves for *D. frontalis* at intervals along the infested bole are illustrated in Figure 14.8.

Of the many parasites, predators, and associates of *D. frontalis, Monochamus titillator* appeared to be of major significance in reducing within-tree populations of *D. frontalis.* The role of this insect as an interspecific competitor was experimentally verified and quantitatively defined (Coulson et al., 1976d).

The processes of within-tree colonization, re-emergence of parent adults, and emergence of brood adults are currently being described and developmental rate constants, measured in both the laboratory and field, have been determined. This information, in conjunction with a series of plant models developed as part of the ESPBRAP, is being used in the synthesis of a general population dynamics model for *D. frontalis.* The quantitative sampling plans for estimating within-infestation populations will be used to test the model under variable site and stand conditions.

In conclusion, we have briefly outlined the approaches and accomplishments of the research on *D. frontalis* initiated in the IBP-IPM program and continuing in the ESPBRAP. As with other insects, research directed to the goal of management of *D. frontalis* is complex, costly, and involves the skills and expertise of academic disciplines far removed from entomology alone.

ACKNOWLEDGMENTS

We are indebted to our colleagues in the WPB and MPB projects of the IPM program for sharing with us the benefit of their counsel, guidance and experience, and to Drs. C. L. Newton and H. E. Burkhart for their contribution to the investigation of impact.

The work reported herein was funded in part by the National Science Foundation project entitled "The Principles, Strategies, and Tactics of Pest Population Regulations and Control in Major Crop Ecosystems," and in part by the U.S. Dept. of Agriculture program entitled "The Expanded Southern Pine Beetle Research and Applications Program," grant numbers CSRS G6163 and CSRS 516-15-58. The findings, opinions and recommendations expressed herein are those of the authors and not necessarily those of sponsoring agencies.

LITERATURE CITED

Coulson, R. N., T. L. Payne, J. E. Coster, and M. W. Houseweart. 1972. The southern pine beetle, *Dendroctonus frontalis* Zimm. 1961–1971. *Texas For. Serv. Publ.* 108.

Coulson, R. N., F. P. Hain, J. L. Foltz, and A. M. Mayyasi. 1975. Techniques for sampling the dynamics of southern pine beetle populations. *Texas Agric. Exp. Stn. Misc. Publ. 1185.*

Coulson, R. N., J. L. Foltz, A. M. Mayyasi, and F. P. Hain. 1975a. Quantitative evaluation of frontalure and cacodylic acid treatment effects on within-tree populations of the southern pine beetle. *J. Econ. Entomol.* 68:671–687.

Coulson, R. N., A. M. Mayyasi, J. L. Foltz, and F. P. Hain. 1976a. Resource utilization by the southern pine beetle. *Can. Entomol.* 108:353–362.

Coulson, R. N., P. E. Pulley, J. L. Foltz, and W. C. Martin. 1976b. Procedural guide for quantitatively sampling within-tree populations of *Dendroctonus frontalis*. *Texas Agric. Exp. Stn. Misc. Publ. 1267.*

Coulson, R. N., A. M. Mayyasi, J. L. Foltz, and P. E. Pulley. 1976c. Production flow system evaluation of within-tree populations of *Dendroctonus frontalis*. *Environ. Entomol.* 5:375–387.

Coulson, R. N., A. M. Mayyasi, J. L. Foltz, F. P. Hain, and W. C. Martin. 1976d. Interspecific competition between *Monochamus titillator* and *Dendroctonus frontalis*. *Environ. Entomol.* 5:235–247.

Coulson, R. N., P. E. Pulley, J. L. Foltz, W. C. Martin, and C. L. Kelly. 1977. Generation survival models for within-tree populations of *Dendroctonus frontalis* (Coleoptera: Scolytidae). *Can. Entomol.* 109:1971–1977.

Daniels, R. F., and H. E. Burkhart. 1975. Simulation of individual tree growth and stand development in managed loblolly pine plantations. FWS-5-75, Division of Forestry and Wildlife Resources, Virginia Polytechnic Inst. and State Univ., Blacksburg, Va.

Dixon, J. C., and E. A. Osgood. 1961. Southern pine beetle. A review of current knowledge. *U.S. For. Serv. S. E. For. Exp. Stn. Pap. No. 128.*

Foltz, J. L., A. M. Mayyasi, F. P. Hain, R. N. Coulson, and W. C. Martin. 1976a. Egg-gallery length relationship and within-tree analyses for the southern pine beetle, *Dendroctonus frontalis* Zimm. (Coleoptera: Scolytidae). *Can. Entomol.* 108:341–352.

Foltz, J. L., A. M. Mayyasi, P. E. Pulley, R. N. Coulson, and W. C. Martin. 1976b. Host–tree geometric models for use in southern pine beetle population studies. *Environ. Entomol.* 5:640–643.

Foltz, J. L., P. E. Pulley, R. N. Coulson, and W. C. Martin. 1977. Procedural guide for estimating within-spot populations of *Dendroctonus frontalis*. *Texas Agric. Exp. Stn. Misc. Publ. 1316, 27 pp.*

Leuschner, W. A., and C. M. Newton, 1974. Benefits of forest insect control. *Bull. Entomol. Soc. Am.* 20:223–227.

Leuschner, W. A., and C. M. Newton, and R. B. Neal. 1974. Impact of the southern pine beetle in E. Texas—1971 & 1972, in T. L. Payne, R. N. Coulson, and R. C. Thatcher (eds.), Proceedings of the southern pine beetle symposium. Texas Agric. Exp. Stn.— U.S. For. Serv., Texas A&M University, College Station, Texas, pp. 22–25.

Leuschner, W. A., H. E. Burkhart, G. D. Spittle, I. R. Ragenovich, and R. N. Coulson. 1976. A descriptive study of host and site variables associated with the occurrence of *Dendroctonus frontalis* Zimm. in East Texas. *The Southwestern Entomol.* *1*:141–149.

Leuschner, W. A., T. G. Matney, and H. E. Burkhart. 1977. Simulating southern pine beetle activity for pest management decisions. *Can. J. For. Res.* *7*:138–141.

Mayyasi, A. M., R. N. Coulson, J. L. Foltz, and F. P. Hain. 1976a. The functional distribution of within-tree larval and progeny adult populations of *Dendroctonus frontalis* (Coleoptera: Scolytidae). *Can. Entomol.* *108*:363–372.

Mayyasi, A. M., P. E. Pulley, R. N. Coulson, D. W. DeMichele, and J. L. Foltz. 1976b. Mathematical descriptions of within-tree distribution of the various developmental stages of *Dendroctonus frontalis* Zimm. Coleoptera: Scolytidae. *Res. Popul. Ecol.* *18*:135–145.

Pulley, P. E., A. M. Mayyasi, J. L. Foltz, R. N. Coulson, and W. C. Martin. 1976. Topological mapping to estimate numbers of bark-inhabiting insects. *Environ. Entomol.* *5*:714–719.

Pulley, P. E., J. L. Foltz, A. M. Mayyasi, R. N. Coulson, and W. C. Martin. 1977a. Sampling procedures for within-tree attacking adult populations of the southern pine beetle, *Dendroctonus frontalis* Zimm. *Can. Entomol.* *109*:39–48.

Pulley, P. E., R. N. Coulson, J. L. Foltz, and W. C. Martin. 1977b. Sampling intensity, informational content of samples, and precision in estimating within-tree populations of *Dendroctonus frontalis*. *Environ. Entomol.* *6*:607–615.

Pulley, P. E., J. L. Foltz, R. N. Coulson, and W. C. Martin. 1977c. Evaluation of procedures for estimating within-spot populations of attacking adult *Dendroctonus frontalis*. *Can. Entomol.* 1325–1334.

Stark, R. W. 1975. Forest insect pest management, in R. L. Metcalf, and W. H. Luckman (eds.), *Introduction to Insect Pest Management*. Wiley, New York, pp. 509–528.

Stark, R. W., and D. L. Dahlsten. 1970. Studies on the population dynamics of the western pine beetle, *Dendroctonus brevicomis* LeConte (Coleoptera: Scolytidae). Univ. Calif. Press, Berkeley.

Thatcher, R. C. 1960. Bark beetles affecting southern pines: A review of current knowledge. *U.S. Dept. Agric. For. Serv., South, For. Exp. Stn. Occas. Pap. 180.*

Waters, W. E. 1974. Systems approach to managing pine bark beetles, in T. L. Payne, R. N. Coulson, and R. C. Thatcher (eds.), Proceedings southern pine beetle symposium, Texas Agric. Exp. Stn.—U.S. For. Serv., Texas A&M University, College Station, Texas, pp. 12–14.

15

SUMMARY

W. E. Waters

Department of Entomological Sciences
and
Department of Forestry and Resource Management
University of California, Berkeley, California

C. B. Huffaker

Division of Biological Control
University of California, Berkeley, California

L. D. Newsom

Department of Entomology
Louisiana State University, Baton Rouge, Louisiana

The foregoing chapters show clearly that significant progress has been made in achieving the primary objective of the NSF-EPA Integrated Pest Management Program, namely the development of improved strategies and tactics for managing major pests, or complexes of such pests, affecting productivity in the six crop ecosystems involved. The specific accomplishments thus far in each crop ecosystem are well described in the respective chapters (3–14), and the main features of both the research conducted and the results obtained are highlighted. The rationale and approach taken by the different sub-projects, and the consequent priorities, have necessarily differed somewhat because of the nature of the specific crop-pest ecosystems concerned, the relevant scientific knowledge and technology available at the start, and the particular objectives of production and marketing relevant to each case.

However, there are fundamental commonalities of concept, objectives, and organization that have been discerned and developed, together with an increasing commitment to goals agreed upon by individual researchers with diverse disciplinary interests and experience.

In this context, several major advances have been made that will serve as a landmark for future research and development of operational integrated pest management systems. The work has entailed a concentration on the insect pests; as a first major effort that seemed necessary. A broadening of such efforts to include other kinds of pests and, indeed, the whole gamut of crop protection, is the logical next step. The general word "pest" is used here insofar as some attention was given to plant diseases, and especially to mites, and because the general advances made and the technologies developed should in general apply to management of whole complexes of pests. The following advances can be enumerated:

1. A focusing of attention on the crop plant as the hub of the system— rather than the *pests* per se.

2. An awareness that the pest management system must be conceptually and operationally linked to the total crop production system. Reciprocally, the optimization of pest managment strategies is dependent to a degree on variables in other components of the production system and on some external conditions and factors affecting the whole system. Thus, the program has taken a major step from the concept and practice of integrated pest *control* as a specialized and more or less independent activity to a higher order of integration, that is, total crop protection systems as an integral part of the total production process.

3. A recognition that the research process, to be efficient, must have an organizational framework that relates to a model structure of the pest management system. The model structure identifies the primary components (i.e., the working groups) and their linkages (i.e., information flows) and indicates the specific kinds of research needed and the priorities in space and time. It is the skeleton, and perhaps the endocrine system, on which the entire multidisciplinary effort is hung.

4. Extensive use of models of many kinds in all aspects of the program— from research planning and execution to actual operations. Descriptive or analytic models are needed to gain an understanding of the complex biological and economic inter-relationships and, ultimately, to reduce the whole to the really significant processes and relationshps and to express these in quantitative terms that are compatible with the management model structure. Modeling in the research phase must be augmented by the use of systems analysis (in the broad sense, from statistical methods to the newest,

most advanced techniques of operations research), and by development of adequate computerized data management resources. Both of these tools are needed in the operational program, where predictive models are the primary mechanism of information flow to decision and action. The latter provide the capability to anticipate and forecast, at least for short term intervals, the occurrence, abundance, population trends, and potential damage of pests, singly or jointly. Such models are needed also to predict the outcomes of alternative control treatments and strategies as a basis for evaluation of benefits and costs and for input to the overall management model. The effective coupling of predictive models in a systems framework for operational use presents problems; it is a still developing aspect of pest management research and its application.

5. Finally, and most important to further advances in pest management research, it has been demonstrated that a large-scale multidisciplinary program involving researchers and technical specialists from both the academic and governmental sectors can be organized and conducted effectively toward a common goal, albeit a broadly defined one. The payoff is obvious and real, in terms both of program and individual achievement. The magnitude and depth of information produced is greatly increased, and the time span to practical applications is significantly shortened, compared to the random walk process of individual research. For the individual participants such a program can provide added resources and incentives which, to those so motivated, can increase productivity and personal recognition.

Objectively, we can say that the NSF-EPA Integrated Pest Management Program has provided both new tools and the stimulus for progress toward the goal of total crop protection systems. It has not provided the complete package for any of the crops under study. In the final year of the program, special emphasis will be given to the synthesis of information gained and reporting this out through a variety of channels to disciplinary peers and to user groups in both the private and public sectors. Similarly, the specific needs for continuing research and implementation in the six crop ecosystems and the justification for expansion to other agricultural and forest resource systems will be spelled out—hopefully, to a listening audience of decision makers. This bequest has already been accepted, in part at least, as exemplified by the ongoing USDA Combined Forest Pest Research and Development Program (Ketcham & Shea, 1977). The challenge and the promise is clear.

LITERATURE CITED

Ketcham, D. E., and K. R. Shea. 1977. USDA combined forest pest research and development program. *J. Forestry* 75:404–407.

Appendix

MANAGEMENT, COORDINATION, AND SPECIAL SERVICE OF OVERALL PROGRAM

Director:

Carl B. Huffaker
University of California, Berkeley

Associate Director:

Ray F. Smith
University of California, Berkeley

Modeling Coordinators:

Christine A. Shoemaker
Cornell University, Ithaca, New York

William G. Ruesink
University of Illinois, Urbana

Not all listed participated for all years of the project. The Editors wish to acknowledge all contributors and apologize for any oversights or omissions. (Affiliations shown here are those at time of initial participation.)

Staff:

Joe C. Ball
Dale G. Bottrell
Donald J. Calvert
M. Jane Clarkin
Nettie M. Mackey
University of California, Berkeley

Steering Committee:

C. B. Huffaker, Chmn. (*1972–1977**), *Univ. Calif., Berkeley*
P. L. Adkisson (*1972–1977*), *Texas A&M Univ.*
J. L. Apple, *North Carolina State Univ. at Raleigh*
E. J. Armbrust (*1976–1977*), *Univ. Illinois*
D. M. Baker, *Clemson Univ.*
S. D. Beck (*1972*), *Univ. Wisconsin*
J. R. Brazzel, *USDA, APHIS*
G. A. Carlson, *North Carolina State Univ. at Raleigh*
J. J. Cartier, *Canada Dept. Agric. Res. Br.*
B. A. Croft (*1975–1977*), *Univ. Michigan*
P. DeBach, *Univ. Calif., Riverside*
D. W. DeMichele, *Texas A&M Univ.*
T. W. Fisher, *Univ. Calif., Riverside*
E. H. Glass (*1973–1977*), *New York State Agric. Expt. Sta.*
J. M. Good, *USDA, Extension Service*
A. P. Gutierrez (*1975–1977*), *Univ. of Calif., Berkeley*
G. E. Guyer (*1972*), *Michigan State Univ.*
J. C. Headley, *Univ. Missouri*
C. S. Holling, *Univ. Br. Columbia*
S. C. Hoyt, *Washington State Univ.*
W. Klassen, *USDA, ARS*
E. F. Knipling, *USDA, ARS*
E. G. Le Roux, *Canada Dept. Agric.*
M. F. Martignoni, *USDA, For. Sci. Lab.*
L. D. Newson (*1973–1977*), *Louisiana State Univ.*
R. L. Rabb, *North Carolina State Univ. at Raleigh*
L. A. Riehl (*1975–1977*), *Univ. Calif., Riverside*
R. C. Riley, *USDA, CSRS*

*Indicates member Executive Committee for years noted.

R. C. Scott, *USDA, Extension Service*
W. C. Shaw, *USDA, ARS*
R. F. Smith (*1972–1977*), *Univ. Calif., Berkeley*
R. W. Stark (*1972–1973*), *Univ. Idaho*
W. E. Waters (*1974–1977*), *USDA, Forest Service*
W. H. Whitcomb, *Univ. Florida*

Technical Panel:

L. G. Brown, *Mississippi State Univ.*
J. J. Cartier, *Canada Dept. Agric. Res. Br.*
C. B. Craig, *Univ. Notre Dame*
D. W. DeMichele, *Texas A&M*
B. Ewing, *Univ. Calif., Berkeley*
G. W. Fick, *Cornell Univ.*
A. P. Gutierrez, *Univ. Calif., Berkeley*
J. C. Headley, *Univ. Missouri*
C. M. Ignoffo, *USDA, ARS*
R. D. Lacewell, *Texas A&M Univ.*
F. R. Lawson, *USDA, ret.*
M. E. Martignoni, *USDA, For. Sci. Lab.*
W. L. Myers, *Michigan State Univ.*
R. B. Norgaard, *Univ. Calif., Berkeley*
B. C. Pass, *Univ. Kentucky*
W. E. Robbins, *USDA, ARS*
G. A. Rowe, *Univ. Calif., Berkeley*
W. G. Ruesink, *Univ. Illinois*
R. I. Sailer, *USDA, ARS*
R. C. Scott, *USDA, Extension Service*
C. A. Shoemaker, *Cornell Univ.*
R. W. Skeith, *Univ. Arkansas*
A. R. Stage, *U.S. Forest Service*
W. J. Turnock, *Canada Dept. Agric. Res. Br.*

ALFALFA SUBPROJECT

Subproject Director:

E. J. Armbrust, *University of Illinois, and Illinois Natural History Survey*
(*1972–1977*)

Principal Investigators:

University of California
> C. G. Summers (*1973–1977*)
> V. M. Stern (*1972*)

Cornell University
> R. G. Helgesen (*1973–1977*)
> G. G. Gyrisco (*1972*)

University of Illinois and Illinois Natural History Survey
> E. J. Armbrust (*1972–1977*)
> W. G. Ruesink (*1973–1976*)

University of Kentucky
> B. C. Pass (*1972–1976*)

University of Nebraska
> D. G. Hanway (*1972–1976*)

Utah State University
> D. W. Davis (*1972–1976*)

Virginia Polytechnic Inst. and State Univ.
> R. L. Pienkowski (*1972–1976*)

Participants and Cooperators:

University of California: J. Andrews, J. Christensen, W. Cothran, R. L. Coviello, F. David, L. Etzel, C. E. Franti, D. Gonzalez, A. Gutierrez, W. Jordan, C. Koehler, K. Lee, W. Lehman, W. L. Loew, W. D. McClellan, C. M. Merritt, G. Moore, R. Norgaard, U. Regev, V. Sevacherian, R. Sheesley, V. Stern, R. Stoltz, C. Summers, and R. van den Bosch.

Cornell University, New York: C. Bremer, N. Cooley, B. Dethier, S. Ditmar, G. Fick, G. G. Gyrisco, R. Helgesen, B. Lin, C. Lowe, R. Murphy, C. Shoemaker, G. Smith, M. Tauber, I. Vignarajah, and M. J. Wright.

University of Illinois and Illinois Natural History Survey: E. Armbrust, D. Bartell, J. Bouseman, R. Bula, S. G. Carmer, R. Cherry, S. Cowen, F. Frosheisier, R. Giese, G. Godfrey, D. W. Graffis, H. Hildebrand, C. N. Hittle, R. Huber, B. Irwin, J. A. Jackobs, M. Kogan, R. Kotek, E. Lichenstein, J. Maddox, D. Miller, M. Nichols, R. Pausch, B. Peterson, P. W. Price, S. Roberts, M. Rose, W. Ruesink, L. Stannard, E. Swanson, C. Taylor, C. E. White, M. Wilson, and M. J. Wright.

University of Kentucky: P. A. Broomal, D. L. Dahlman, S. Diachun, M. Latheef, G. Mitchell, V. Montgomery, W. Morrison, J. Parr, B. Pass, T. Taylor, W. Templeton, C. Thompson, Jr., R. Thurston, R. Tucker, W. Wiggins, A. Young, and K. V. Yeargan.

University of Nebraska: E. Dickason, D. Hanway, W. Kehr, S. Kindler, L. Klostermeyer, J. Kugler, G. Manglitz, R. Ogden, L. Palmer, and J. K. Skinner.

Purdue University: R. Bula, G. Miles, and M. C. Wilson.

Utah State University: K. Allred, G. Bohart, W. A. Brindley, D. W. Davis, W. G. Dewey, B. A. Haws, T. H. Hsiao, L. Jensen, D. R. McAllister, W. P. Nye, M. W. Pedersen, and N. N. Youssef.

Virginia Polytechnic Institute and State University: R. L. Pienkowski, and D. D. Wolf.

U.S. Department of Agriculture: W. Day, M. Pederson, P. F. Torchio, and R. Ratcliffe.

CITRUS

Subproject Director:

L. A. Riehl, *University of California, Riverside* (*1977*)

L. A. Riehl, *University of California, Riverside*
and
W. H. Whitcomb, *University of Florida*
  Co-directors: 1975, 1976

T. W. Fisher, *University of California, Riverside*
and
Paul DeBach, *University of California, Riverside*
  Co-directors: 1974

Paul DeBach, *University of California, Riverside* (*1972, 1973*)

Principal Investigators:

University of California
  Paul Debach (*1972–1974*)
  T. W. Fisher (*1973–1976*)

R. F. Luck (*1977*)
L. A. Riehl (*1975, 1976*)

University of Florida

J. C. Allen (*1977*)
R. F. Brooks (*1973–1976*)
W. H. Whitcomb (*1972, 1974–1976*)

Texas A&M University

H. A. Dean (*1972–1973, 1975–1976*)

Participants and Cooperators:

University of California: D. Baasch, J. Black, R. Brewer, R. Burns, E. Calavan, F. David, M. Clark, A. Deal, P. DeBach, R. Doutt, C. Elder, H. Elmer, W. Ewart, T. Fisher, D. Flaherty, H. Francis, K. Greene, K. Hagen, K. Hench, M. Huftile, L. Jeppson, L. Jordan, C. Kennett, D. Lewis, R. Luck, J. McMurtry, D. Mead, A. Mostafa, K. Opitz, R. Orth, G. Oster, J. Pehrson, R. Platt, J. Quezada, L. Riehl, V. Sevacherian, J. C. Shaw, J. Snyder, S. Warner, W. White, and W. Willey.

University of Florida: C. Albriego, J. Allen, R. Brooks, F. Collins, C. McCoy, H. Niggs, W. Menke, R. Townsend, M. Vittelli, and W. Whitcomb.

Texas A&M University: J. Bush, C. Cintron, H. Dean, H. Hartley, N. Maxwell, A. Seils, and B. Villalon.

U.S. Department of Agriculture: D. Chambers, H. Dulmage, W. Hart, D. Moreno, D. Reed, A. Selhime, and J. Shaw.

Church of Jesus Christ of Latter-Day Saints: T. Stay, and M. Whitlock.

Citrus Advistory Commission: R. McLean.

Corona College Heights Packing Association: C. Livingston.

Exeter Packing House: F. J. Stetson.

Fillmore-Piru Packing Association: G. E. Nehrig.

Fillmore Citrus Protec. Dist.: N. Pennington.

Gilkey Farms: D. Riddle.

Grandview Heights Citrus Association: W. C. Orr.

Perma Rain Corp.: R. Walters.

Pest Management Associates: J. Gordon.

Rincon-Vitova Insectaries: E. Dietrick.

Shiells Ranch Co.: J. Shiells, G. Wren.

Sunkist: D. Avis.

Superior Farming Co.: H. Chavez, W. Duncan.

COTTON

Subproject Director:

P. L. Adkisson *Texas A&M University* (*1972–1977*)

Principal Investigators:

University of Arkansas
   J. R. Phillips (*1972–1977*)

University of California
   A. P. Gutierrez (*1975–1977*)
   T. F. Leigh (*1974*)
   H. T. Reynolds (*1972, 1973*)

Mississippi State University
   L. G. Brown (*1976, 1977*)
   F. A. Harris (*1974, 1975*)
   F. G. Maxwell (*1972, 1973*)

Texas A&M University
   P. L. Adkisson (*1972–1977*)
   D. G. Bottrell (*1974*)

Participants and Cooperators:

University of Arkansas: R. Allen, L. Carter, C. Lincoln, W. Nicholson, J. Phillips, W. Sabbe, R. Skeith, J. Slosser, N. Tugwell, B. Waddle, and W. C. Yearian.

University of California: E. J. Butterfield, K. Byerly, R. Cave, F. David, W. Denton, J. DeVay, L. E. Ehler, L. A. Falcon, D. Grimes, A. P. Gutierrez, D. Gonzalez, K. Hagen, D. Hogg, D. Hueth, R. Jones, K. Lee, T. F. Leigh, W. B. Loew, R. Norgaard, U. Regev, H. T. Reynolds, A. Sorensen, R. van den Bosch, D. Westphal, and L. T. Wilson.

Mississippi State University: G. Andrews, T. Brooks, L. Brown, H. Chambers, R. Creech, A. Douglas, W. Fox, J. Frazier, F. Harris, L. Hepner, V. Hurt, M. Laster, E. Lloyd, F. Maxwell, R. McClendon, J. Megahee, M. Parker, D. Parvin, H. Pietre, M. Schuster, K. Shaunak, P. Sikorowski, E. Thredgill, and C. Wilson.

Texas A&M University and Texas Agricultural Extension Service: P. Adkisson, R. Billingsley, L. Bird, I. Borosh, D. Bottrell, H. Burke, D. Calvert, B. Casey, J. Cate, G. Curry, D. DeMichele, R. Hanna, J. Harding, S. Harp, L. Keeley, R. Lacewell, S. Nemec, G. Niles, F. Plaff, R. Powell, P. Sharpe, W. Sterling, H. Talpaz, J. Thomas, J. Walker, Jr., and L. Wilkes.

U.S. Department of Agriculture: D. Baker, G. Butler, R. Colwick, W. Cross, P. Fryxell, J. Hamil, D. Hardee, A. Hartstack, J. Hollingsworth, A. Hyer, J. Jenkins, E. Lloyd, D. Martin, R. McLaughlin, W. Parrott, R. Ridgway, E. Stadelbacher, and J. Witz.

PINES (BARK BEETLES)

Subproject Director:

W. E. Waters, *University of California, Berkeley* (*1974–1977*)
R. W. Stark, *University of Idaho* (*1972, 1973*)

Principal Investigators:

University of California
D. L. Wood (*1972–1977*)

Duke University
D. O. Yandle (*1972*)

University of Idaho
R. W. Stark (*1972–1977*)

Texas A&M University
R. N. Coulson (*1972–1975*)

Virginia Polytechnic Institute and State University
W. W. Leuschner (*1974–1975*)
C. M. Newton (*1972–1973*)

Washington State University
A. A. Berryman (*1972–1977*)

Participants and Cooperators:

University of California: F. Cobb, Jr., D. Dahlsten, B. Ewing, J. Helms, J. Parmeter, P. Rauch, W. Waters, and D. Wood.

University of Idaho: D. Adams, L. Anderson, E. Godfrey, E. Michalson, D. E. Olson, A. Partridge, A. Rigas, J. Schenk, R. Stark, K. Stoszek, J. Thomas, and C. Warnick.

Texas A&M University: J. Coster, R. Coulson, G. Curry, D. DeMichele, J. Foltz, F. Hain, T. Harney, W. Martin, A. Mayyasi, T. Payne, P. Sharp, and J. Zolnowski.

Virginia Polytechnic Institute: H. Burkhardt, H. Heikkenen, W. Leuschner, C. Newton, M. Reynolds, and A. Sullivan.

Washington State University: A Berryman, R. Bruce, D. Burnell, R. W. Dingle, C. Ditweiler, R. Gilkerson, B. Hyman, L. V. Pienaar, H. Rigas, R. Shew, and C. White.

U.S. Forest Service: G. Amman, S. Barras, W. Bedard, W. Bennett, D. Cole, R. Dalleske, G. Deitschman, C. DeMars, A. Drooz, M. Furniss, M. Galbraith, R. Gustafson, D. Hamilton, R. Heller, W. Klein, K. Knauer, C. Leaphart, P. Lorio, Jr., J. Lotan, C. Lyon, W. McCambridge, J. Moser, J. Myers, N. Norick, P. Orr, W. O'Regan, R. Pfister, D. Pierce, J. Rauchensberger, A. Roe, C. Sartwell, W. Schmidt, R. Schmitz, R. Shearer, A. Stage, R. Stevens, and R. Thatcher.

POME AND STONE FRUITS

Subproject Director:

B. A. Croft, *Michigan State University* (*1974–1977*)
S. C. Hoyt, *Washington State University* (*1972–1973*)

Principal Investigators:

University of California
L. E. Caltagirone (*1972, 1973*)
R. E. Rice (*1974–1976*)

Michigan State University
B. A. Croft (*1962–1977*)

New York State Agric. Expt. Station
E. H. Glass (*1972–1976*)

Pennsylvania State University
Dean Asquith (*1972–1976*)
W. M. Bode (*1977*)

Washington State University
E. C. Burts (*1972–1976*)
S. C. Hoyt (*1973–1977*)

Participants and Cooperators:

University of California: N. Akesson, J. Ball, W. Batiste, D. Brown, L. Caltagirone, A. Davidson, C. Davis, J. Dibble, L. Etzel, B. Ewing, L. Falcon, A. Gutierrez, K. Hagen, J. Ogawa, G. Okumura, G. Oster, C. Pickel, R. Rice, H. Riedl, W. Roelofs, and G. Rowe.

Mitchigan State University: B. Croft, S. Harsh, D. Haynes, A. Howitt, S. Hoying, J. Hull, A. Jones, H. Koening, R. Meyer, M. Nakashima, A. Putnam, H. Riedl, J. Rodriguez, W. Thompson, L. Tummala, and S. Welsh.

New York State Agricultural Experiment Station: S. E. Lienk, J. Tette, K. Trammel, C. M. Watve, and H. R. Willson.

Pennsylvania State University: D. Asquith, D. Baugher, W. Bode, R. Coleburn, D. Daum, P. Dress, G. Greene, C. Herbert, K. Hickey, L. Hull, B. Kelly, R. Koene, F. Lewis, W. McCarthy, P. Mowery, and S. Smith.

Washington State University: E. W. Anthon, N. Benson, J. Bruner, E. Burts, R. Covey, J. Eves, R. Harwood, S. Hoyt, D. Ketch, J. Logan, K. Olsen, and L. Tanigoshi.

U.S. Department of Agriculture: J. Adams, W. Butt, R. Goodwin, H. Hoffitt, and E. Stahly.

SOYBEAN SUBPROJECT

Subproject Director:

L. D. Newsom, *Louisiana State University* (*1972–1977*)

Principal Investigators:

University of Arkansas
F. D. Miner (*1972–1977*)

Clemson University
  S. G. Turnipseed (*1972–1977*)

University of Florida
  W. G. Whitcomb (*1972–1977*)

University of Illinois and Illinois Natural History Survey
  Marcos Kogan (*1972–1977*)

Louisiana State University
  L. D. Newsom (*1972–1977*)

North Carolina State University
  R. L. Rabb (*1972–1977*)

USDA, ARS
  D. F. Martin (*1973–1976*)

Participants and Cooperators:

University of Arkansas: F. Miner, A. Mueller, H. Scott, N. Tugwell, H. Walters, W. Yearian, and S. Young.

Clemson University: G. Carmer, E. Hartwig, J. Maxwell, J. Schillinger, B. Shepard, M. Sullivan, and S. Turnipseed.

University of Florida: G. Allen, L. Buschman, P. Callahan, G. Greene, R. Hemenway, L. Kish, N. Leppla, J. Nickerson, D. May, W. Menke, and W. Whitcomb.

University of Illinois: E. Armbrust, J. Bouseman, M. Erwin, B. Ford, G. Godfrey, M. Kogan, J. Maddox, M. Miller, S. Roberts, W. Ruesink, L. Stannard, and G. Waldbauer.

Louisiana State University: W. Birchfield, B. Farthing, F. Gilman, J. Graves, D. Herzog, A. Hammond, N. Horn, R. Jensen, A. Larson, F. Martin, R. McPherson, L. Newsom, R. Rogers, W. Rudd, V. Srinivasan, F. Wiegman, and C. Williams.

North Carolina State University: J. Bradley, C. Brim, W. Brooks, W. Campbell, G. Carlson, F. Guthrie, T. Herbert, E. Hodgson, P. Miller, H. Neunzig, R. Rabb, J. Rawling, T. Sheets, J. Ross, and J. Van Duyn.

U.S. Department of Agriculture: R. Bernard, W. Birchfield, C. Brim, E. Hartwig, and D. Martin.

INDEX